Invasive Species Management

Techniques in Ecology and Conservation Series

Series Editor: William J. Sutherland

Bird Ecology and Conservation: A Handbook of Techniques
William J. Sutherland, Ian Newton, and Rhys E. Green

Conservation Education and Outreach Techniques
Susan K. Jacobson, Mallory D. McDuff, and Martha C. Monroe

Forest Ecology and Conservation: A Handbook of Techniques
Adrian C. Newton

Habitat Management for Conservation: A Handbook of Techniques
Malcolm Ausden

Conservation and Sustainable Use: A Handbook of Techniques
E.J. Milner-Gulland and J. Marcus Rowcliffe

Invasive Species Management: A Handbook of Principles and Techniques
Mick N. Clout and Peter A. Williams

Invasive Species Management

A Handbook of Principles and Techniques

Edited by
Mick N. Clout
and
Peter A. Williams

OXFORD
UNIVERSITY PRESS

Great Clarendon Street, Oxford OX2 6DP

Oxford University Press is a department of the University of Oxford.
It furthers the University's objective of excellence in research, scholarship,
and education by publishing worldwide in

Oxford   New York

Auckland   Cape Town   Dar es Salaam   Hong Kong   Karachi
Kuala Lumpur   Madrid   Melbourne   Mexico City   Nairobi
New Delhi   Shanghai   Taipei   Toronto

With offices in

Argentina   Austria   Brazil   Chile   Czech Republic   France   Greece
Guatemala   Hungary   Italy   Japan   Poland   Portugal   Singapore
South Korea   Switzerland   Thailand   Turkey   Ukraine   Vietnam

Oxford is a registered trade mark of Oxford University Press
in the UK and in certain other countries

Published in the United States
by Oxford University Press Inc., New York

© Oxford University Press, 2009

The moral rights of the authors have been asserted
Database right Oxford University Press (maker)

First published 2009

All rights reserved. No part of this publication may be reproduced,
stored in a retrieval system, or transmitted, in any form or by any means,
without the prior permission in writing of Oxford University Press,
or as expressly permitted by law, or under terms agreed with the appropriate
reprographics rights organization. Enquiries concerning reproduction
outside the scope of the above should be sent to the Rights Department,
Oxford University Press, at the address above

You must not circulate this book in any other binding or cover
and you must impose the same condition on any acquirer

British Library Cataloguing in Publication Data

Data available

Library of Congress Cataloging in Publication Data

Data available

Typeset by Newgen Imaging Systems (P) Ltd., Chennai, India
Printed in Great Britain
on acid-free paper by
CPI Antony Rowe, Chippenham, Wiltshire

ISBN 978–0–19–921632–1 (Hbk.)
ISBN 978–0–19–921633–8 (Pbk.)

10 9 8 7 6 5 4 3 2 1

Introduction

Invasive alien species are now one of the main threats to biodiversity worldwide. The transport of organisms by humans since earliest times, and now through increased levels of trade and tourism, has led to the widespread breaching of natural biogeographic barriers at historically unprecedented rates. The consequences for native biota and natural ecosystem processes can be severe, especially in previously isolated ecosystems. Global climate change further exacerbates the spread of alien species, as climatic zones shift and potential ranges expand.

This book focuses on those alien species, spread inadvertently or deliberately by humans, which invade natural or semi-natural ecosystems. Such species are agents of ecological change, which includes extinction or decline of vulnerable endemic species, alteration of the structure and composition of communities, loss of ecosystem services, and disruption of successional pathways. Alien species causing ecological change can be termed 'invasive alien species', or merely 'invasive species'. These invaders are also given many other names, including pests, weeds, exotics, aliens, introduced species, or non-indigenous species.

Biological invasion is a staged process whereby, to become a successful invader, a species must cross a series of spatial, environmental, and biological barriers. The main stages are transport, establishment, and spread. Different sets of species attributes are likely to confer potential success at these different stages, (e.g. attributes necessary for transoceanic dispersal may be quite different from those favouring population establishment at a new location). Managing biological invasions therefore involves identifying pathways or vectors and the species attributes that confer success at each barrier, and using appropriate methods or strategies to prevent, eradicate, or control the species of concern.

To say that alien invasions are a human construct is both a tautology and a fact. But because we are the principal cause of invasions, we are also, paradoxically, the principal source of the solution (McNeely 2006). The purpose of this book is to introduce the reader to the underlying principles necessary for the successful prevention and control of biological invasions. Although concerned primarily with conserving biodiversity, many of the principles are applicable to a broader range of sectors. This applies particularly to the first eight chapters. Invasions occur across the full spectrum of taxonomic groups, as shown in Chapters 9–14, and they often interact, so that managing the system as a whole becomes necessary (Chapter 15).

The role of humans in the spread of invasive species probably began with the introduction of dogs and edible plants by hunter-gatherers, before the advent of agriculture. Many prehistoric invasions are so ancient, such as the introduction of dogs (dingoes) to Australia and the Polynesian rat throughout the Pacific, that the main evidence we have of their initial impacts are the bones of extinct species.

Arguably the most culturally influential biological invasion was the spread by Roman trade of the black rat through Europe. These animals became the carriers of the bubonic plague that decimated the European population in the years immediately following 1348. This led to such restructuring of western society that it laid the foundations for the Renaissance (Benedictow 2004). This in turn, led to the intensive study of nature and so, it could be argued, to the concept of nature conservation itself, and ultimately to the concept of alien species and their impacts *on* nature.

The modern era of alien species dispersal and introductions began only c. 500 years ago with the advent of square-rigged ships in the 15th century and the discovery of the New World. Now we live in a world where anything can be transported anywhere, often overnight, providing a myriad of vectors and pathways, some purposeful and others not, for everything from microbes or mussels, to mice and mambas. As a consequence, the degree of international trade into a country is now the strongest predictor of the number of invasive alien species in that country (Westphal *et al.* 2008). The dependence of human societies worldwide on introduced crops and livestock has led to many introductions, and trade has fuelled other global changes, including human population growth.

Biological invasions have had impacts on many aspects of human society, including agriculture, horticulture, aquaculture, and human health (Pimental 2002). They also have had major impacts on biological diversity and ecosystem services. Examples of invasive alien species with severe impacts range across the whole globe. Avian malaria, which is carried by alien mosquitoes, has virtually eliminated native birds below 1300m in Hawaii. Accidental introductions of predatory vertebrates, such as rats or the brown tree snake, have decimated vulnerable native wildlife in many parts of the world, especially in previously isolated island ecosystems. Introductions of plants for pleasure or profit, including via well-intentioned aid projects, have seen whole ecosystems transformed, e.g. through the spread of acacia in South Africa. Invasions in fresh and salt water, of both plants and animals, have been equally devastating. Examples include the reduction of aquatic biodiversity and devastation of local economies by water hyacinth covering tropical lakes. Many similar examples, and the pathways through which they have occurred, are summarized by Wittenberg and Cock (2001); and a sound classification scheme is suggested by Hulme *et al.* (2008).

As human society has tried to adapt to biological and environmental changes, even potential solutions have sometimes led to yet more devastating invasions. An example is the historical liberation of mustelids in New Zealand as biological control agents for previously introduced rabbits—leading to decline of native wildlife species, but no effective control of rabbits.

The concurrent increase in global consumption of resources, especially fossil fuels, has lead to human-induced climate change, which is already affecting the range and behaviour of invasive species. A current fear, generated by diminishing resources and the 'fuel crisis', is that one proposed solution, in the form of crops grown for biofuel, will spawn yet more weed invasions. The pressure on biological systems is relentless. Although public awareness and research into biological invasions has increased dramatically since the book by Elton (1958), there has not been a proportional investment in the management of invasive alien species (Hulme 2006).

Those working with alien species on a daily basis must nevertheless continually be aware we do not all have the same value systems and consequently not everyone accepts such species as necessarily undesirable. Some claim that the drive to reduce the spread and impacts of alien species is displaced xenophobia and is lineally descended from the native plant-loving Nazis (see Simberloff 2003a). Although this argument has been deconstructed by Coates (2006), we confess to some disquiet over the frequent use of martial metaphors by invasion biologists to describe alien (or non-native) species. Nevertheless, we are aware that there is no universally accepted set of terms to describe an alien species, as it progresses from a benign species, to a potential threat, to becoming invasive and almost universally unwanted in a new area. For this reason, we have not attempted to standardize terms amongst the contributors to this volume.

Although biological invasions are not everyone's primary concern in life, and care must be taken in the terms we use to communicate about them, there is now tremendous popular support in many places for their effective management. Obtaining this support, and directing it appropriately, is a crucial step in controlling invasions. This can be achieved, as Boudjelas explains (Chapter 7), only by harnessing the values that people already have towards natural systems, and showing how invasive species interfere with these. As explained by De Poorter (Chapter 8), at the international level there are now several laws and conventions that provide support for the management of alien species and oblige governments to prevent their introduction and spread and minimize their impacts.

Some of the strongest international legislation aims to prevent the introduction of alien species from one country to another, and began to be enacted over 100 years ago (Maynard and Nowell, Chapter 1). These authors go on to explain the multitude of pathways by which alien species may enter new areas and the monitoring systems and procedures that must be put in place to prevent *introduction*. They also explain that predicting which species and pathways are likely to be problematic is important for designing quarantine systems, and even for locating facilities. Stohlgren and Jarnevich (Chapter 2) explain what we need to know to estimate, and predict the risks associated with invasive organisms—there are net bioeconomic benefits to doing this (Keller *et al.* 2007)

In the end, some alien species are always going to slip through the border, perhaps because neither the organism nor its pathway was identified as a threat. At this point, future invasions can be prevented only if effective detection and early warning systems are in place and are backed up by ready action plans as described by Holcombe and Stohlgren (Chapter 3). These authors point out that the window of opportunity here may be quite narrow before the species becomes fully *naturalized*, (i.e. maintaining a self-sustaining population in the wild). Once this happens, the problems change to considering whether or not it is possible to eradicate such populations. There are significant economic and environmental benefits of *eradication*, as opposed to perpetual control. However, unless they are properly planned and executed, eradication programmes often fail, for reasons detailed by Parkes and Panetta (Chapter 4). Key factors in eradication success are proper planning, commitment to persevere until the last individual is removed, and ensuring that reinvasion is prevented. If eradication is not possible, the issue then becomes one of *containment*, to prevent the invasive species expanding to saturate all potential habitats as differentiated by Grice (Chapter 5). Containment is a valid strategy only for those situations where it is possible to curb or halt the rate of range expansion. In reality, there is seldom a 'permanent' barrier to range expansion (Grice, Chapter 5) and the vast majority of resources, are expended on *controlling* the populations to minimize their impacts in particular places. Biological control usually has this effect because one organism is seldom able to exterminate another, as Murphy and Evans explain (Chapter 6). 'Classical' biological control has mostly been for the control of arthropod and plant invaders, but efforts have been made on invasive mammals, invasive marine organisms, and other species.

The principles outlined in Chapters 1–8 are applicable to most invasive species in a range of habitats. However, since there are particular management challenges associated with different groups of invasive organisms, the second part of the book (Chapters 9–15) consists of individual chapters devoted specifically to these. Firstly, in a chapter on terrestrial plants, Holt (Chapter 9) explains how their particular characteristics, such as high genetic plasticity and the tremendous dispersal potential of propagules that can 'hide' in seed banks, raise particular problems at all stages of invasion. Aquatic plants have many similar characteristics, but, as described by Coetzee and Hill (Chapter 10), they also have some vitally different ones relevant to their management—such as reduced accessibility to the foliage for entirely submerged species. However, in the case of still waters, the distribution of aquatic invaders is often confined to discrete areas, with the result that infestations of aquatic plants have frequently been eradicated. Techniques ranging from biocontrol to total modification of the environment (e.g. drainage) have been employed to eradicate aquatic invasive plants. Invertebrates are arguably the most pervasive and widespread group of invaders around the planet, transported (usually inadvertently) by a variety of vectors and pathways. The management of invasive invertebrates presents special challenges, as described by Green and O'Dowd (Chapter 11) in their illuminating case study of the yellow crazy ant on Christmas Island. Prevention or early detection and eradication are the best policies for invertebrate invaders, as they are for other organisms. These strategies are likely to be successful on a landscape scale only when the biology of the species is properly understood and the necessary resources are committed for the duration of the programme.

Terrestrial vertebrate pests are arguably the best-studied and theoretically the most tractable group of invaders, since they do not have dormant life stages and, with the notable exception of commensal rodents, are rarely transported by accident. However, there are some special challenges in managing terrestrial vertebrates, as described by Parkes and Nugent (Chapter 12). Terrestrial vertebrates have been introduced to many locations as domestic livestock, as pets, or for hunting, and are often valued by people. Opposition by some sectors of the community to their eradication or control is therefore more common than for most other groups of organisms. Other challenges in the management of terrestrial vertebrate pests stem from their relatively complex behaviour.

In aquatic ecosystems, fish are the major group of vertebrate invaders. Many fish species have been deliberately transported around the world for sport, aquaculture, or as aquarium pets. Like other groups of organisms, some are relatively benign, whereas others are highly invasive. Ling (Chapter 13) describes methods and approaches that have been successfully used to prevent, control, or eradicate invasive fish from a range of aquatic habitats around the world, and some of the special difficulties that are inherent in their management.

The management of invasive species in marine environments poses challenges that are not prevalent in freshwater and terrestrial systems. As pointed out by Piola and colleagues (Chapter 14), marine environments are expansive, inter-connected, and often only partially accessible. Although the successful management of marine invaders must clearly focus on effective prevention, there are nevertheless some tools available for post-border management of marine pest species. Effective marine biosecurity should consist of vector management, surveillance, incursion response, and control measures that target particular pests or suites of functionally similar ones, coupled with generic approaches to reduce human-mediated transport.

Finally, it is increasingly evident that managers must be concerned not only with the effects of invasive species on native species in the ecosystems that they have invaded, but also with the interactions between invasive species. Bull and Courchamp (Chapter 15) use examples of eradications of invasive vertebrates from islands to illustrate this; highlighting phenomena such as hyper-predation, mesopredator release, competitor release, and the release of invasive plants from introduced herbivores. Understanding the functional relationships within invaded ecosystems is a significant challenge, but is important for the restoration of native species and natural ecosystem processes. This underscores the point that the effective management of invasive species requires underpinning by sound ecological science.

The subject of invasive species management is so extensive that a single book cannot possibly prescribe detailed techniques for every species or situation. The first aim of this book is therefore to describe strategies for managing invasive species at different stages of the invasion process. The second aim is to describe the general tools and approaches that are recommended for the successful management of particular groups of invasive organisms. We hope that this handbook will be useful to a range of readers, including invasive species managers, legislators, students, and the broader community concerned with biological conservation.

Mick N. Clout
and
Peter A. Williams

Acknowledgements

We thank all of the contributers to this book, the staff of OUP for their professional help, and Carola Warner for her invaluable assistance in the completion of this work.

Contents

Contributors   xxi

1 Biosecurity and quarantine for preventing invasive species   1

Glynn Maynard and David Nowell

1.1 Introduction	1
1.2 Invasiveness and impacts	1
1.3 Legislative frameworks	3
1.3.1 International framework	3
1.3.2 National frameworks	3
1.4 Pathways	4
1.4.1 Natural spread and host range extension	5
1.4.1.1 Natural disasters	5
1.4.2 Accidental introductions	5
1.4.2.1 Trade	5
1.4.2.2 Traditional movement of people and goods	6
1.4.2.3 Emergency food, disaster relief, and development aid	6
1.4.3 Deliberate introductions	7
1.4.3.1 Biological control	7
1.4.3.2 Plant introductions	8
1.4.3.3 Smuggling	9
1.5 Actions	9
1.5.1 Pre-entry	10
1.5.2 Entry (border)	11
1.5.3 Emergency actions	14
1.6 Summary	16
1.7 Acknowledgements	18

2 Risk assessment of invasive species   19

Thomas J. Stohlgren and Catherine S. Jarnevich

2.1 Introduction	19
2.1.1 Why do we need a formal approach to invasive species risk assessment?	20
2.1.2 Current state of risk assessment for biological invaders	21
2.1.3 The ultimate risk assessment challenge	22

2.2 Components of risk assessment for invasive species   23
  2.2.1 Information on species traits   23
  2.2.2 Matching species traits to suitable habitats   25
  2.2.3 Estimating exposure   27
  2.2.4 Surveys of current distribution and abundance   29
  2.2.5 Understanding of data completeness   30
  2.2.6 Estimates of the 'potential' distribution and abundance   31
  2.2.7 Estimates of the potential rate of spread   31
  2.2.8 Probable risks, impacts, and costs   32
  2.2.9 Containment potential, costs, and opportunity costs   33
  2.2.10 Legal mandates and social considerations   33
2.3 Information science and technology   34
2.4 The challenge: to select priority species and priority sites   34
2.5 Acknowledgements   35

3 Detection and early warning of invasive species   36

Tracy Holcombe and Thomas J. Stohlgren

3.1 Introduction   36
  3.1.1 Fire as a metaphor for invasion   37
  3.1.2 Definitions   37
3.2 Early detection and rapid assessment   38
3.3 Guiding principles for early detection and rapid assessment   40
  3.3.1 Data and information management   41
  3.3.2 Global and regional invasive species databases   41
  3.3.3 Species reporting requirements   45
3.4 Conclusions   45

4 Eradication of invasive species: progress and emerging issues in the 21st century   47

John P. Parkes and F. Dane Panetta

4.1 Introduction   47
4.2 From scepticism to positive consideration   47
4.3 Feasibility   48
4.4 Advances in eradication of vertebrate pests   50
4.5 Advances in eradication of weeds   52
4.6 Emerging issues   55
  4.6.1 Eradication on mainlands   55
  4.6.2 Does scale count?   56
  4.6.3 Delimiting boundaries and detecting survivors and immigrants   56
  4.6.4 A particular problem with weeds—seed banks   57

4.6.5 Tricky species	58
4.6.6 Institutional commitment	58
4.6.7 Local elimination	59
4.7 Conclusions	59
4.8 Acknowledgements	60

5 Principles of containment and control of invasive species — 61

Tony Grice

5.1 Introduction	61
5.2 Control and containment—strategies without an end-point	62
5.2.1 When to contain, when to control	63
5.2.2 Feasibility of containment	63
5.2.3 Elements of a containment strategy	65
5.2.4 To control or not to control	65
5.3 Principles of containment and control	67
5.3.1 Evaluate impacts of invasive species	67
5.3.2 Assemble knowledge of species' biology, ecology, and responses to management	67
5.3.3 Map distribution and abundance	68
5.3.4 Set priorities for species and places	68
5.3.5 Coordinate management of multiple, functionally similar invasive species	69
5.3.6 Take action early in the invasion process	69
5.3.7 Direct effort where benefit: cost ratio is high	70
5.3.8 Direct containment effort at the periphery of an expanding distribution	70
5.3.9 Exploit natural barriers to range expansion	70
5.3.10 Exploit times when invasive species' populations are low	71
5.3.11 Acquire continuing commitment	71
5.3.12 Resolve conflicting interests	72
5.3.13 Monitor the consequences	72
5.4 Examples	73
5.4.1 Containment of rubber vine in northern Australia	73
5.4.2 Containment of leucaena—a commercially grown fodder shrub in Australia	74
5.4.3 Control of invasive mammalian predators in New Zealand	74
5.4.4 Invasive pasture grasses in Australia	75
5.5 Conclusions	76
5.6 Acknowledgements	76

6 Biological control of invasive species — 77

Sean T. Murphy and Harry C. Evans

- 6.1 Introduction — 77
- 6.2 Why classical biological control is an appropriate tool for managing invasive species — 78
- 6.3 The practice of classical biological control — 79
  - 6.3.1 *Early history and development* — 79
  - 6.3.2 *Biological control projects against invasive species in natural ecosystems* — 80
  - 6.3.3 *Success, failures, and the economics of biological control* — 83
- 6.4 Modern methods of biological control — 86
  - 6.4.1 *The characteristics of efficacious agents* — 86
  - 6.4.2 *Issues related to ecological risks* — 87
- 6.5 Constraints to the implementation of biological control — 91
- 6.6 Conclusions — 92

7 Public participation in invasive species management — 93

Souad Boudjelas

- 7.1 Introduction — 93
- 7.2 Why involve the public? — 93
  - 7.2.1 *Ethics* — 93
  - 7.2.2 *Compliance* — 93
  - 7.2.3 *Effectiveness* — 95
    - 7.2.3.1 Locally relevant — 95
    - 7.2.3.2 Maximize the resource effort — 95
    - 7.2.3.3 Public support — 98
    - 7.2.3.4 Part of the problem; part of the solution — 99
- 7.3 How to successfully involve the public — 99
- 7.4 Conclusions — 106

8 International legal instruments and frameworks for invasive species — 108

Maj De Poorter

- 8.1 Introduction — 108
- 8.2 Scope and types of international instruments — 110
- 8.3 Invasive species and global instruments for conservation of biological diversity — 112
  - 8.3.1 *The Convention on Biological Diversity (CBD)* — 112
  - 8.3.2 *The Convention on Wetlands (Ramsar)* — 113
  - 8.3.3 *The Convention on International Trade in Endangered Species (CITES)* — 114
  - 8.3.4 *Convention on the Conservation of Migratory Species of Wild Animals (CMS)* — 114

  8.3.5 The UN Convention on the Law of the Sea (UNCLOS) 115
  8.3.6 The Code of Conduct for Responsible Fisheries 115
 8.4 Invasive species and regional instruments for conservation of biological diversity 115
  8.4.1 The International Council for the Exploration of the Sea (ICES) Code of Practice 115
  8.4.2 United Nations Environment Programme (UNEP) Regional Seas Programme 116
  8.4.3 Other agreements 116
 8.5 Invasive species and instruments relating to phytosanitary and sanitary measures 117
  8.5.1 The International Plant Protection Convention (IPPC) 117
  8.5.2 Other regulations 117
 8.6 Invasive species and instruments relating to transport operations 118
  8.6.1 International Maritime Organization (IMO) 118
  8.6.2 The International Civil Aviation Organization (ICAO) 119
 8.7 Relationship with multilateral trading systems 119
 8.8 International instruments and approaches relevant to invasive species 120
 8.9 Relation between invasive species and sustainable development programmes 121
 8.10 Regional strategies and plans 122
  8.10.1 South Pacific Regional Environment Programme (SPREP): invasive species strategy for the Pacific Island region 122
  8.10.2 European Strategy (Council of Europe) 122
  8.10.3 European Union 122
  8.10.4 Pacific Invasives Initiative 122
  8.10.5 Pacific Ant Prevention Programme (PAPP) 123
 8.11 International programmes and organizations 123
  8.11.1 The Global Invasive Species Programme (GISP) 123
  8.11.2 The International Union for Conservation of Nature and Natural Resources (IUCN) 124
  8.11.3 The Invasive Species Specialist Group (ISSG) 124
  8.11.4 The Global Invasive Species Database (GISD) and Global Invasive Species Information Network (GISIN) 124
 8.12 Conclusions 125

9 Management of invasive terrestrial plants 126

Jodie S. Holt

 9.1 Introduction 126
 9.2 Classification of weeds and invasive plants 126
 9.3 Plant characteristics important in management 127

Contents

9.4 Management of terrestrial invasive plants	128
9.4.1 *Principles of prevention, eradication, containment, and control*	128
9.4.2 *Physical methods of invasive plant control*	132
9.4.2.1 Hand pulling and using manual implements	132
9.4.2.2 Fire	133
9.4.2.3 Using machines for invasive plant control	134
9.4.2.4 Mulching and solarization	136
9.4.3 *Cultural methods of invasive plant control*	136
9.4.3.1 Prevention	136
9.4.3.2 Competition	137
9.4.4 *Biological control*	138
9.4.5 *Chemical control*	138
9.4.6 *Integrated weed management*	139

10 Management of invasive aquatic plants   141

Julie A. Coetzee and Martin P. Hill

10.1 Introduction	141
10.2 Plant characteristics important in management	141
10.3 Modes of introduction and spread	145
10.4 Management of aquatic invasive plants	146
10.4.1 *Utilization*	146
10.4.2 *Manual/mechanical control*	146
10.4.3 *Herbicidal control*	148
10.4.4 *Biological control*	148
10.4.5 *Integrated control*	151
10.5 Prevention, early detection, and rapid response	152

11 Management of invasive invertebrates: lessons from the management of an invasive alien ant   153

Peter T. Green and Dennis J. O'Dowd

11.1 Introduction	153
11.2 History	155
11.2.1 *The yellow crazy ant as a pantropical invader*	155
11.3 YCA invasion of Christmas Island	156
11.3.1 *The interim response*	157
11.3.2 *The aerial control campaign*	161
11.3.2.1 Legislative approval	162
11.3.2.2 The helicopter	162
11.3.2.3 Dispersion of Presto®01 ant bait	163
11.3.2.4 Mapping supercolonies	163
11.3.2.5 Trial of aerial baiting	164

	11.3.2.6 Measuring the success of the island-wide operation	164
	11.3.2.7 Non-target impacts	165
	11.3.3 Evaluation and lessons learned from the aerial campaign	167
11.4	Conclusions	170
11.5	Acknowledgements	171

12 Management of terrestrial vertebrate pests — 173

John P. Parkes and Graham Nugent

12.1	Introduction	173
12.2	Tools to prevent new species arriving	174
12.3	Tools to manage established wild populations	175
	12.3.1 Detection tools	175
	12.3.2 Exclusion	176
	12.3.3 Control tools	176
	12.3.3.1 Snares and traps	176
	12.3.3.2 Shooting	178
	12.3.3.3 Poisoning	179
	12.3.3.4 Biocontrol	181
	12.3.3.5 Fertility control	182
12.4	Conclusions	183

13 Management of invasive fish — 185

Nicholas Ling

13.1	Introduction	185
13.2	The role of humans	186
13.3	Risk assessment	187
13.4	Economics of eradication and control	188
13.5	Marine versus freshwater	188
13.6	Indigenous fish as invasive species	188
13.7	Routes of introduction and spread	189
	13.7.1 Ballast water and vessel hull transport	189
	13.7.2 Live fish importation and sale	189
	13.7.3 Aquaculture for the aquarium trade	191
	13.7.4 Aquaculture for food	191
13.8	Eradication and control	192
	13.8.1 Early response	192
	13.8.2 Response tools	192
	13.8.2.1 Preventing spread: physical barriers, electrical barriers, interstate/interisland biosecurity barriers	192
	13.8.2.2 Chemical control	193
	Rotenone	195
	Antimycin-A (Fintrol®)	196

	Natural saponins	196
	TFM and niclosamide	197
13.8.2.3	Biocontrol measures	197
	Predatory fish	197
	Pheromones	197
	Fish pathogens	197
	Habitat modification and restoration	198
	Immunocontraceptive control and genetic modification	198
13.8.2.4	Physical removal	199
13.8.3	Case studies in the effectiveness of physical removal	200
13.8.3.1	Nile perch in Lake Victoria	201
13.8.3.2	Common carp in Lakes Crescent and Sorrell, Tasmania	203
13.9	Conclusions	203

14 Marine biosecurity: management options and response tools — 205

Richard F. Piola, Chris M. Denny, Barrie M. Forrest, and Michael D. Taylor

14.1	Introduction	205
14.2	Pre-border management	207
14.2.1	Human-mediated invasion pathways	207
14.2.2	Management of human-mediated pathways	210
14.3	Post-border management	211
14.3.1	Early detection and rapid response	212
14.3.2	Response tools	214
14.3.2.1	Physical removal	214
14.3.2.2	Wrapping and smothering	221
14.3.2.3	Physical treatment	224
14.3.2.4	Chemicals	224
14.4	Discussion	227
14.5	Acknowledgments	231

15 Management of interacting invasives: ecosystem approaches — 232

Leigh S. Bull and Franck Courchamp

15.1	Introduction	232
15.2	Cases when removal of alien species does not lead to ecosystem recovery	234
15.2.1	When the alien species has an important functional role	234

15.2.2	When the alien species has a long lasting effect	234
15.2.3	When the alien species interacts with other aliens	235
	15.2.3.1 Interactions resulting from conspicuous aliens	
	Hyperpredation	235
	15.2.3.2. Interactions resulting from inconspicuous aliens	
	Release from introduced herbivores	238
	The mesopredator release effect	241
	The competitor release effect	242
15.3 Mitigating actions		243
15.3.1	Pre-eradication studies	244
15.3.2	Exclosure experiments	244
15.3.3	Control strategies	245
	15.3.3.1 Hyperpredation	245
	15.3.3.2 Mesopredator release	245
	15.3.3.3 Competitor release	245
	15.3.3.4 Post-eradication monitoring	246
15.4 Conclusions		247

References	249
Index	295

Contributors

Souad Boudjelas
Pacific Invasives Initiative
School of Biological Sciences
The University of Auckland
Auckland, New Zealand

Leigh S. Bull
Universite Paris-Sud XI
Laboratoire Ecologie, Systématique et
Evolution
Orsay, France

Mick N. Clout
Centre for Biodiversity and Biosecurity
School of Biological Sciences
The University of Auckland
Auckland, New Zealand

Julie A. Coetzee
Rhodes University
Department of Zoology and Entomology
Grahamstown, South Africa

Franck Courchamp
Universite Paris-Sud XI
Laboratoire Ecologie, Systématique et
Evolution
Orsay, France

Maj De Poorter
Invasive Species Specialist Group
Centre for Biodiversity and Biosecurity
The University of Auckland
Auckland, New Zealand

Chris M. Denny
Cawthron Institute
Nelson, New Zealand

Harry C. Evans
CAB International Europe (UK)
Silwood Park
Ascot, UK

Barrie M. Forrest
Cawthron Institute
Nelson, New Zealand

Peter T. Green
La Trobe University
Department of Botany
Bundoora, Australia

Tony Grice
CSIRO Sustainable Ecosystems
Aitkenvale, Australia

Martin P. Hill
Rhodes University
Department of Zoology and Entomology
Grahamstown, South Africa

Tracey Holcombe
US Geological Survey
National Institute of Invasive Species Science
Fort Collins Science Centre
Fort Collins, CO, USA

Jodie S. Holt
University of California
Riverside, CA, USA

Catherine S. Jarnevich
US Geological Survey
National Inst of Invasive Species Science
Fort Collins Science Centre
Fort Collins, CO, USA

Nicholas Ling
The University of Waikato
Hamilton, New Zealand

Glynn Maynard
Office of the Chief Plant Protection Officer,
Department of Agriculture, Fisheries &
Forestry
Canberra, Australia

Contributors

Sean T. Murphy
CAB International Europe (UK)
Silwood Park
Ascot, UK

David Nowell
International Plant Protection Convention (IPPC)
Rome, Italy

Graham Nugent
Landcare Research NZ Limited
Lincoln, New Zealand

Dennis O'Dowd
Australian Centre for Biodiversity
School of Biological Sciences
Monash University
Victoria, Australia

F. Dane Panetta
Department of Natural Resources and Mines
& CRC for Australian Weed Management
Alan Fletcher Research Station
Sherwood, Australia

John P. Parkes
Landcare Research NZ Limited
Lincoln, New Zealand

Richard F. Piola
Cawthron Institute
Nelson, New Zealand

Thomas J. Stohlgren
US Geological Survey
National Institute of Invasive Species Science
Fort Collins Science Centre
Fort Collins, CO, USA

Michael D. Taylor
Cawthron Institute
Nelson, New Zealand

Peter A. Williams
Landcare Research NZ Limited
Nelson, New Zealand

1

Biosecurity and quarantine for preventing invasive species

Glynn Maynard and David Nowell

1.1 Introduction

The saying 'prevention is better than cure' applies to the entry of invasive species but this is difficult to achieve, particularly in the absence of physical or ecological barriers to the movement of invasive species, or where human activities and vectors provide pathways for their entry. For the purposes of this chapter, the term 'invasive species' applies to those that enter and establish in a new area and have the ability to spread aggressively, to intrude or overwhelm other organisms. This can apply to organisms affecting human food safety, human health and culture, and agricultural, natural terrestrial, and aquatic ecosystems. *Biosecurity* includes all policies and measures that a country implements to minimize these harmful affects, ranging from preventing the entry of unwanted species into an area to their management if they do enter. Biosecurity is a broader concept than *quarantine* (system), but at times these terms are used interchangeably. Together they are usually integrated measures that cross over all sectors that relate to the protection of the environment in general.

Internationally the term *quarantine* is used in several ways. In the broad sense it refers to all activities aimed at preventing the introduction, and/or spread, of a species of concern. In a narrower sense, it is the official confinement of organisms that have a risk of invasiveness (FAO 2007a). In this chapter, the broader sense of quarantine refers to a quarantine system and the narrower sense refers to a quarantine facility, quarantine procedure, or quarantine measure. All of these systems or measures are tools used to reduce the likelihood of entry of invasive organisms.

We stress that accurate identification of the species involved is critical to all aspects of biosecurity and invasive species management to enable appropriate decisions or actions.

1.2 Invasiveness and impacts

The impacts of invasive species range from negligible to extremely high and they can be difficult to understand. Certain components are clearly quantifiable, such as the loss of human lives (e.g. West Nile disease) and financial losses (e.g. direct loss of agricultural production or increased cost of control measures). Many other impacts are less easily quantified, including environmental impacts (e.g. loss or change of biodiversity), impingements on human lifestyle, and amenity losses. In general, if invasive species can be prevented from establishing in an area, the resources used in prevention are usually significantly lower than those needed for eradication, containment, long-term control, or the consequences of doing nothing. Hence, where an invasive species does enter and is detected, it is essential to have well organized and implemented emergency management procedures to minimize the risk of widespread establishment and the subsequent need for an eradication campaign. Eradication can be difficult to achieve and often entails fairly severe measures that may need to be maintained over lengthy periods (see Chapter 4).

Many organisms that enter a new or endangered area either do not establish or necessarily become pests even if they do establish. Other species will establish but do not appear to have a significant impact, at least initially, because the populations are small in size and initially not problematic. However, some species, after several to many generations—which may take months to years—the population can reach sufficiently high levels to become problematic. Invasiveness may increase with a change in conditions, ranging from broad-scale climate change resulting in more available habitat, to a single event such as the introduction of a more efficient plant pollinator resulting in greater seed set. Such changes may result in previously benign species becoming invasive with unacceptable impacts. In addition to exotic species, previously benign native species can invade, through habitat modification such as the introduction of new nutrient resources (e.g. new hosts), to such an extent that they require control.

A critical aspect of all aspects of biosecurity, particularly at the border, is the great difficulty of predicting which species will be invasive. There is no single set of characteristics that determine if a species will be invasive, although certain characteristics increase the likelihood that a species is more likely to be successful in establishing and possibly becoming invasive. A general formula for invasiveness = [Nutrition availability (food or niche availability) + capacity to spread + sensitivity (ecological and human) of ecosystem to change/impact] × Constraints (climate requirements, parasites, disease, reproductive constraints etc.)

There are many texts on this subject with greater or lesser details with many predictive models available that have varying degrees of usefulness. The major requirement for a species to establish and invade in an area is the availability of suitable nutrition, which ranges from host organisms (e.g. parasites) to niches on the landscapes. Once established, other constraints that determine the extent of impact include: its dispersal capacity either by its own means such as flight, or mediated by other means such as passive wind dispersal or attachment to bird feathers or mammal coats; tolerance of different environments; lack of competition at the site of invasion; presence/absence of predators, disease, or parasites; abundance of high value nutrition; host suitability; reproductive capacity; climate; coincidence of climate and host life stages; niches availability and disturbance levels (e.g. change of landscape by humans).

1.3 Legislative frameworks

1.3.1 International framework

When considering the establishment of quarantine systems/measures there are many issues to be taken into account. At the broadest level there are obligations to international conventions and intergovernmental organizations to which governments are members or contracting parties. The principle conventions and intergovernmental organizations that deal with invasive species are the International Plant Protection Convention (IPPC), World Organisation for Animal Health (OIE), Convention on Biological Diversity (CBD), Codex Alimentarius, Ballast Water (International Maritime Organisation, IMO), and the Convention on International Trade in Endangered Species of Wild Flora and Fauna (CITES). These international agreements provide a framework of principles to guide a country in developing mechanisms/measures to reduce the threats from invasive species (see Chapter 8).

The agriculture sector has been developing procedures, methodologies, and tools to lower the likelihood of entry of organisms of concern for many years. The first international plant health agreement was signed in Bern, Switzerland in 1881, as a response to spread of a plant pest, *Phylloxera*, on grape vine planting material. It was called the Convention on measures to be taken against *Phylloxera vastatrix* (now renamed *Daktulosphaira vitifoliae*). This agreement was the forerunner of the current IPPC. Over the past 12 years international phytosanitary standards (ISPMs) have been developed within the framework of the IPPC, to provide guidance on a range of plant quarantine issues covering plant pests—under IPPC definitions the term plant pest includes plant pathogens, insect pests, and weeds (FAO 1997, 2004a–c, 2005a, 2007a,b; IPPC 1997) These standards are generally adaptable across all sectors. The risk assessment standards have been unofficially adapted for use by some scientists for invasive species in areas other than agriculture, e.g. aquatic species.

1.3.2 National frameworks

These agreements, international standards, and frameworks are often administered through different national mechanisms, usually implemented by different sectors of government, and hence not necessarily applied in a unified or coordinated manner.

National issues that can impact on a country's capacity to implement quarantine systems/measures include the national economic status, effectiveness of governance, social and political stability, and the well-being of a populace. If major social problems prevail—such as poverty, famine, civil unrest, or war—then the prevention of entry or deliberate introduction of potential invasive species is likely to have a relatively low national priority. That is, the short-term imperatives of human issues are likely to have higher priority than establishing quarantine measures despite the potential longer-term consequences of not doing so.

1.4 Pathways

Apart from the biological characteristics of invasive organisms, many other factors need to be considered during the implementation of quarantine measures. In particular are the pathways via which organisms can enter new areas. There are three broad categories of pathways of introduction:

1) Species that spread naturally either passively by water or wind, including extreme events such as cyclones, or actively flying, crawling, or swimming.
2) Species that are accidentally introduced hitch-hiking or vectored on/by something for example,: during trade; movement of material during emergency relief or conflicts; traditional movement of people; movement of plants, animals, or soil; scientific materials; traveller's personal effects; movement of contaminated agricultural, military, or industrial equipment; ships, including ballast water.
3) Species that are deliberately introduced, e.g. new genetic stock, biological control, hunting, pets, or ornamental trade. These may be introduced legally or illegally (smuggled) into an area. A further layer of complexity arises when one community considers an organism invasive and another community (often in the same country) considers the species beneficial.

The volume of human-facilitated movement of goods and organisms, and people travelling around the globe is huge and increasing every year. These movements provide pathways that are possibly the most significant sources of potential invasive organisms. Every traveller and item of goods that are imported into a country potentially provides a pathway for an invasive species. The following describes some data of known movements of people and goods in two countries with relatively well-controlled borders, the USA and Australia. They give some idea of the magnitude of the task of minimizing the risk of invasive species entering a country. Both the USA and Australia have invested heavily in the prevention of entry of invasive organisms, as well as in management of pest incursions. The USA has a mix of long land borders and sea borders. Australia is entirely surrounded by sea with only one area in close proximity to another country.

The USA intercepted about 325,000 pests between 1991–96, and inspected over 315,000 ships. In 1996, they inspected over 66 million passengers (APHIS web facts). Australia intercepted about 140,000 pests between 1993–2003. In the year 2006–07 there were 1.6 million sea cargo containers inspected; rising to 1.8 million in 2007–08—an increase of over 10% in one year. Currently 12 million air passengers are screened each year and around 45,000 items of quarantine concern are seized every month and about a quarter of these are undeclared—there are 3,300 staff (inspection and non-inspection) in the agency that manages the international border. Hence, to complement the effectiveness of the work done by government agencies, there needs to be a high level of cooperation by a greater part of the populace

1.4.1 Natural spread and host range extension

The Queensland fruit fly (*Bactrocera tryoni*) exemplifies the combination of natural spread and host range extension resulting from human modification of the environment. It is native to south east Queensland, Australia, where it originally lived on native fruit. When exotic horticultural species of fruit and vegetables were introduced into its native range, *B. tryoni* was exposed to a novel host range to which it readily adapted. It subsequently expanded its distribution to wide areas of eastern Australia where it has severe consequences for some crops. High levels of control and monitoring are required to ensure that major production areas are kept free from this native, invasive species.

1.4.1.1 Natural disasters

Natural disasters affect the entry of invasive species. Cyclones (hurricanes) can result in the movement of organisms over abnormally long distances. Similarly, large-scale disturbance of landscapes can create conditions in which invasive organisms can establish, e.g. the spread of insect vectors or large-scale destruction of land cover creates opportunities for establishment of weeds. In addition, natural disasters often generate emergency relief actions and the rapid importation of largely uncontrolled goods. The associated quarantine risks often lead to accidental introductions (see Section 1.4.2).

1.4.2 Accidental introductions

There are many historical examples of accidental introductions, e.g. rodents via ships, weeds in fodder, and European woodborer in furniture. More recent examples include aquatic diseases introduced through ornamental fish and hyperparasites introduced with biological control agents. The impacts of such accidental introductions can rival those of deliberate introductions. For example, there are many documented examples of rodents causing at least local extinction of certain bird species.

Most accidental introductions enter via contaminants of commodities or organisms. They can also result from the deliberate introduction of another species upon which they are parasites, e.g. parasitic mites of bumblebees introduced into Japan for pollination purposes where the parasitic mites moved from the imported species to native species where they have had a negative impact (Goka *et al.* 2006).

1.4.2.1 Trade

Trade is the major pathway for short- and long-distance movement of small-to-large quantities of materials and goods. Over recent decades, there have been vast improvements in the capacity to rapidly move greater volumes, and an increased variety, of material. Organisms associated with trade goods no longer need to survive for as long a period as they previously needed to in order to arrive in a viable condition at a destination. Hence, there is a significant increase in the potential for trade to be a source of invasive species. As the primary objective of trade is profit or economic benefit, this often involves moving goods as fast as possible from source to outlet. This need is frequently in conflict with quarantine measures designed to lower the likelihood of entry of invasive species, because the measures may be seen to impede the speed of movement of goods. This can result in a tendency to avoid these measures. Therefore, there is a need to develop pragmatic quarantine measures, in conjunction with stakeholders, which are commensurate with the threats posed by the trade. If possible, these measures should be undertaken as close to the entry point into a country and as efficiently as possible.

If the quarantine measures are integrated into normal practices, there is a greater probability of a high level of compliance than if the measures cause change to normal practices. However, there are circumstances where changes to normal practices are unavoidable to achieve the required level of protection. It is important to remember that not only the goods themselves are potential pathways for invasive alien species, but also the conveyances by/on which they are transported, e.g. wood borers in packing crates.

1.4.2.2 Traditional movement of people and goods

Traditional movement of people and goods are pathways that have existed for a considerable time, particularly between countries with land borders. These pathways can allow the entry of invasive alien species, including both accidental introductions such as cattle diseases, and purposeful introductions such as food crops with weed potential. When establishing a quarantine system these movements should be given consideration but with an understanding that some may be difficult to manage. The objective is not necessarily the complete exclusion of traditional movements, which often encourages illegal trade. Rather, the need is to look at what goods (animals, foodstuff, plant material, and conveyances) are involved, the areas through which they pass, the risks involved, and the available management options. Communities should be engaged to increase awareness of the risks and develop appropriate quarantine measures to reduce the risks.

1.4.2.3 Emergency food, disaster relief, and development aid

Emergency food and disaster relief, and development aid, can be a source for the introduction of invasive species, e.g. stored product pests such as larger grain borer (*Prostephanus truncatus*), aquatic species, and weed seeds. Such aid often involves the rapid deployment of people, vehicles, goods, and products, and in many cases quarantine or preventative measures are intentionally or unintentionally ignored. Often there are few, if any, controls enforced when people, vehicles, and goods enter a country. Political interference can over-ride justified safety measures. Thus, the risks of the introduction of invasive species are often substantially increased during such a process. It is predicted that the need for food aid is likely to increase with the effects of global warming with a resulting increase in risk of invasions. The application of controls in source countries probably has the greatest potential to lower the likelihood of entry of invasive species into recipient countries.

1.4.3 Deliberate introductions

Deliberate introductions of alien species for any purpose, including pasture and genotype improvement, new crops, biological control, land rehabilitation (e.g. for erosion control or post-mining activities), leisure activities (e.g. gardening), the pet trade, hunting, research, agricultural or horticultural purposes can have wide reaching, and often unexpected, consequences. Hence, there should be careful consideration of impacts beyond those of the immediate focus of the introduction programme when undertaking a risk analysis before importation.

A large number of invasive species have been human assisted at least in the first instance, a process that has been going on for thousands of years. During the colonization of the New World, especially in the 1800s, there were proactive moves to introduce a greater range of species into new areas for food or utility, for ornament, as pets, or for hunting (acclimatization societies) to make the new lands 'feel like home'. Birds were often introduced to control insect outbreaks and later on, mammalian predators were introduced to control outbreaks of the previous 'useful' introductions. A few of those introduced to Africa, Australia, and New Zealand include common mynahs (Aves: Sturnidae: *Acridotheres tristis*), rabbits (Mammalia: *Oryctolagus cuniculus*), foxes (Mammalia: Canidae: *Vulpes vulpes*), stoats (Mammalia: Mustelidae: *Mustela erminea*), weasels (Mammalia: Mustelidae: *Mustela nivalis*), European starlings (Aves: Sturnidae: *Sturnus vulgaris*), sparrows (Aves: Passeridae: *Passer domesticus*), deer (Mammalia: Cervidae: *Cervus* spp.), lantana (Magnoliopsida: Verbenaceae: *Lantana camara*), prickly pear (Magnoliopsida: Cactaceae: *Opuntia* spp.), pasture grasses (Liliopsida: Poaceae), goats (Mammalia: Bovidae: *Capra hircus*), and pigs (Mammalia: Suidae: *Sus* spp.). Many of these introductions, especially food plants, have been critical to the new colonies. Some introductions, when undertaken with appropriate consideration of the potential off-target impacts, can be highly beneficial, such as biological control agents.

A major trap for those proposing to import new organisms is that most species are normally not considered invasive in their home range. When introduced to a new range, these same species, by adaptation to new hosts or niches, have resulted in drastically altered habits and ecosystems, including the extinction of some native species. The consequences of many of the above introductions still have major impacts today.

1.4.3.1 Biological control

Biological control usually involves the deliberate introduction and release of new organisms into areas, often in a repetitive manner. The objective of quarantine systems is to prevent the entry of new organisms that can have a negative impact and, as such, the release of biological control agents is contrary to most quarantine measures. Just because an organism is labelled as a biological control agent does not automatically mean that it will be safe or beneficial in all circumstances. Even though an organism may have been used successfully and safely as a biological control agent in a particular target area, it does not necessarily mean it will be safe in another ecological area. An example of this is with the highly successful use of the prickly pear moth, *Cactoblastis cactorum* in Australia to control prickly pear (Dodd 1940), where, even 70 years after initial release, it is still controlling prickly pear in that country. This same species of moth has accidentally entered the USA where it is an invasive pest (Hight *et al.* 2002; Vigueras and Portillo 2002; Zimmermann *et al.* 2002), and threatens species of cacti in their area of origin and livelihoods of subsistence farmers. This does not discount from the usefulness of this species as a safe biological control agent, provided appropriate non-target testing has been undertaken as well as consideration of the possible affects of this species in areas outside of the release areas, i.e. ability of this species to disperse (or be transported) to other geographic areas. Hence, when releasing biological control agents, if a country has land borders with another country or is in close proximity to another country, then potential impact of the biological agent in those areas should be considered.

1.4.3.2 Plant introductions

Plants have been introduced around the world in an attempt to improve productivity in many areas. However, significant numbers of these plants have either not been particularly useful or become invasive weeds. For example, in northern Australia, 463 pasture species of legumes and grasses were introduced for pasture improvement between 1947 and 1985; only four species proved useful with no invasive consequences, 17 species were proved useful as well as having weedy characteristics that caused problems for some sectors, and a further 60 species are considered invasive and as having no useful characteristics (Lonsdale 1994).[1] Other examples of species introduced for commercial reasons include pines, acacias, and eucalypts, which form the basis of commercial plantation industries in many countries. However, in some areas outside of their native ranges, these introduced species have become invasive. For example, in South Africa in climatically suitable areas all three groups have become invasive (Rouget *et al.* 2002).

Many ornamental plants that have been deliberately introduced into various areas around the world have become highly invasive. Some continue to be promoted in various areas as ornamental species. Part of the problem is that the combination of hardiness and attractiveness makes them desirable in horticulture because they require little maintenance to produce sometimes spectacular floral displays. One

---

[1] These findings eventually resulted in a weed risk assessment system for the introduction of plants into Australia (Pheloung *et al.* 1999).

such species is *Lantana camara*. This species originally came from Central and South America, it now occurs throughout most tropical and subtropical areas in the world (e.g. Asia, Australia, the pacific islands, Africa). *Lantana camara* has attractive flowers and thrives in a broad range of temperatures, soil types, and rainfall ranges. It is a major weed of disturbed areas including agriculture and areas of significant environmental value. It is tolerant of slashing and chemical controls and hence is difficult to control. It has been the subject of biological control efforts for many years; these have, at best, only achieved partial success. It continues to expand its range and become denser in areas where it already exists. It is difficult to eradicate and takes considerable persistence to do so, and vigilance to prevent reinfestation.

The recent trend for introducing biofuel crops raises the spectre of new weed invasions, especially onto marginal land in the tropics or the encouragement of the clearing of virgin forests to plant alien species, hence creating significant areas of ecological disturbance.

1.4.3.3 Smuggling

Deliberate illegal introductions can have significant consequences, not only from the species smuggled, but also other pests and diseases entering with the material. This can be because of ignorance of the consequences or deliberate avoidance of extremely strict quarantine regulations. Hence, to lower the likelihood of such instances there is a need to facilitate movement of material wherever possible, and where this is not possible to work with people and provide reasons and information as to why it is so. Therefore, the engagement of populace and good communication is critical to a robust quarantine system and indeed all aspects of biosecurity.

Another factor that complicates the issue of quarantine or biosecurity is the possibility of malicious or deliberate introduction of species in an attempt to create damage or fear e.g. the possible release of zoonotic diseases or widespread distribution of crop diseases. This risk should be considered when conducting the overall risk analyses of threats to a country.

1.5 Actions

The types of action that can be taken to reduce the risk of entry/establishment of invasive species fall into three broad categories. These are: pre-border actions—those actions taken outside a country or region; border actions; and post-border emergency actions. Each category has two main components—physical (e.g. infrastructure, materials, finance) and human (e.g. legislation, procedures, skills). It should be noted that to establish and maintain quarantine system takes the ongoing commitment of physical and human resources.

The objective of quarantine systems is usually to lower the risk or prevent the entry of identified unwanted organisms. When a quarantine system is developed, various aspects should be considered. These include understanding of the degree of natural biological isolation of the country, infrastructure capacity, the legal and political situation, available technical expertise, communication capacity, and personnel capabilities. There are generic similarities between most quarantine systems, but every quarantine system should be developed specifically for the particular circumstances that prevail within a country. In the case of plant quarantine systems, ISPM 20 (FAO 2004b) provides general guidance for the elements of an import system for plants—these are applicable to most systems and can be adapted to lower the risk of entry of invasive species. See Fig. 1.1 for a diagram of assessment and management of risks associated with invasive species

In quarantine systems, actions with regards to particular species can take place before the border (pre-entry), at the border (entry), or as a reaction to the detection of an invasive species (emergency actions).

1.5.1 Pre-entry

There is a need to identify the organisms or groups of organisms that pose risks and assess their potential impacts. This will enable appropriate guidance and contingency resources for detecting or controlling them should they enter or escape. When undertaking a risk analysis or assessment, issues that are useful to consider include those mentioned in Section 1.1, in particular the pathway/s via which an organism is most likely to arrive. Control of pathways of entry of invasive organisms provides the best opportunity to prevent the entry.

Risk mitigation measures prior to entry include: pre-export inspection; pre-export treatments; field treatments; selection of material from areas free of the invasive species or areas where there is low populations of the invasive species of concern; or treatment of goods that may provide a pathway for the target invasive species at discharge at the airport/port of entry.

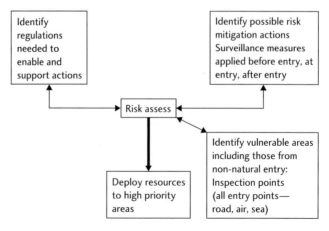

Fig. 1.1 Assessment and management of risks.

1.5.2 Entry (border)

Risk mitigation measures include: inspection of goods; checking that goods meet conditions of entry; treatment if required (heat/cold treatment, chemical treatment) including holding in a containment facility for observation; and either release or destruction or re-export if unable to meet conditions of entry. Quarantine facilities should be built to cope with the highest risk organism likely to be handled by the facility (Boxes 1.1 and 1.2). This may be a highly specialized facility or laboratory, or general facility that houses a wide range of organisms likely to be vectors for invasive species. Items possibly needing such facilities include, for example, agricultural crops such as cuttings, insect pathogen vectors (plants or animals), farm animals or pets carrying diseases or weed seeds, pets, and potential biological control agents.

Box 1.1 Notes on inspection facilities and processes

Inspection processes have inherent risks arising from various areas, but with careful management and competent inspections they can be minimized. Risks arise in several areas such as: lack of appropriate facilities to contain potential risks; poor staff knowledge and skills; poor procedures or lack of appropriate application of procedures such as opening sealed containers.

Actions to mitigate risks

Inspection facilities should be sited on appropriate terrain largely immune from natural disasters such as earthquakes. They should be located near the country entry point to minimize movement of material and the potential for escape. They also need to be as far as possible from suitable habitats/hosts such that in the event of an escape the likelihood of establishment is lowered.

Potentially invasive organisms (and risk goods) need to be examined in a secure area where they cannot escape. The quarantine containment rooms or laboratories need to be perceived, and function, as though biologically isolated from the surrounding country, with a system of physical and procedural barriers that lead out into the country. The physical isolation of the facility must be complemented and reinforced by operational procedures and systems established to run it.

Facility components

The floor plan of a facility and installation of equipment should make operational procedures flow smoothly and have efficient practices. Otherwise, staff may adopt practices that potentially compromise the integrity of the quarantine facility/system. The design and operation of the facility should focus on controlling pathways of biological organisms into and out of the facility (including by hitch-hiking on humans or materials used in the facility). This type of design often contradicts standard design and construction used for other types of buildings or facilities where the focus is often economy, aesthetics, and/or people-friendly functionality. To have a facility that operates well, it is imperative to have highly competent staff, otherwise there is a high likelihood of system failures. A minimum standard should be a fully enclosed room with solid floor, walls, and ceilings made of impervious materials and sealed joints, as well as with sealed or screened vents, and appropriately sealed water supply, drainage, waste disposal, and power supply.

The safeguards/measures used in a facility should be appropriate to its function and be independent from each other such that each safeguard/measure has a different mode of action. Hence if one fails, all safeguards are not compromised (e.g. a series of sealed doors with vestibules in-between and one with a light at one end to attract organisms back towards it). The degree of security needs to reflect the acceptable level of risk of material likely to be held. All procedures need to reflect a similar degree of security.

Particularly vulnerable areas of quarantine facility rooms/buildings need biologically-secure structural designs, and procedures include: specialized doors, non-opening windows, appropriate air vents and drains, and sealed light fittings. The materials of ceilings, floors, walls, and all joints should be impervious without crevices where organisms can lodge. All surfaces should be capable of being decontaminated. Facilities should have contingency procedures to contain escapes within the facility or emergency action taken if something is detected outside a facility in close proximity.

Traceability

Traceability is critical to control and rectification of problems. It requires the ability to track goods backwards to the point of origin or likely area(s) of contamination, and any subsequent points where the contaminated goods may have released unwanted organisms. This includes knowledge of the source and country of origin or production of the goods, where they entered the country, and their subsequent fate and destinations. If there is an incursion, this information can assist in enabling the implementation of containment controls, and in preventing further introductions. Components of tracking systems often have to have regulatory and administrative elements to ensure compliance.

Disposal of hazardous waste

Disposal of hazardous waste requires the ability to prevent a potential invasive species escaping. Hence the need to have a system that enables the safe disposal of quarantine wastes. It is preferable that the waste is not transported over long distances as this is a vulnerable point in the system and introduces factors that increase biosecurity/quarantine risk.

Procedures and work instructions

These should describe how to check identity of goods and organisms; check the exterior of conveyances for contamination; examine goods and organisms for contaminants, parasites, disease. They need to include information about all of the data collection required.

Biosecurity and quarantine for preventing invasive species | 13

Box 1.2 Some of the tools and equipment for general bench inspections

Inspections are undertaken in many situations. In general a large quantity of commercial goods (or a sample taken from a container of goods) or personal goods involve examination at the border on a bench or table. However, other situations include inspections in ship holds, aircraft holds, of animals in holding areas, or field inspections prior to entry. The following is a guide to tools and equipment to enable adequate bench inspections. Some of these tools and equipment are used in other forms of inspection—all require adequate lighting for inspection of goods.

- Inspection bench dimensions (for one-person work area): height 900mm, width 1 metre and length 2 metres; with an impervious, white surface. The bench should be sturdy, level, and stable, and have adequate fluorescent lighting along the entire length. The surface of the bench should lend itself to easy cleaning and sterilization where necessary. Power outlets should be located in close proximity to the inspection bench to enable the use of microscopes and supplementary lighting.
- Microscope (binocular dissecting microscope with a minimum magnification of 30×)—if possible mounted on a flexible arm to enable greater access to material to be inspected.
- Microscope light (preferable cold light—however, need to consider ability to obtain replacement lamps).
- Tweezers/forceps (fine).
- Specimen jars or tubes for collection of samples in a range of sizes e.g. 30mL, 60mL, 100mL with some of larger size—the jars or tubes need to be leak-proof and not be affected by the preservation fluid. There should also be the ability to apply some form of labelling; this could be comprised of a label written with alcohol (preservation fluid) -proof pen inserted into the tube with the specimens or an external label applied to the jar/tube such that it will remain attached without the writing being affected by the preservation fluid.
- Preserving medium (often 70% ethanol) with appropriate handling facilities if preserving fluids are toxic.
- Capacity to kill samples or specimens effectively and efficiently where necessary, such as access to a freezer (−18°C or below) or other methods.
- Sealable plastic bags for material preserved by sealing or needing to be sent off-site for further diagnostic examination or testing (it is preferable the material is a non-viable state).
- Brushes (fine bristle or camel-hair brushes size 00 for collecting very small organisms; larger size 15 or 20 for brushing down items to detect specimens).
- Hand lens/magnifier (for example, folding 10× and 20×).

- Magilamp (magnifying lens at least 4× with associated lighting) on adjustable arm.
- Sheets of white paper (for disposal after each inspection) or large white trays on which to work.
- Large flat, white inspection trays.

1.5.3 Emergency actions

Knowing how to respond if an invasive species is detected can minimize the impact that the species has on an area and maximize the potential to control or eradicate it. Hence the development of *contingency response plans* should be part of the overall risk assessment process. Contingency plans can be developed as overall generic plans with supporting documents for specific situations.

Organisms are usually detected after the national border via systems of inspection (including regular inspection of fields by farmers), surveillance, monitoring, or sample testing. Systems need to be in place to enable effective and efficient reporting of the incident to the persons responsible for initiating control actions. The options that are available on detection of an invasive species are: emergency management to prevent establishment; containment; eradication; or no action.

It is useful to undertake full-scale exercises/simulations on a theoretical or potentially real invasive species threat to: test the ability of the functionality of the components of a system; train personnel as to their roles and responsibilities in the event of a real detection of a pest/invasive species; and make necessary adjustments should they be found not to be effective or practical. The overall management of an incident must focus on the outcome(s) and ensure that that procedures and processes support the effective and efficient achievement of that outcome. To this end, a response plan should contain an explanation of why these actions are needed and the goals that are desirable to achieve. It should identify the individuals responsible for particular activities, and the regulations that enable these activities. It should clearly state at what point the response plan will be activated. The response plan should also:

- Detail the scope of the actions, work plan, and lines of communication and expected interactions between whom, and when, or at what points in a incident.
- Detail the hierarchy of reporting, and information flow, and define the types of documentation needed through all stages of the incursion.
- Identify the resources and equipment required to execute the plan and who/which organizations are to provide them (this includes technical advice and support).
- Define the types and numbers of personnel needed and allow for appropriate rotation of personnel to avoid fatigue.

Biosecurity and quarantine for preventing invasive species | 15

- Include the training schedules for all staff, and records of the training undertaken.
- Stakeholders need to be identified and their responsibilities stated where appropriate.
- As the response progresses a large amount of information will accumulate and the plan needs to detail its management.
- There needs to be mechanisms for review and assessment of the plan, both during an incident and following an incident. The plan itself should state how and when changes are to be implemented.

A minimum set of (at least generic) equipment should always be accessible and available—otherwise delays to the initiation of actions may occur.

Administrative support for emergency actions is essential for efficient functioning of the plan and it is useful to identify the following: who is going to provide the administrative support for the process; what degree of confidentiality is required of the personnel involved in response; what potential conflicts of interest could arise. As part of the overall plan it should contain strategies of how, when, and why stand down of actions should occur. Triggers for stand down could include:

- Eradication of the invasive species has been achieved as per pre-agreed definition of level on non-detection.
- A move into a containment phase.
- Or, if neither eradication nor containment are achievable with resources available, then move to management of the invasive organism as it spreads.

Such plans should be developed taking into account a country's or authority's capacity to take actions, including the ability to make and enforce exclusion zones; carry out treatments, and elimination or species—these could be because of legal limitations, resource limitations, or physical (environmental) barriers to undertaking actions. There is no single way of doing this, and authority to undertake actions can vary between states, provinces, or regions within national borders.

An example that provides guidance on overall structure of documentation and possible processes can be found in AS ISO 10013–2003 (Australian Standard on Guidelines for quality management system documentation). The plans that have been based on this structure are Aquavet (http://www.daff.gov.au), Plant plan (http://www.planthealthaustralia.gov.au), and Ausvetplan (http://www.animalhealthaustralia.gov.au). The following is a brief summary of stages of a response:

- Investigation phase—determine if there is a possible problem.
- Alert phase—provide concerned parties with information that a potential situation exists.
- Activation—initial meeting/communication of all decision makers as per contingency plan; summaries of information of current situation; prognosis for immediate and long term provided; clarification of objectives of response plan; execution of actions as per plan; coordination of plan; communication of incident.
- Stand down—response concluded as per plan, review of plan, incident, and actions.

1.6 Summary

Organisms can be introduced into an area many times, but never become established or invasive, yet for others a single event is all that is needed. The establishment of an invasive species is therefore the end point of a combination of many coinciding events and conditions.

Islands that are isolated by significant stretches of water (sea) and areas isolated by significant biological or geographic barriers, such as deserts, have distinct advantages in their ability to control the entry of terrestrial invasive organisms. Under such circumstances the prevention of entry of invasive species can be a realistic option. However, where there is limited or no biological or geographic isolation then to have a system totally focused on the prevention of the entry of invasive species is fraught with practical difficulties. Under such circumstances, prevention in combination with other strategies, such as very good early detection and management systems, are more likely to lead to a more robust system that lowers the number of invasive species establishing and becoming problematical. This combination of strategies implemented in both non-biologically isolated and biologically isolated areas will provide the greatest opportunity to minimize the impacts of species that do enter. Measures that involve early detection, with contingency plans for eradication/containment of invasive species that can be efficiently implemented, are more likely to have a higher cost/benefit outcome.

The same situation exists for aquatic or marine invasive species—where there are significant biological barriers or isolation, there is the potential to develop workable quarantine measures. However, complicating factors in aquatic systems are that there is little biological separation when water is the vector and by practices such as the discharge of ballast water by vessels and the fouling of hulls of vessels. A history of invasiveness and patterns for specific species in the Pacific region can be found on the PIERS database for the Pacific region (http://www.hear.org/pier/index.html).

Invasive species occur in most taxa, a few examples of species in various taxa and broad categories of introduction are given in this chapter. Table 1.1 summarizes general problematic organisms associated with broad categories.

Table 1.1 Summary of potential quarantine issues associated with organisms.

Organism type	State	Potential quarantine/invasive issues
Terrestrial plants	Live	• Invertebrate pests • Diseases • Weed potential • (Soil associated with plants carry further risks)
	Dried foliar (non-viable)	• Disease
	Dried stem (or thicker plant parts)—non-viable	• Diseases • Invertebrate pests

Table 1.1 (Con't.)

Organism type	State	Potential quarantine/invasive issues
Aquatic plants	Seeds	• Weed potential • Diseases • Invertebrate pests—particularly insects
	Live	• Invertebrate pests • Diseases • Weed potential
	Dried foliar (non-viable)	• Diseases
	Dried stem (or thicker plant parts) non-viable	• Diseases • Invertebrate pests
	Seeds	• Diseases • Invertebrate pests • Weed potential
Terrestrial and aquatic invertebrates	Live	• Pest potential • Diseases • Parasites
	Dried/dead (non-viable)	• Stored product pests • Diseases
Terrestrial and aquatic vertebrates	Live	• Pest potential • Diseases • Parasites
	Dried/dead (non-viable)	• Diseases • Stored product pests
Fungi	Live	• Invertebrate pests • Parasites (other fungi and diseases) • Spores—propagative material—invasiveness potential of species • Hyperparasites
	Dried (non-viable)	• Invertebrate pests • Parasites (other fungi and diseases) • Spores—propagative material—invasiveness potential of species
Biological control agents		• Potentially all types of organisms—hence issues associated with all other organism
Bacteria		• Contaminants
Viruses		• Contaminants

1.7 Acknowledgements

Comments and discussions with various colleagues, including Mike Grimm, Mike Robbins, Robert Langlands, Bill Crowe, Paul Pheloung, Bill Magee, and Tony Callan were much appreciated.

2
Risk assessment of invasive species

Thomas J. Stohlgren and Catherine S. Jarnevich

2.1 Introduction

Risk assessments have long been used for the analysis of human health risks associated with chemical contaminants and other hazards (National Academy of Sciences 1983). For chemical hazards, risk assessment has been defined as 'A set of formal scientific methods for establishing the probabilities and magnitudes of undesired effects resulting from the release of chemicals. Risk assessment includes quantitative determination of both exposure and effects' (Society of Environmental Toxicology and Chemistry 1987). Humans often were the target species of concern. Assessments were typically restricted to hazard identification, dose-response assessments, exposure assessments, and human health risk characterization. The effect of the pesticide DDT on a variety of bird species is a classic example of risk assessment (Ratcliff 1967).

Risk assessment for biological invasions is somewhat similar to those other types of hazards. For example, evaluating chemical spills requires basic information on where a spill occurred; exposure level and toxicity of the chemical agent; knowledge of the physical processes involved in its rate and direction of spread; and potential impacts to the environment, economy, and human health relative to containment costs. However, unlike typical chemical spills, biological invasions can have long lag times from introduction and establishment to successful invasion, they reproduce, and they can spread rapidly by physical and biological processes. We can view potentially harmful, non-native species (i.e. species foreign to the ecosystem in which they are now found) as biological hazards (Stohlgren and Schnase 2006). And, borrowing from the physical sciences (Society of Environmental Toxicology and Chemistry 1987), we can define risk assessment of invasive species as 'A set of formal scientific methods for establishing the probabilities and magnitudes of undesired effects resulting from the introduction of non-native (or non-indigenous) biological organisms. Risk assessment includes quantitative determination of the current and potential abundance and distribution of the organisms and their economic, environmental, and human-health effects'.

2.1.1 Why do we need a formal approach to invasive species risk assessment?

We need a formal approach to invasive species risk assessment because of the rapidly growing costs associated with harmful invasive plants, animals, and diseases around the globe. Pimentel *et al.* (2005) estimate that in the United States alone, the economic costs associated with invasive organisms exceeds $120 billion/year in lost production, maintenance, eradication efforts, and direct health costs. These same authors suggest that up to 80% of endangered species worldwide could be adversely affected by competition or predation by invasive species. For example, the introduction of the brown tree snake in Guam led to the direct extinction of a dozen species of birds (Jaffe 1994). Meanwhile, the costs to human health are obvious from the notorious examples of plague, West Nile virus, and the potential effects of Asian bird flu.

Thus, risk assessment for invasive species may expand the number of target species being considered to any and all species in an ecosystem, including humans. Typical target species, in addition to humans and charismatic animal species, include threatened and endangered species, rare native species assemblages, and selected ecosystem processes such as competition and predation (e.g. Connell 1983), ecosystem services (Gross 2006), and fire frequency and intensity (Freeman *et al.* 2007), as seen by the increase in wildfires aided by non-native annual grasses in the western United States.

Formal approaches to ecological risk assessment are not new. In the 1990s, assessments of 'ecological risks' expanded data requirements for complete and accurate risk analyses by recognizing the inherent complexity of ecosystems. The stressors to ecosystems have grown to include climate change, genetically modified organisms, disturbance, and natural disasters such as earthquakes, floods, and wildfires. This led Lipton *et al.* (1993) to suggest that information is needed on 'the biotic components and organization of the system, as well as assessing the distribution of the stressor within biotic components' including 'risk cascades' and 'biological, ecological, and societal relevance'. Despite these general, well-recognized needs, specific strategies, methods, and the costs and difficulty of acquiring detailed information on all relevant ecosystem components and processes relative to complex stressors such as multiple air or water pollutants or climate change, remain elusive (Stohlgren and Schnase 2006).

In this century, the challenges of risk assessment must take another astronomical leap as consideration extends from abiotic, chemical, and climatic threats to invasive non-native organisms. There are thousands of species of plants, animals, and diseases that have invaded the United States from other continents—species that can cause harm to the environment, our economy, and to human health (Mack *et al.* 2000). Notorious examples in the United States include zebra mussels, cheatgrass, West Nile virus, the brown treesnake, plague, kudzu, salt cedar, the Argentine fire ant, yellow star thistle, sudden oak death, hydrilla, Burmese pythons, and Dutch elm disease, to name a few. It is discomforting that no county in the United States is free of invasive species (Stohlgren *et al.* 2006), and more are arriving all the time (see http://www.invasivespecies.gov). Thus, a formal approach to risk assessment is essential to detect, evaluate the spread and effects, respond to, and monitor harmful invasive species.

2.1.2 Current state of risk assessment for biological invaders

We conducted a cursory review of recent literature that is not exhaustive, but rather symptomatic of the types of risk analysis approaches that readers are likely to find. Many current 'case studies' of risk analysis for invasive species fall into three categories:

1) Species-specific risk assessments;
2) Habitat-specific risk assessments; or
3) Species and habitat risk assessments.

One of many possible examples of species-specific risk assessments includes predicting the introduction of West Nile virus to the Galápagos Islands (Kilpatrick *et al.* 2006). The authors devised a predictive model for the virus by evaluating the likely 'pathways' such as avian migration, transportation of day-old chickens, infected humans, mosquitoes in cargo containers, etc., to assess risk. The probabilities of spreading the disease by various pathways were estimated to develop prevention strategies. In another example, Pemberton and Cordo (2001) evaluated the risk of biological control on *Cactoblastis cactorum* in an attempt to control the escaped cactus moth that decimates Opuntia cacti. The release of biological control organisms also assumes elements of risk.

In other cases, several species are evaluated simultaneously to rank species for prevention, screening, or early detection and control. For example, Tassin *et al.* (2006) ranked 26 of 318 introduced woody species on Réunion Island in the Indian Ocean as more serious invaders, based on historical records.

Habitat risk assessments attempt to describe and map the suitability or vulnerability of various habitats to invasion. In this way, early detection efforts might be guided to a subset of potential habitats. For example, riparian zones and mesic habitats appear more prone to invasion than xeric habitats in arid landscapes (Stohlgren *et al.* 1998).

Risk analysis also must consider limits to, and the connectivity of, the potential habitat of organisms. For example, Bossenbroek *et al.* (2001) developed a deterministic model to estimate zebra mussel (*Dreissena polymorpha*) distributions using a distance coefficient (i.e. connectivity), Great Lakes boat-ramp attractiveness, and colonization potential. This 'gravity model' constrained the risk assessment to successfully forecast zebra mussel dispersal into inland lakes of Illinois, Indiana, Michigan, and Wisconsin. Thus, species and habitat risk assessments can be combined to set priorities for management based on species traits and habitat characteristics (Chong *et al.* 2006), greatly narrowing the number of species considered and confining assessments to priority habitats, thereby reducing the costs of managing invasive species.

Significant hurdles remain in assessing risks for multiple species and multiple habitats and regions. Holt *et al.* (2006) provide a quantitative approach to averaging scores based on establishment potential, host range, dispersal potential, economic impact, environmental impact, and pathway potential (and trading partners), where categorical scores from one to three are subjectively determined for each factor. Then, the upper limits and lower limits are considered along with average rankings across species. Obviously, such determinations are greatly affected by the completeness and accuracy of the information about each species.

2.1.3 The ultimate risk assessment challenge

A tremendous challenge remains in documenting, mapping, and predicting the establishment and spread of biological organisms in space and time (Chong *et al.* 2001; Schnase *et al.* 2002b). Imagine the often difficult case of predicting generally large chemical spills, collecting basic information on where a spill occurred, the toxicity and amount of the chemical, knowledge of physical dispersion processes involved in the rate and direction of spread, and the potential impacts and costs to the environment, economy, and human health relative to containment costs. Now, imagine the difficulties in detecting the initial establishment of tiny, often cryptic organisms that can have long lag times from introduction and establishment to successful invasion; that can reproduce and spread rapidly by physical and biological processes; and that may have leap-frog like re-introductions by human transportation and trade. Many species that arrive in a country are intentionally introduced (via seed trade, horticulture, released unwanted pets, etc.). For example, 50% of the alien plant species in China were intentionally introduced as pasture, food or forage, ornamental plants, textile, or medicinal plants, while 25% of alien invasive animals were intentionally introduced (Xu *et al.* 2006a). In addition, many species are introduced unintentionally as 'hitchhikers' (i.e., pathogens, ballast water species).

Because there is a continuing threat of intentional and unintentional introductions of harmful species, a formal approach to risk assessment is needed to:

- Guide prevention and screening efforts;
- Plan an early detection and rapid response programme;
- Help select priority species and priority areas from control and restoration; and
- Iteratively improve and integrate all these aspects of invasive species management to reduce the costs and effects of current and future invasions.

We outline the general components of risk assessment for invasive species, along with examples from the field and laboratory.

2.2 Components of risk assessment for invasive species

What do we need to know to understand, estimate, and predict the risks associated with invasive biological organisms? Risk assessment for biological organisms requires information on the invading species, vulnerability of habitats to invasion, modelled information on current and potential distributions, and the costs associated with containing (or failing to contain) harmful species (Table 2.1).

We summarize information needs and a general strategy for risk assessment for invasive species based on the ultimate risk assessment challenge. The most important aspect of the process is that it is an iterative approach, improving risk assessments as new information and modelling become available. The initial process requires:

- Detailed information on invading species traits.
- Matching those traits to suitable habitats for the invading species.
- Estimating exposure (or propagule pressure).
- Surveys of current distribution and abundance.
- An understanding of data completeness.
- Estimates of the 'potential' distribution and abundance of the species.
- Estimates of the potential rate of spread (and pathways, corridors, and barriers to invasion).
- The probable risks, impacts, and costs of the invading species to the environment, economy, and human health.
- The containment potential, costs, and opportunity costs.
- The legal mandates and social considerations in containing and controlling the species (Table 2.1).

2.2.1 Information on species traits

Some species are better invaders than others, and classifying potentially harmful species is our first difficult task in risk assessment of biological hazards. Plant biologists have long tried to identify an 'ideal' invader species based on traits of successfully colonizing species (Table 2.2) (Bazzaz 1986; Roy 1990; Thompson et al. 1995). For example, Grotkopp and Rejmánek (2007) clearly showed that high seedling relative growth rate and specific leaf area are traits of invasive woody angiosperm species in Mediterranean climates. Many general traits (Baker 1965; Lodge 1993) and strategies (Grime 1974; Newsome and Noble 1986) are associated somewhat with increased invasion potential (Table 2.2), but an exclusive set of invader traits has not emerged (Newsome and Noble 1986; Crawley 1987; Roy 1990), hampering the ability to predict responses of individual species (Hobbs and Humphries 1995; Reichard and Hamilton 1997; Lee 2001). Sometimes, species, taxonomic, and behavioural traits do help identify and rank invaders (Panetta and Mitchell 1991; Lee 2001). This is based on observations that particular species in

Table 2.1 Information needed for risk assessment for invasive species (adapted from Stohlgren and Schnase 2006), and selected examples.

Information needed	Description or examples
1. Information on species traits	Propagule size, number, and mode of dispersal (e.g. wind, water current, animal-assisted); growth rate, age to reproduction, competitive capabilities, predation prowess, etc.
2. Matching species traits to suitable habitats	Affinity to disturbed areas and per cent disturbed areas in a county, available phosphorus in lakes and aquatic plant productivity, agricultural (corn) fields and non-indigenous bird abundance.
3. Estimating exposure	The number, distribution, and virility of propagules, or the frequency and intensity of their arrival by various pathways.
4. Surveys of current distribution and abundance	Current presence and abundance, and 'absence' in areas; museum records; known barriers and corridors to invasion, pathways of invasion.
5. Understanding of data completeness	The geographical extent of the data; sampling intensity; temporal extent of surveys and monitoring; data on associated species and habitats; key gaps in information.
6. Estimates of the 'potential' distribution and abundance	Usually modelled from point or polygon data from a few known locations extrapolated to larger, un-surveyed (or poorly surveyed) areas. These models should be presented with a quantified level of uncertainty.
7. Estimates of the potential rate of spread	Rates of spread may not be simple dispersion models given complex pathways by wind, birds, human trade and transportation. Spread may be multi-directional, with areas of contraction and extirpation.
8. Probable risks, impacts, and costs	Economic costs (containment costs and opportunity costs), environmental costs (competition with or predation of native species, altered ecosystem services, or disturbance regimes), and costs to human health.
9. Containment potential, costs, and opportunity costs	Some species are more easily contained than others. Long-term restoration costs must be considered. Actual costs include the 'costs of doing nothing' and opportunity costs associated with selecting some species for control while others continue invading and spreading.
10. Legal mandates and social considerations	Legal directives may restrict choices for invasive species management. Likewise, social considerations affect management decisions.

Table 2.2 Some general traits of successful invaders adapted and summarized from the studies cited in the text.

Trait	Example
1. Exceptional dispersal characteristics (e.g. by wing, water, animals, zoospores, pelagic stages, etc.)	Wind-blown seed of dandelions, many bird species carrying West Nile virus.
2. Rapid establishment and growth to reproductive age	Mediterranean woody angiosperms, annual grasses in California, New Zealand mud snail.
3. Few natural enemies or predators in the new environment	Mongoose in Hawaii, brown treesnake in Guam.
4. Ability to sequester underused resources	Shade-tolerant Japanese honeysuckle, zebra mussels.
5. Copious reproduction	All organisms mentioned above.

selected regions have had more predictive success. Species' life history traits also are important determinants of invasion potential (Rejmánek 1996; Rejmánek and Richardson 1996; Reichard and Hamilton 1997).

Unfortunately, obscure species traits may be particularly important for some invaders. For example, European wild oats has awns that self-bury, allowing greater resilience to wildfire with a plentiful seed bank. Plant pathogens such as white pine blister rust had the plasticity to find alternate hosts and target species after arrival in the United States in the early 1900s. Thus, there are many exceptions to the generalizations in Table 2.2. Not all invaders have all the successful traits, and some species have many of the successful traits, but are not good invaders, at least not yet! Residence time (i.e. the time since the species was introduced) may be an important 'attribute' of a species indirectly affecting genetic diversity or adaptations of the species (Wilson *et al.* 2007).

2.2.2 Matching species traits to suitable habitats

All species require suitable habitats to establish, reproduce, and spread (Fox and Fox 1986; Panetta and Mitchell 1991; Hobbs and Huenneke 1992). So, invasion also depends on environmental characteristics that may predispose a habitat to invasion (Table 2.3) (Tyser 1992; Robinson *et al.* 1995; Lee 2001). As with species characteristics, generalizations of habitat vulnerability to invasion have also been slow to emerge (Usher 1988; Lodge 1993; Lonsdale 1999).

The quantity and quality of available resources also may be important in assessing the vulnerability of an ecosystem to invasion. In some cases, an invading species may take advantage of under-used resources in an ecosystem. For example, *Bromus tectorum* (cheatgrass) in some regions is said to benefit from fall or winter precipitation, while many native plant species are senescent (Bates *et al.* 2006). In addition,

Table 2.3 Some general characteristics of successfully invaded habitats and poorly invaded habitats.

Characteristic of successfully invaded habitats	Examples
1. Low diversity habitats with available resources	Islands in proximity to the mainland (Elton 1958; Rejmanek 1996).
2. High diversity habitats with available resources	Local areas of regions with high light, soil nutrients, water, and warm temperatures (Stohlgren et al. 1999, 2005, 2006).
3. Disturbed habitats	Post-fire habitats (Fox and Fox 1986).
Characteristic of poorly invaded habitats	**Examples**
1. Areas with a limiting resource	Some tropical forests (low light in the understory), some mangrove forest (anoxic soil conditions).
2. Areas with extreme climates	Extreme deserts, tundra, deep oceans.
3. Areas with atypical environments	Serpentine outcrops, hot springs.

cheatgrass invasion is strongly linked to fire regimes (Keeley and McGinnis 2007). This species illustrates how temporal changes in resource availability may be very important to invasion success (Davis et al. 2000).

Thus, identifying invasive species hazards requires an understanding of the receptor ecosystem (genotypes, species, populations, resource availability, and disturbance regime), and information on the invading species' traits (Table 2.2). Invasion is possible only when a vulnerable habitat meets with a species whose traits allow for establishment, growth, and spread (although lag times between introduction and spread are common; Mack et al. 2000).

Suitable habitats might be assessed with climate matching approaches (Chicoine et al. 1985; Panetta and Mitchell 1991; Venevski and Veneskaia 2003). Climate matching requires knowledge of the climatic conditions in the original home range of the non-indigenous species and the abundance and distribution of the species (or genotypes) throughout its range (Morisette et al. 2006). However, many non-indigenous plant species are found in higher and lower latitudes than the same species in its home range, suggesting a possibility of an expanded range in the receptor country (Rejmánek 1996). This pattern may be due to many interacting forces (reduced competition, predators, or pathogens) in the receptor country, greater dispersal (perhaps aided by more wind or birds), or different levels of disturbance in the newly invaded country (Fox and Fox 1986; Hobbs and Huenneke 1992; Burke and Grime 1996) or flooding (DeFerrari and Naiman 1994; Planty-Tabacchi et al. 1996).

Risk assessment is especially challenging in natural ecosystems since the interactions of many species and ecosystem processes are poorly understood, quantified, and mapped at a high enough resolution to effectively manage patches or small populations of invaders. Habitats are usually classified and mapped based on a few dominant species, regional climate factors, or a few environmental gradients (e.g. precipitation, temperature, water depth or pH), so we have little knowledge of the distributions and abundance of most species that respond to micro-habitats that may span several coarse-scale vegetation classifications.

We recognize that matching species traits (Table 2.2) to micro-environments over large areas of potential invasion will not be easy. Wainger and King (2001) found that only two of 13 invasion assessment methods incorporated species traits and habitat characteristics into the decision analysis. Perhaps because many species possess some or all invasive traits (Table 2.2), many ecologists are focusing on a habitat approach to understand invasion patterns (Williamson and Fitter 1996b, Davis *et al.* 2000, Stohlgren *et al.* 2002).

Given the challenges above, an important step in risk assessment of invasive species is creating a tractable problem from a seemingly intractable one (Lee 2001; Chong *et al.* 2006). For example, Chong *et al.* (2006) evaluated 34 non-native species found in 142 plots of 0.1ha in 14 vegetation types within the Grand Staircase–Escalante National Monument, Utah. A species invasive index, based on frequency, cover, and number of vegetation types invaded, showed that only seven of 34 plant species were highly invasive (Fig. 2.1). This analysis reduced the number of species needing further assessments. Furthermore, a plot invasion index, based on non-native species richness and cover, showed that only 16 of 142 plots were heavily invaded. Modelling non-native species richness reduced the risk assessment to a subset of habitats in the Monument (Fig. 2.2). Together, the study showed that even with such a modest sampling intensity (<0.1% of the landscape), managers could quickly create a tractable problem from a seemingly hopeless task in the 850,000 ha Monument (Chong *et al.* 2006).

However, it is also clear that the invasion process may be as 'individualistic' (Crawley 1987; Hobbs and Humphries 1995) as the species themselves or the habitats they invade. There may be species-specific 'invasion windows' in time and space (Johnstone 1986; Mack *et al.* 2000). New invasions occur and species adapt and spread over time. Thus, an iterative process is needed, making improvements to risk assessment immediately as new information becomes available (Stohlgren and Schnase 2006).

2.2.3 Estimating exposure

Estimating exposure requires detailed information on propagule pressure (the number of organisms invading), the viability of the organisms, and the habitat susceptibility to invasion. Even after an 'invasion window' opens, exposure assessments will be difficult for moving organisms. General pathways may be clearly identified, but very poorly quantified. For example, many aquatic organisms have arrived in

28 | Invasive species management

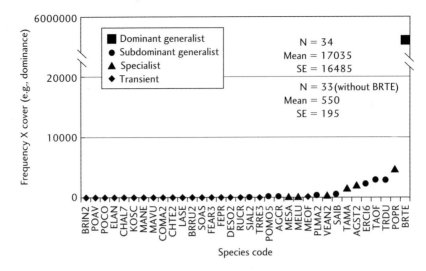

Fig. 2.1 An example of selecting priority species of non-native plants in the Grand Staircase–Escalante National Monument, Utah, based on frequency and foliar cover data from 351 plot locations in the Monument. BRTE, *Bromus tectorum* (cheatgrass), was the most dominant weed, followed by POPR, *Poa pratensis* (Kentucky bluegrass), and TRDU, *Tragapogon dubius* (western goatsbeard).

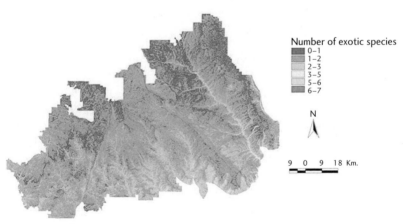

R^2 based on OLS and regression trees = 42%

Fig. 2.2 Modelled distribution of exotic (non-native) plant species richness based on 303 vegetation plots (1000 m² plots with minimum mapping unit of 10 m²) in the Grand Staircase–Escalante National Monument (GSENM), Utah, showing 'hot spots' of invasion along riparian zones in the 850,000 ha Monument.

estuaries, rivers, and lakes from ballast water, with small organisms (larvae, eggs, pelagic stages, etc.) being stored on board in the ships home port before transportation and release in a receptor port (Lodge 1993). Still, even rough estimates of the abundance, viability, and condition of arriving organisms are unknown—for invasive or less-invasive species. Likewise, small seeds of non-native annual weeds contaminate native forage and crops. It is very costly to examine and purify every large bag of seeds. Shipping manifests rarely describe organic hitchhikers in sufficient detail to accurately assess exposure.

The quantification of 'propagule pressure' for biological organisms remains elusive in risk assessment. For aquatic organisms (e.g. invasive fish, zooplankton), propagule pressure might be quantified as the number and viability of reproductive units arriving at a given location (Lodge 1993). However, many invasive species do not have large, obvious, easily counted propagules, and quantifying propagule pressure over large areas is problematic. Many pathways and corridors to invasion are poorly understood. Corridors may include the matrix of roads and riparian zones which may facilitate the spread of invasive riparian plants such as purple loosestrife or tamarisk. Railroads also are linear, disturbed, habitats of invasion for many non-native plants species. Can any country say how many seeds, spores, and pelagic stages arrive undetected? How many establish in each habitat? As difficult as this task seems, some estimates are possible for some species (based on trade and transportation volumes and patterns, surveys, and rudimentary models). We need more 'practice' estimating exposure.

2.2.4 Surveys of current distribution and abundance

Surveys in the early stages of most invasions are made difficult by small population size, patchy distributions, and the cryptic nature of many initially-rare species in complex landscapes and waterways. Cost and efficiency of information gain are major considerations because only a small percentage of any area can be affordably surveyed. Completely random survey techniques may be unlikely to detect new cryptic invaders, especially if costs constrain sampling intensity and completeness. However, probabilistic, iterative sampling (e.g. surveys guided to particularly vulnerable habitats) may be an important tool in risk assessments (Crosier and Stohlgren 2004; Stohlgren and Schnase 2006).

The first step in risk surveys is augmenting initial opportunist or subjective survey information with more systematic, less-biased, and more comprehensive surveys in an iterative approach (Fig. 2.3). In the initial phase, only a few established individuals or populations are known to investigators. They may add a few other observations nearby in similar habitats (highly probable strata) with some locations in less-probable habitats to get a conceptual, first-approximation model of the species distribution and abundance (i.e. a species presence and absence model and map). Upon the initial sightings, the proper authorities are alerted for rapid response, containment, and restoration efforts. Species affinities to habitat types are noted, as are information gaps such as un-surveyed habitat types or areas.

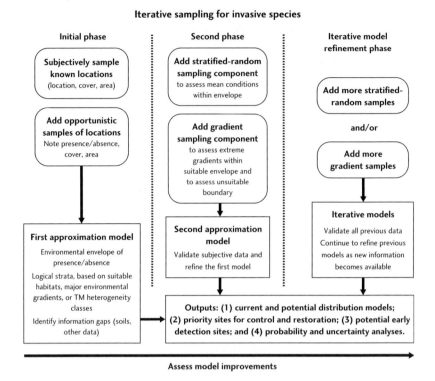

Fig. 2.3 An iterative sampling approach for documenting, mapping, and predicting the abundance, distribution, and spread of invasive species.

The second phase of surveys integrates unbiased stratified-random sampling with gradient sampling designs for robust spatial statistical models (Fig. 2.3). This provides much needed information on the probability of occurrence in different habitat types, and preliminary information on the environmental tolerances of the target species. Note that the actual current and potential distributions of a species are very difficult to determine from limited surveys—but it's a start. Statistical and spatial interpolation models based on stratified random and gradient analysis techniques allow for a second approximation of species' distribution and abundance. New survey data from opportunistic sampling, stratified random sampling, and gradient sampling further improve and validate the distribution maps over time. Since species migrate, adapt, hybridize, expand, and contract in population size, risk analysis surveys for invasive species must be an iterative process (Fig. 2.3).

2.2.5 Understanding of data completeness

Another critical feature of risk analysis for invasive species is an understanding of the taxonomic, geographic, and temporal completeness of data in the region of concern. Most biotic inventories in natural areas are woefully incomplete (Stohlgren *et al.* 1995). In most areas, only one or a few species or genotypes might have been surveyed (low taxonomic completeness; Crall *et al.* 2006). Within a landscape, county, or region, large areas may be poorly surveyed for target species (low geographic completeness). In addition, large areas of a site or region may have been surveyed only once or very few times (low temporal completeness). Maps of the distribution or abundance of invasive species should be accompanied by information on the various levels of completeness throughout the study area so that managers would understand the limitations of the maps for risk assessment.

2.2.6 Estimates of the 'potential' distribution and abundance

Estimating the potential distribution and abundance of an invasive species requires the information needs outlined above (Table 2.1; needs 1–5), integrated with a geographic information system (GIS), remote sensing, and fairly sophisticated mathematical models. In many cases, the resulting product is a 'habitat suitability map', a map or model that describes the vulnerability of habitat to target species' invasion (e.g. Chong *et al.* 2001; Schnase *et al.* 2002b; Venevski and Veneskaia 2003). These models are generally based on a few climate, topographic, or soil variables. Maps and models of target species' abundance are rare because abiotic and biotic factors must be carefully quantified relative to species population estimates, and because of the plasticity and adaptive potential of the target species and genotypes. Patterns of habitat invasibility have been slow to come, let alone mechanisms explaining these patterns (Mack *et al.* 2000), and the complexity of this task should not be underestimated.

It is well recognized that 'potential distribution' models have several limitations. They are based on only a few predictive factors and are affected by the selection of the factors, scale, resolution, and accuracy of spatial data inputs (Morisette *et al.* 2006). The models do not include information on more than one biological species (the target organism), so they do not include the complex of interspecies interactions (e.g. competition, herbivory, predation). The environmental factors in the models are all held constant, and the local disturbances (e.g. fires, floods) and processes such as grazing and mortality of competing species are generally presumed constant, which is rarely or never the case. Species–habitat relationships and species mapping (except for humans) is in its infancy. Yet, developing these capabilities is paramount to the next difficult challenge in risk assessments of invasive organisms—predicting rates of spread of invasive species.

2.2.7 Estimates of the potential rate of spread

Mathematical models predicting the spread of invasive species are essential. Managers may want to set priorities for control based partly on the priority pathways of spread, the area of potential habitat, the effects of the species throughout its range, and rate at which the species could spread from its current distribution to its full potential distribution (Pemberton and Cordo 2001). There are many models being explored for predicting rates of spread from simple dispersion or deterministic spatial models to stochastic models (see Hastings *et al.* 2005 for review).

Spread models may have more limitations than potential distribution models because they often are heavily dependent on complete information on the distribution and abundance of the target species, and the predictions of the establishment, growth, reproduction, and migration of metapopulations in complex environments. Moderately sessile organisms such as plants might provide simple cases to begin developing estimates of species spread. There may be a link between establishment success and invasion success (i.e. frequency and abundance) for many species. However, accurate monitoring of the distribution, abundance, and spread of metapopulations, species, and genotypes remain rare in the ecological literature (e.g. Harrison 1991).

2.2.8 Probable risks, impacts, and costs

The costs associated with invading species may be environmental, economic, or impacts to human health. Assessing environmental risks includes potential losses or declining populations (or genotypes) of native species, declines in ecosystem services (e.g. water quality, fire prevention), or undesirable effects on ecosystems processes (e.g. increased fire frequency). There are many examples of native species declines including the effects of Dutch elm disease caused by the fungus (*Ophiostoma ulmi*) on elm trees (*Ulmas* spp.) or the loss of native populations of *Phragmites* due to invasive non-native genotypes of same species. About 42% of the species listed on the United States Threatened and Endangered species list, are listed because of threats from non-native species (Wilcove *et al.* 1998). Direct and immediate losses of native species might be rare, but are exemplified by the loss of 12 native species of birds on Guam due to the voracious invading brown treesnake (Fritts and Rodda 1998). Quantifying reductions in populations of native species, loss of native genetic diversity, and extinctions requires non-market valuations (Stohlgren and Schnase 2006).

Invasive species can degrade habitat quality for native species, affect nutrient cycling, and promote disturbances such as wildfire (Mack *et al.* 2002). These impacts may be slow and chronic, such as the salinization of soils invaded by salt cedar (*Tamarix* spp.), or they may be cataclysmic such as the rapid spread of aquatic weeds such as hydrilla (*Hydrilla verticillata*) in the southeastern United States, or the spread of sudden oak death in California (*Phytophthora ramorum*).

Total annual costs have been mentioned, but we often lack site-specific costs and valuations for individual species, partly because we have poor maps of the distribution and abundance of species, much less damage estimates throughout their ranges.

Invasive species can directly and indirectly affect human health. Direct affects are seen by the 100 human deaths due to West Nile virus in the United States between 1999 (when it arrived) and 2004 (http://www.cdc.gov/ncidod/dvbid/westnile/surv&controlCaseCount04_detailed.htm). Several human deaths have resulted from plague, Africanized honey bees, Argentine ants, and other invaders. Indirect effects on human health include secondary effects of pesticides, herbicides, and allergic reactions, bites, and unknown long-term effects from, say, coating the skin with harsh chemicals to avoid mosquito bites and West Nile virus.

2.2.9 Containment potential, costs, and opportunity costs

Selecting priority species to control depends on the potential effectiveness of control and restoration efforts relative to costs (Stohlgren and Schnase 2006). For example, cheatgrass (*Bromus tectorum*) is widespread in many states in the United States, but there are no cost-efficient techniques for manual, chemical, or biological control over large areas, and the threat of re-invasion is high after fire. There are readily available biological control agents for several non-native thistles, but the thistles can persist in small populations and as scattered individuals to re-populate controlled areas later. The effects of control agents on non-target species must also be considered. Still, containment potential, relative to costs and potential for long-term success are important considerations when setting priorities for control. There are more than a dozen invasive species ranking systems available on the Internet. Many of these systems use a mix of quantitative and qualitative data on ease or cost of control relative to effectiveness. However, many are designed for local or regional use for select taxa (plants mostly).

Opportunity costs should also be considered. If you choose to spend funds and effort on containing widespread species A, will species B, C, and D take the opportunity to expand unchecked? Conversely, attacking species B, C, and D while their populations are small may be more cost-efficient in the long run compared to species A. Obviously, such decisions would benefit from predictive modelling of potential rates of spread linked to environmental, economic, and human-health costs. Predictive models should be linked to data management systems to improve our ability in selecting priority species for control (Graham *et al.* 2007). Future systems will need to integrate economic analyses to better quantify impacts of invading species.

2.2.10 Legal mandates and social considerations

Priority species for control will be influenced by legal mandates, county regulations, and a sense of 'urgency' based on other social considerations. For example, in the United States, some states and counties are legally bound to address weeds classified as 'noxious' (often poisonous) regardless of the abundance, spread potential, and other impacts of other weeds in the area. Other social considerations include threats to listed threatened or endangered species or habitats, private property rights, or unfairly distributed economic costs for control. In some cases, legal mandates and social considerations are complex in terms of national and international policies because current policies are at odds with cost effective prevention and early detection programs for invasive species (Lodge *et al.* 2006).

2.3 Information science and technology

The challenge of risk assessments for invasive species is compounded by the demanding requirements it places on information science and technology. Information must be immediately available on species locations and abundance, the ecology, and natural history of a species, and on the characteristics of favourable habitats. These are non-trivial information management problems (Graham *et al.* 2007). Humans play a crucial role in the assembly and filtering of this type of information, but technological advances are helping to synthesize available information from the Internet, make available downloadable datasets, and making data ready for easy input into analysis tools. This process of creating higher-order understanding from dispersed datasets is a fundamental intellectual process in any strategy for risk assessment.

The challenge of understanding taxonomic, geographic, and temporal completeness of data in a region of concern translates into a requirement to systematically catalogue 'metadata' knowledge about the information used in analyses. These metadata are a crucial aspect of all scientific databases.

Risk assessments for globally invasive species will require an unprecedented level of integration of 'living maps' of harmful species along with field-based environmental measurements and new remote sensing data products (Morisette *et al.* 2006). New geostatistical modelling approaches are needed at landscape- or watershed-scales to regional and continental scales, requiring the use of high-performance computing. Estimating the potential rate of spread of an invasive species, and the probable risks, impacts, and costs associated with that species, may require entirely new approaches to predictive modelling in space and time—models that combine temporal, spatial, stochastic, mechanistic, socioeconomic, and scenario-based approaches. Comprehensive risk assessment for invasive species remains largely uncharted territory. Finally, aggregating this information in ways that allow decision-makers to systematically evaluate containment potential, costs, opportunity costs, and make reasoned trades against legal mandates and social considerations will require a new generation of decision support environments tailored to the needs of invasive species risk analysis (Graham *et al.* 2007).

2.4 The challenge: to select priority species and priority sites

Managers responsible for invasive species often ask two simple questions: where is it, and how do I kill it? The underlying challenge is to select priority species for control in a constantly changing triage approach to risk assessment (Fig. 2.1). A first cut is to target frequently occurring, highly abundant invasive species. At present, some widespread species for which there is little hope of containment or control might have to be put on the back burner, while currently easily contained species get our attention. Local and regional decisions and priorities on species and sites will be set based on a mix and match of the criteria outlined above—hopefully in cooperation with other local entities since propagules and species cross boundaries with trade, transportation, wind, or with animals as vectors.

Sharing data, modelling tools, and expertise is the first step towards effective risk assessment. Coordination and cooperation are the keys. Integrated teams of taxonomists, survey and monitoring specialists, economists, landscape ecologists, modellers, remote sensing specialists, and information technology experts are needed to meet invasive species challenges at local to international scales. Increased public awareness and involvement (e.g. volunteer networks) may be needed to populate databases to map the abundance and distributions of many invasive species.

Finally, predictive modelling and synthesis will become increasingly important in risk assessment (Fig. 2.2). Targeting the few areas with clusters of invasive species may cost-effective. New species are likely entering every country each week. Even if only a small fraction of these species become established, and a fraction of those spread and cause harm (e.g. like zebra mussels have since the 1970s and West Nile virus has since 1999 in the United States), then the potential high cost of ignoring or not containing invasive species must be considered. Interdisciplinary scientists and modellers will need to work closely with agencies, non-government organizations, and communities to reduce the costs associated with harmful invasive species.

2.5 Acknowledgements

Several members of our research team contributed ideas to this manuscript. Michele Lee, Yuka Otsuki, and Geneva Chong were instrumental in literature review. The U.S. Geological Survey and National Aeronautics and Space Administration (NRA-03-OES-03) provided funding for the research and modelling aspects of the paper. Logistical support was provided by the USGS Fort Collins Science Center, Colorado State University's Natural Resource Ecology Laboratory, and NASA Goddard Space Flight Center. Sunil Kumar and Paul Evangelista provided helpful comments on an earlier version of the paper. To all we are grateful.

3
Detection and early warning of invasive species

Tracy Holcombe and Thomas J. Stohlgren

3.1 Introduction

It is well known that invasive species are a problem of epidemic proportions around the world, causing economic losses of up to $120 billion per year in the USA alone (Pimentel *et al.* 2005). As trade and travel across international boundaries increase, so do invasions (Mack and Lonsdale 2001). Early detection and rapid assessment are effective strategies to minimize the impacts that invasive species have on economies and on ecosystems that they invade (Rejmanek and Pitcairn 2002). Because the task of invasive species control can sometimes be daunting, managers need to be able to set priorities for prevention and control of these organisms (Byers *et al.* 2002). It is important to obtain accurate assessments of location and abundance of invasive species so that managers can set these priorities and have the information to quickly and effectively combat the invaders. It is also important to identify barriers to invasion and habitats where an invasive species cannot persist or cause much harm.

To be informed in the initial stages of a species on the way to becoming a successful invader, we need early detection. Early detection is a very low probability event that is critically dependent on adequate surveillance. It involves sampling strategies sufficiently rigorous to detect incursions at sufficient frequency and, assuming a response programme, to influence the chance of establishment and spread.

In our quest for early detection it is important to remember that invaders can be any type of organism—from microbes to mammals—and come in many forms. We need to be aware of plants, animals, insects, pathogens, and parasites that can all be invasive or be vectors for invasions. Examples of these include plague (*Yersinia pestis*), West Nile virus (*Flavivirus* spp.), gorse (*Ulex europaeus*), common cord-grass (*Spartina anglica*), nutria (*Myocastor coypus*), and sudden oak death (*Phytophthora ramorum*). There are many well-known examples of invasive plants and animals, including feral pigs (*Sus scrofa*), miconia (*Miconia calvescens*), red imported fire ant (*Solenopsis invicta*), and starlings (*Sturnus vulgaris*). Fish examples include western mosquito fish (*Gambusia affinis*), carp (*Cyprinus carpio*), brown trout (*Salmo trutta*), and Nile perch (*Lates niloticus*). These notorious examples would have benefited from early detection and rapid assessment.

3.1.1 Fire as a metaphor for invasion

A metaphor that could be applied to invasive species is wildfire. Wildfires sometimes grow large by sending out sparks that start small spot fires in places where conditions are right for fire to spread. Wildland firefighters know this and try to extinguish spot fires expediently, preventing the fire from growing larger. Even if the fire is already fairly large, wildland firefighters will make spot fires a priority over the large burning mass that may be too large to slow under current conditions. It is always best to detect the fire early and prevent it from spreading. This model of movement also applies well to invasions. Invasive organisms put out progeny—similar to sparks—which may move far from the parent, furthering its invasion potential. If invasive species managers, with limited resources, focus first on these smaller invasions this may do more to slow the spread of an invasive species than trying to tackle large well-established invasions (Rejmanek and Pitcairn 2002). The challenges are how to find these small invasions of cryptic species, and to assess the risks and threats of each invader.

3.1.2 Definitions

The terms 'early detection' and 'rapid response' were defined by Worall (2002):

- **Early detection**, as applied to invasive species, is a comprehensive, integrated system of active or passive surveillance to find and verify the identity of new invasive species as early after entry as possible, when eradication and control are still feasible and less costly. It may be targeted at areas where introductions are likely (such as near to pathways of introduction) and in sensitive ecosystems where impacts are likely to be great or invasion is likely to be rapid.
- **Rapid response** is a systematic effort to eradicate, contain, or control invasive species while the infestation is still localized. It may be implemented in response to new introductions or to isolated infestations of a previously established, non-native organism. Preliminary assessment and subsequent monitoring may be part of the response. It is based on a system and infrastructure organized in advance so that the response is rapid and efficient.

Whilst this chapter will focus on rapid assessment more than rapid response to invasive species, there are two important points that are made in this rapid response definition. One is that preliminary assessment must be part of the response. This is a crucial step to take once a species has been detected. If the situation is not quickly inventoried, patches or individuals may be missed and the opportunity of catching the species while the invasion is small may be lost. The second important point in the Worall (2002) definition of rapid response is that of having the infrastructure already in place. Early detection and rapid assessment require frequent monitoring, which requires effort, a strategy, and funding. It may be possible to organize a group of volunteers to conduct monitoring, but in many cases someone needs to be hired to carry out this task. In general this is much less expensive than the alternative. The mimosa tree (*Mimosa pigra*) in Australia illustrates this well

(Cook *et al.* 1996). A small stand of mimosa trees were found in Kakadu National Park (KNP) in 1983. The staff at KNP immediately sent out a team to find any mimosa trees in the park and intervene. There are now occasional reports of a tree found in the park that is quickly eliminated, but no large stands. The programme costs the park about $2 per hectare per year. In a nearby floodplain called Oenpelli a stand of about 200ha was found at about the same time. The response was not as swift and by the year 1990 the infestation covered about 8200 hectares of the floodplain. A control effort was finally undertaken and a very large aerial spray operation was carried out. The spray programme cost $220 per hectare per year for 5 years to get the tree under control. Now they, like KNP, spend about $2 per hectare per year for maintenance. This is a clear example of the costs associated with neglecting rapid response.

There are two distinct types of invasion that should be recognized when discussing early detection (Fig. 3.1). The first type of invasion is one in which a native species moves within its own native country, state, or habitat. If it moves to an area where it did not previously exist it can be considered as invading that area. For example, many game fish species in the western USA are non-native transplants. These fish disturb the native ecology of the western lakes, yet they have remained in their country of origin.

The second type of invasion crosses international borders and oceans, often moving between similar ecological zones. Tamarisk (*Tamarix* spp.), which comes from very arid regions of the Middle East and Asia, exemplifies this. It has invaded the southwestern USA in a climate similar to its native range. It is possible for invasions such as this one to be intercepted at the borders of the country (Lodge *et al.* 2006).

Land managers could benefit from accurate maps showing current distributions and local and sub-regional models of potential habitats of invaders to address both of these types of invasion. Knowing current species distribution would help land managers concentrate on the frontier of invasion and control small invasions in new areas separate from larger invasions. Identifying these small, isolated areas would be beneficial because the most effective time for control is when an invasion is small (Rejmanek and Pitcairn 2002). Determining the potential distribution of invasions would help managers focus on the areas at a high risk of being invaded, aiding in early detection/rapid assessments of new areas being invaded (Stohlgren and Schnase 2006).

3.2 Early detection and rapid assessment

Basic components of an early detection and rapid assessment (EDRA) programme include:

1) Access to current and reliable scientific and management information.
2) Ability to identify species quickly.
3) A functional risk assessment plan.
4) Mechanisms in place to coordinate a control effort.

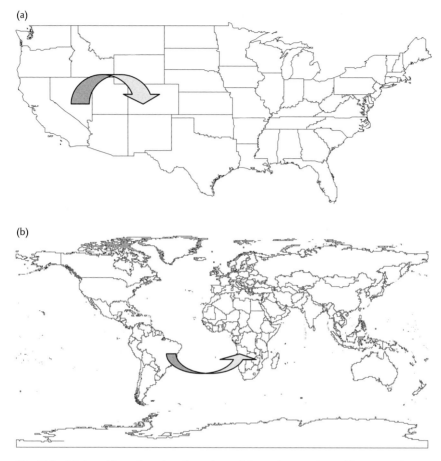

Fig. 3.1 (a) Local invasions involve a harmful species moving within a single country, state, or county to a new area within that country, state or county. (b) Global or intercontinental scale invasions pertain to a harmful species moving between countries, often over an ocean.

5) Providing adequate technical assistance (e.g. quarantine, monitoring, information sharing, research and development, and technology transfer) and rapid access to stable funding for accelerated research of invasive species biology, survey methods, and eradication options. The system's success will depend in part on public participation in efforts to report and respond to invasions.

Each of these elements, particularly points 1 and 2, are important to an early detection and rapid response programme. One tool that can aid in those specific processes is 'watch lists'. Watch lists are list of species that are either nearby, or known to invade similar habitats to the area of the list. For example, caulerpa seaweed (*Caulerpa taxifolia*), an aggressive invader introduced to the Mediterranean around 1984, was placed on the US Federal Noxious Weed List in 1999 by the Southern California Caulerpa Action Team, a watch group for early detection of this detrimental organism (Anderson 2005). When it was found offshore of California, USA in June 2000 there was already an infrastructure in place and action was taken to eradicate the plant within 17 days of its discovery. This was probably due to a well-prepared action and assessment team and the fact that the plant was on a watch list before it even entered the country. It is important to know about the biology of an invader before it arrives so that when it appears you can be prepared with strategies and have already determined potential habitat and effects. Ideally, countries would share information about their invaders with each other, but unfortunately official reporting (e.g. to the UN Convention on Biodiversity) is very limited, and the global-scale invasive species information exchange systems that collect and share this information do not receive sufficient financial support.

Westbrooks (2003) defines the essential attributes of an EDRA programme similarly as: including aids for species identification; authentication/verification of new field observations; reporting records; maintaining a database of species' occurrences and locations; alerting appropriate officials and rapid response teams; and monitoring management actions. Simple EDRA programmes have been developed for selected taxa in some local areas using these principles. For example, the state of Wisconsin in the USA has an early detection programme for purple loosestrife (*Lythrum salicaria*) whereby public service announcements prompt television viewers to call in purple loosestrife locations to a hotline with awaiting weed coordinators. This system incorporates adding new information to a database as soon as a specimen is found with alerting the appropriate officials so that a response team can be notified.

Additional aspects of EDRA components have recently been added for: user ID and validation; reporting; expert verification; occurrence database; and rapid assessment (Simpson *et al.* 2006). This paper highlights the importance of a centralized data-sharing system. The authors also mention the importance of species profiles on the Internet for quick identification of new invaders, biological and ecological information, global distribution with details about instances of invasion, and information about management options, including case studies of early detection and rapid response.

3.3 Guiding principles for early detection and rapid assessment

Suggested guiding principles are as follows:

- An early detection programme must be fully integrated into a comprehensive, science-based, research and management programme that coordinates aspects of prevention, EDRA, research, surveys, and monitoring, and outreach and reporting.

- The database of observations must remain in the public domain with free and open access to unclassified, peer-reviewed data.
- Because many aspects of an EDRA programme require extensive research and development (e.g. integrating millions of field observations with remotely sensed information and new forecasting tools; greatly improved information technologies; and high-performance computing), basic research and a scientific method must underpin the design, testing, and phased implementation of the programme and these programmes must be developed prior to new invasions.
- The long-term success of any national or international EDRA programme is dependent on a long-term commitment of funding, personnel, and equipment of all key components in the system, plus the continued cooperation of many government and non-government organizations, engaged volunteers, and public acceptance.
- It would be impossible to create a comprehensive EDRA programme for the thousands of species on Earth. For any country (or region within a country) it might be more realistic to focus preliminary efforts on those top priority species that are identified as serious potential invaders. Once the system is more fully tested, it could be expanded to cover more species.

3.3.1 Data and information management

Data and information management represent the single greatest challenge of an effective EDRA programme. Information is needed on probable and current species distribution and abundance, habitat suitability, and containment strategies and techniques. High resolution maps and models of current and potential spread of harmful species and their effects—which are being used in developed countries to assess and manage invasive species problems—can be used to provide insights into invasion ecology and to develop guidelines for response options for those facing similar problems in other parts of the world. Based on US surveys of resource managers and the public, there is an unprecedented need for a 'comprehensive, integrated system' for early detection, and 'a systematic effort to scope the severity of the issue' for rapid assessment (Stohlgren and Schnase 2006).

3.3.2 Global and regional invasive species databases

Biological invasion is a global problem so it is clear that global-scale clearinghouses that share data from all over the world are a crucial component of any effective response. Existing global-scale systems include the Global Invasive Species Database (GISD http://www.issg.org/database), which has comprehensive information on more than 500 of the world's worst invasive species; the Global Register of Invasive Species (GRIS), which provides the names and full taxonomy of all known invasive species, along with geographic records of introduction and invasion; and the Global Invasive Species Information System (GISIN), which is developing a system for the exchange of invasive species data and information between local, national, regional, and international databases over the Internet.

The Global Organism Detection and Monitoring system (GODM) of the US National Institute of Invasive Species Science (NIISS, http://www.niiss.org), and the International Nonindigenous Species Database Network (NISbase, http://www.nisbase.org) include global information and CAB International (http://www.cabi.org) launched the first phase of their Invasive Species Compendium in 2008. Regional information systems include Delivering Alien Invasive Species Inventories for Europe (DAISIE, http://www.europe-aliens.org) and I3N, the invasive species thematic network of the Inter-American Biodiversity Information Network (IABIN, http://i3n.iabin.net/).

All of these databases provide data free to the public, but have limited access to those contributing to the system to ensure data quality. Websites such as these are a great benefit to early detection and rapid assessment. They have the potential to form a global network of information on all harmful invasive plants, animals, and pathogens (especially if geographic information gaps are addressed and if they become providers of data to the GISIN). Here we specifically outline the components and potential uses of the GODM system to illustrate how databases can be used in EDRA. All of these components may not be available in each system but they all represent great future potential.

- **User ID/tracking:** this first step involves users that may contribute data registering with the website and entering contact information that includes their name, email address, location, and level of expertise (specifically regarding the information about to be entered). This is important so that the information entered can be tracked to its source and checked for reliability.
- **Verify records and 'first alert':** only a limited number of well-trained users and coordinators may enter data into the system. The user may wish to exclude suspect data in analyses, mapping, and modelling by selecting data that is confidently identified. Location data are matched with other known reported locations and modelled distributions—this step allows for detection of novel, urgent species establishments in new habitats, ecosystems, counties, or states. After taxonomic identifications are verified, novel/urgent observations of occurrences can be sent to officials or agencies responsible for sending specific alerts.
- **Taking in new information:** new records are systematically added. Metadata need to accompany all data. Ancillary data (e.g. soil texture, land use characteristics, etc.) should be available for any data points collected in the field. All data are screened for quality (measures within acceptable ranges), stamped with a 'certainty-level', and then served on the Internet for public consumption or download.
- **Rapid assessment, data synergy, invasive species forecasting system:** this point illustrates the real power of having multiple databases on the Internet. Datasets from one database can be linked with datasets from other inventory and monitoring programmes to map the current distribution and abundance of a target species or multiple species. Simple, 'first approximation maps'

derived from a choice of several commonly-used species distribution models (e.g. multiple logistic regression, classification trees) using multiple datasets can be created. For some very common, less-harmful species (e.g. dandelion (*Taraxacum* spp.), ladybug beetle (*Coccinella* spp.)), distribution maps may be all the 'modelling' that is needed. For newly detected species, species on 'noxious' or 'invasive' lists, or watch-list species, more advanced modelling could be performed. Potential distributions can be modelled from occurrence and abundance data, ancillary data, and remotely sensed data to produce maps of probable/potential distribution and abundance, habitat vulnerability analysis, and uncertainty analysis (Stohlgren and Schnase 2006). Modelled information and species attribute data can be used to create 'second approximation models' of potential rates of spread, and corridors and barriers to invasion. The current distribution and abundance data can be overlaid on the model outputs and habitat maps to identify priority survey, control, and restoration sites. All data and model outputs can be served on the Internet, and all data and metadata associated with selected models can be archived.
- **Rapid response and monitoring effectiveness:** based on new reports and modelled outputs of distribution and abundance, habitat vulnerability, potential spread rate, and risks, alerts can be targeted to authorities and to groups of concerned citizens where appropriate. Typically, 'exotic invasive management teams' can be provided with a location and a method of extermination. We suggest a more sophisticated use of rapid response teams where far more information is provided to the team to maximize efficiency (Table 3.1).

Table 3.1 Suggested information provided to rapid assessment teams.

Information provided	Reasoning
Species identification aids.	To effectively target cryptic invasive species rather than look-alikes.
Accurate location data of known occurrences and predictive models of target species, information on other highly invasive species in the local area, and high probability sites nearby and along the route to the primary site.	Improves cost-effectiveness of rapid assessment efforts, while reducing propagule pressure and source populations nearby.
Comparable (standardized) monitoring protocols.	To help quantify 'what works, where', share success stories, and document performance goals.
Instructions to upload data into a distributed database to share information on what techniques work best in different habitats under a variety of conditions.	Improve accountability and data sharing for better predictive modelling, early detection, and restoration.

44 | Invasive species management

A first critical step here is to serve, store, and share monitoring data to use in an adaptive management framework when combating invasive species. Initial control efforts may not be successful, and vulnerable habitats may be quickly re-invaded from seeds, propagules, or source populations nearby. Thus, rapid assessment is an iterative process improved by careful monitoring and information sharing (Stohlgren and Schnase 2006).

A second critical step is to use predictive spatial models to revise maps of current and potential species distributions and abundance to select the next highest priority control sites in a strategic manner. This step may include isolating source populations from vulnerable habitats by concentrating on corridors of invasion or two-pronged attacks on both well-established source populations and newly invading sub-populations (Fig. 3.2). Each habitat must be prioritized and acted upon according to the priority it is assigned. A key feature here is documenting all management actions to better understand the invasion process and to be able to

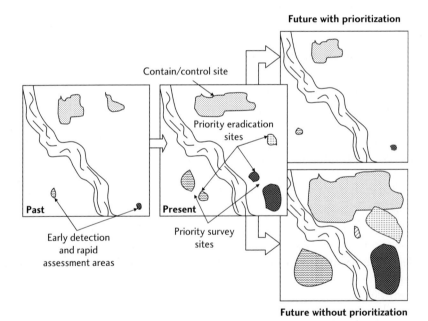

Fig. 3.2 In this conceptual model of invasion the past shows where a species may have been introduced. The present shows where the species is when it is first found. Priority survey sites are areas between two close invasions, contain/control sites are large patches, and priority eradication sites are areas of small populations. If these sites are prioritized expediently the damage can be minimized and money saved. Without prioritization, species will continue to spread and cause more ecological and financial burden. The concept of this figure applies equally well to plants, animals, and diseases.

extrapolate successful actions to additional species and habitats. This will improve costs of future control and restoration efforts, while tracking performance measures and overall cost effectiveness.

Researchers and modellers must also track the accuracy and utility of modelling capabilities for EDRA and document the economic and environmental savings by using modelling products. Likewise, we must document customer satisfaction in the use of modelling products to improve decision support.

Determining the spatial extent and severity of invasions is of utmost importance (Simberloff *et al.* 2005). Unfortunately, ground surveys of each invasive species require large amounts of time and funding, and most managers do not have the resources required to complete the task. Statistical techniques linked to targeted field surveys may achieve fairly accurate measurements of potential distributions in large areas. These models produce maps of habitat suitability or barriers to invasion. The information contained in remotely sensed images can be used in these spatial models of habitat suitability (Reich *et al.* 1998, 2004; Crosier 2004; Barnett *et al.* 2007). These models provide information on the potential habitat of an organism with minimal field data on newly invading species. These methods could prove invaluable for targeted early detection surveys.

3.3.3 Species reporting requirements

While most would agree that reporting new locations of harmful invasive species is important, there are a few published recommended data requirements for early detection. Extreme minimum requirements include 'who, what, when, and where' data (Table 3.2, required fields), sometimes referred to as the Dublin Core (See http://www.gisinetwork.org/Documents/GISINProc2004HTML/GISINProc20041.html and http://dublincore.org/). This general advice could be greatly improved by an understanding of the potential to model species distribution and abundance data in space and time. For example, ancillary data on abundance, dominant native species present, other non-native species present, environmental data (e.g. soils or disturbance information for plants, water depth for fish, nest tree species for birds, etc.) and noticeably absent native and non-native species can be extremely important information in predictive modelling (Table 3.2) (Morisette *et al.* 2006).

3.4 Conclusions

EDRA could be the most effective tools that land managers have to stop an invasion before it becomes an ecological nightmare. A relatively modest investment in existing global-scale information exchange systems will provide the world with access to information about all known invaders. 'Watch lists' should be created, maintained, and updated for local areas. When information is obtained about a particular invasive species in a local area, it should be shared on global websites and with local land managers so that others can benefit from this knowledge. It is

Table 3.2 Generic species reporting requirements (* = required field)

Data field	Example	Comments
Recorder Name*	Chuck Darwin	Observer
Date*	July 17, 2005	
Time*	17:35	24-hour clock
Y coordinate*	4405547	Exact UTM Northing or Longitude
X coordinate*	10644277	Exact UTM Easting or Latitude
Species*	Spotted knapweed	Genus, species, or common name
Abundance	10	Count or % foliar cover
Location certainty	±10	m (specify metres or feet)
Area surveyed around point	40	m^2 (specify units as m^2, ft^2, ac, or ha)
Dominant native species present	*Pinus ponderosa*	Genus, species, or common name
Other non-native species present	*Bromus tectorum*	Genus, species, or common name
Other non-native species noticeably absent	Yellow sweet clover	Genus, species, or common name
Comments	Old field	Any helpful ancillary information

important to look at the habitat that surrounds an area and determine what species are possible invaders and survey for them. Probable distribution models and habitat suitability maps should be used, and surveys conducted along corridors and entry points for invasion. EDRA is a very effective tool when used efficiently.

4

Eradication of invasive species: progress and emerging issues in the 21st century

John P. Parkes and F. Dane Panetta

4.1 Introduction

At an international conference on the eradication of invasive species, held in 2001, Simberloff (2002) noted some past successes in eradication—from the global eradication of smallpox (Fenner *et al.* 1988) to the many successful eradications of populations (mostly mammals) from small islands (e.g. Veitch and Bell 1990; Burbidge and Morris 2002). However, he cautioned that we needed to be more ambitious and aim higher if we are to prevent and reverse the growing threat of the homogenization of global biodiversity. In this chapter we review how the management strategy of eradication—the permanent removal of entire discrete populations—has contributed to the stretch in goals advocated by Simberloff. We also discuss impediments to eradication success, and summarize how some of the lessons learnt during this process have contributed to the other strategies (prevention and sustained control) that are required to manage the wider threat posed by invasive alien species. We concentrate on terrestrial vertebrates and weeds (our areas of expertise), but touch on terrestrial invertebrates and marine and freshwater species in the discussion on emerging issues, to illustrate some of the different constraints these taxa and habitats impose on the feasibility of eradication

4.2 From scepticism to positive consideration

A major advance in management of vertebrate pests and, to a lesser extent, weeds has been a general change in mind-set among managers and decision makers. Before the 1980s, many were sceptical about the possibility of eradication, but now eradication is at least seriously considered as the first option to deal with pests—especially for colonizing populations, those with limited or patchy distributions, and island populations. Part of this change has been driven by the successes, but part has been because those proposing eradications have been developing more convincing feasibility plans.

Thus, eradication of pest and weed populations is an attractive option for pest managers because successful removal of all individuals in the target population may (a) forestall any adverse impacts before these manifest themselves (e.g. the prompt removal of wallabies from Great Barrier Island; Eadie *et al.* quoted in Clout and Russell (2006)), and (b) reverse any impacts and restore the system to its previous state or at least set the affected ecosystem along more acceptable trajectories (Fukami *et al.* 2006). Eradication also does not require the complex knowledge and ongoing commitment to efficiently and effectively manage pests under the alternative of a sustained control strategy (Choquenot and Parkes 2001).

4.3 Feasibility

However desirable it might be, attempts at eradication can be counterproductive if they are not feasible. The costs of promoting eradication when it is not possible obviously include failure and, at worst, abandonment of the problem, when a properly planned, sustained control campaign might have addressed the problem. For example, there was a hiatus of a decade in effective management of rabbits in New Zealand between the abandonment of the 'last rabbit' policy in the early 1980s and the implementation of sustained control in priority areas in the early 1990s (Gibb and Williams 1994). The costs of unachievable eradications also include foregone opportunities to act elsewhere and a return to scepticism, especially among funding agencies. This leads to risk aversion exactly when we need some considered risk-taking if we are to aim higher. Thus, as part of good planning, most eradication proposals start with a feasibility study. This is aimed at convincing funding agencies that (a) eradication is needed either to eliminate current impacts or as a precaution; (b) it is possible if obligate conditions can be met and any constraints are either managed or acceptable; and (c) the programme is adequately funded.

There have been several attempts to set criteria to assess whether or not eradication is feasible. The suggested criteria mix essential rules and desirable attributes, which are often overlapping, and some have suffered from definition drift. They have also reflected their authors' backgrounds. For example, Parkes (1990a) focused on obligate biological criteria for vertebrates on islands; Panetta and Lawes (2005) and Cacho *et al.* (2006) stressed detectability and delimitation of infestation for weeds on mainlands; Bomford and O'Brien (1995) included economic and social criteria; Myers *et al.* (2000) and Baker (2006) included organizational commitment as criteria; and Cromarty *et al.* (2002) stressed the planning process. These criteria may be summarised in a more parsimonious set of three essential rules that cover both island and mainland eradications:

- The average annual long-term rate of removal in source populations must be greater than the annual intrinsic rate of increase. This rule covers the older 'all at risk and rates of removal' criteria, and is applicable to situations where we have source-sink populations or where Allee effects have been argued to negate

the 'all at risk' condition (Liebhold and Bascompte 2003). It also implies the funds are available to achieve the rule.
- There is no immigration of individuals that can breed.
- There must be no net adverse effects. Eradication may not be desirable if the adverse affects on non-target species of the control methods available are predicted to be unacceptable and unresolvable, or if the consequences of removal of the pest outweigh the benefits (Courchamp *et al.* 2003).

Non-target problems are generally manageable either by changing the control method (e.g. toxic baits to eradicate ship rats (*Rattus rattus*) on Middle Island were presented in bait stations that excluded the native bandicoots (*Isoodon auratus barrowensis*) (Morris 2002), or by temporarily taking some or all of the non-target populations to a safe place (e.g. protection of the endemic deer mice (*Peromyscus maniculatus anacapae*) during Norway rat (*Rattus norvegicus*) eradication on Anacapa Island (Howald *et al.* 2003).

Predicting and managing adverse effects that result from the removal of the pest(s) is one area that requires further consideration. Removing one pest may lead to an increase in another pest. This may be unexpected, e.g. the eruption of the exotic vine *Operculina ventricosa* following eradication of feral goats and pigs on Sarigan Island (Kessler 2002). However, some effects are predictable, and certainly can be tested experimentally before the eradication is conducted. For example, the eradication of feral goats from Guadalupe Island, Mexico has resulted in an increase the biomass of both native plants and in exotic grasses. The latter has caused an increase in abundance of mice (*Mus musculus*) and probably of feral cats (*Felis catus*) with potential consequences for native biota. The increase in grass biomass has also increased the risk of fire with potential consequences for the regenerating relict stands of the endemic *Pinus radiata guadalupensis* (A. Aguirre pers. comm.). These consequences are manageable (eradicate the mice and cats, and plant the pines in widely dispersed parts of the island), but it raises the question of the order in which pest species should be eradicated.

The range of other issues that may also have to be considered in a feasibility plan may constrain conformity to one or more of these criteria. For example, eradication was not practical for Himalayan thar (*Hemitragus jemlahicus*) in New Zealand because of intractable objections by some landowners (Hughey and Parkes 1996). Not all thar could be put at risk, so the rate of removal and immigration criteria could not be met. Attempts at eradicating hedgehogs (*Erinaceus europaeus*) over a range of about 50,000 ha on the Uist islands, Scotland, also appear unlikely to succeed because animal welfare groups have limited control intervention to a short annual period when the females are active but not lactating (Warwick *et al.* 2006), thereby increasing the likelihood that the rate of removal (without heroic efforts) will not exceed the annual rate of increase.

Feasibility of eradication must therefore be viewed in the context of the minimum effort that can be mustered to meet the obligate rules and overcome constraints

(Rainbolt and Coblentz 1997). If resources are not limited, the upper boundary for a decision on feasibility would be where eradication was no longer the most cost-effective option. The assessment of eradication feasibility cannot be divorced from the hazard posed by the pest, with more effort being justified by pests with greater threats (Panetta and Timmins 2004).

A key lesson learnt in feasibility planning since 2001 is that it is an essential part of eradication but cannot be produced as a mere recipe because every case has a unique mix of issues that have to be considered and managed (Towns and Broome 2003).

Assuming the feasibility plan is accepted and the operation funded, the next phase requires a detailed operational plan to describe when, how, and who will conduct the eradication operations, and include how the outcomes and consequences of the attempt will be measured (e.g. Cromarty *et al.* 2002). This entire planning process is critical if the inevitable failures when aiming higher are not to prove fatal to ongoing eradication attempts. Just as important are funded and planned reviews of progress towards eradication, especially for weeds where projects commonly extend over many years (Mack and Lonsdale 2002) and for animals where eradication is achieved by successive culling events. Such reviews are necessary to determine whether or not the project is on track, or if it is likely to become a *de facto* sustained control project (Bomford and O'Brien 1995; Panetta and Lawes 2007).

4.4 Advances in eradication of vertebrate pests

Eradication of some alien terrestrial vertebrates from islands was a well-established practice before 2001. For example, there were 282 successful eradications (and 25 failures) of rodents reported between 1951 and 2000, and 66 successes (and nine failures) reported since 2001 (Howald *et al.* 2007). The failure rate has not changed ($r^2 = 0.54$, $P = 0.46$) but the size of islands on which rodent eradication has been achieved has increased (Table 4.1). Reviews of other vertebrate eradications have been published for feral goats (Campbell and Donlan 2005) and feral cats (Nogales *et al.* 2004) (Table 4.1).

Other terrestrial vertebrate species that have been routinely eradicated before and after 2001 include feral pigs (*Sus scrofa*) from at least 20 islands with the largest being Santiago, Galapagos (58,465 ha) in 2000 (Cruz *et al.* 2005); rabbits (*Oryctolagus cuniculus*) from at least 20 islands with the largest being St Paul, Kerguelens at 800 ha in 1996 (Lorvelec and Pascal 2005); brushtail possums (*Trichosurus vulpecula*, Fig. 4.1) from 16 islands in New Zealand with the largest being Kapiti at 1970 ha (Cowan 2005); and red (*Vulpes vulpes*) and arctic foxes (*Alopex lagopus*) from 39 Aleutian islands including 93,000 ha Attu Island (Ebbert 2000) and red foxes from many smaller islands in Western Australia (Burbidge and Morris 2002) mostly before 2001. Less routinely, several other vertebrate species occurring on many islands have been eradicated from only a few—sometimes because no one has

Eradication of invasive species | 51

Table 4.1 Summary of islands from which six species of vertebrate pests have been eradicated, before and after 2001. Rodent data from Howald *et al.* (2007); goat (*Capra hircus*) data from Campbell and Donlan (2005); cat (*Felis catus*) data from Nogales *et al.* (2004).

Taxa	Before 2001			After 2001		
	No. islands	Mean area (ha)	Largest island (ha)	No. islands	Mean area (ha)	Largest island (ha)
Rattus norvegicus	51	173	3103	22	695	11300
Rattus rattus	82	60	800	37	63	1022
Rattus exulans	44	158	1965	3	2011	3083
Mus musculus	22	96	710	4	119	200
Capra hircus	102	1421	36100	5	18288	58465
Felis catus	45	1318	29000	7	2843	9700

Fig. 4.1 Invasive arboreal mammals such as the brushtail possum (*Trichosurus vulpecula*) and ship rat (*Rattus rattus*) are key predators of New Zealand birds. This is a primary motivation for attempts to eradicate these invasive species. Photo: David Mudge, Nga Manu Images.

tried but usually because they are difficult. Mongoose (mostly *Herpestes javanicus*) are found and controlled on many islands in the Pacific and Caribbean, but to date have been eradicated, in 2001, only from 120 ha Fajou Island in Guadeloupe in the Caribbean (Lorvelec and Pascal 2005). Coypu (*Myocaster coypus*) were eradicated from East Anglia in England in the 1980s (Gosling and Baker 1989), but have not been eradicated from other places despite massive control efforts in Europe and the USA. North American mink (*Mustela vision*) are also a widespread pest in Europe and were eradicated from many small (Nordstrom *et al.* 2002) and one large island (Hiiumaa at 100,000 ha) in the Baltic in 1999 (Macdonald and Harrington 2003), and from South Uist (32,026 ha) and Benbecula (8203 ha) in Scotland in 2007 (S. Roy, pers. comm.).

Apart from the single mongoose eradication, other advances are also evident in the list of new vertebrate species that have been successfully eradicated since 2001. These include hedgehogs (*Erinaceus europaeus*) from 88 ha Quail Island, New Zealand in 2004 (Clout and Russell 2006); feral donkeys (*Equus asinus*) from 58,465 ha Santiago Island, Galapagos in 2004 (Carrion *et al.* 2007); turkeys (*Melaegris gallopavo*) from 24,900 ha Santa Cruz Island, California in 2007 (S. Morrison pers. comm.); and feral pigeon (*Columba livia*) from their range on part Santa Cruz Island, Galapagos in 2004 (F. Cruz, pers. comm.).

Current actions against alien terrestrial vertebrates are even more ambitious, with plans to attempt to eradicate *Rattus rattus* from Pinzon Island (1815 ha); *Mus musculus* from Gough Island (6500 ha); both rats and mice from Tristan da Cunha Island (9600 ha); rabbits, rats, and mice from Macquarie Island (12 800 ha); feral cats from Guadalupe Island (24,399 ha); North American beavers (*Castor canadensis*) from Tierra del Fuego (7,000,000 ha); ruddy ducks (*Oxyura jamaicensis*) from all of England; and feral goats from Isabela Island (500,000 ha)—the latter almost completed (V. Carrion pers. comm.). Advances in control techniques and how they are specifically applied to achieve eradication of vertebrates are discussed in other chapters in this book.

4.5 Advances in eradication of weeds

As we discuss in section 4.6.4, eradicating weeds presents some particular problems. Rejmánek and Pitcairn (2002) used the data from the California Department of Agriculture on 18 agricultural weed species (a mix of annual and perennial species) in 53 separate infestations to show that eradication was likely for small infestations of <1 ha, but that only 33% of those between 1–100 ha were eradicated and only 25% of those between 100–1000 ha. They suggested that eradication of infestations over 1000 ha was unlikely to succeed with a 'realistic amount of resources'.

Recent advances in weed eradication have been mainly conceptual and theoretical, in an attempt to understand the scale-cost mechanisms behind Rejmánek and Pitcairn's conclusions, and to extend the conclusions to environmental weeds. Panetta and Timmins (2004) have suggested that the effort required to achieve eradication should be a function of both gross infestation area (= the area over which the weed is distributed and must be searched in repeat visits following control treatments which may or may not be applied over the whole area at each event), and what they call 'impedance' (= groups of constraints such as accessibility, detectability, weed biological characteristics, control efficacy).

Two regression approaches have been used to model the costs of eradication against scale of infestation and other attributes of weeds. Cunningham *et al.* (2003) surveyed experts in weed eradication for their views on the probability of success for each of 15 hypothetical weeds with different life histories and status, within eight broad cost ranges. Regressions of the relationships between eradication cost and weed incursion/attributes showed that only four variables (as categorical values) were significant: the total gross area of infestation (a), the number of infestations (b), the ease of access for control (c), and seed longevity (d):

$$\text{Cost} = 1000 \exp[9.43 + (-0.5a) + (-0.63b) + (-0.36c) + (-0.42d)]$$

Woldendorp and Bomford (2004) modelled costs (estimated from actual effort) for 20 weed eradication projects in various stages of completeness from around the world against a variety of attributes. This study showed that the cost of eradication was largely determined by the total area of infestation that was treated, the net area (c.f., the gross area of infestation in Cunningham *et al.* (2003)) with no improvement to the model (Cost = $e^{9.89}$ Area$^{0.66}$) by inclusion of other variables, although these may have been implicit within the original decisions that selected the projects as being feasible. Interestingly, the actual costs for those projects successfully completed were almost always lower than the costs modelled (Table 4.2), which hopefully is not an indication of false claims of success!

Plant eradications generally take longer to achieve than animal eradications, but application of some of the concepts discussed above has suggested that the view of Rejmánek and Pitcairn (2002) about the feasibility of larger-scale eradications might have been too pessimistic, or conversely that agencies have been prepared to invest more resources than considered possible by Rejmánek and Pitcairn. The tumbleweed bush (*Bassia scoparia*) has been eradicated from over 2480 ha in Western Australia, although, since it was introduced deliberately and planted as a forage plant, its total extent was known, making the task easier than it otherwise might have been (Panetta and Lawes 2005). An eradication campaign in the USA against *Striga asiatica*, an obligate root parasite of cereals and legumes, has since the early 1970s reduced the infestation from c. 200 000 ha to less than 1600 ha (Eplee 2001; R.E. Eplee, unpubl. data). Eradication may be possible for this plant because there is an effective means of eliminating soil seed banks, which are proving a major impediment to eradicating another annual parasitic weed, branched broomrape (*Orobanche ramosa*), over 7000 ha in South Australia (Panetta and Lawes 2007).

Table 4.2 Estimated eradication costs for 20 weeds (based on the methods of Cunningham et al. (2003) and Woldendorp and Bomford (2004)), compared with the actual costs expended to 2004 in 10 successful projects (C), three still being monitored (M), seven ongoing projects (O), and one where eradication failed (F). The lowest figure for each case study is given in bold font. Table modified after Woldendorp and Bomford (2004).

Species		Net area (ha)	Costs (AU$)		Actual costs to 2004
			Cunningham et al. (2003)	Woldendorp and Bomford (2004)	
Centaurea trichocephala	C	<1	56,800	20,000	1378
Eichhornia crassipes	C	2.4	287,100	35,600	8800
Eupatorium serotinum	C	0.5	64,700	12,600	9842
Helenium amarum	C	<50	290,000	276,300	72,835
Hieracium pilosella	C	0.005	14,900	600	394
Jatropha curcas	C	0.25	34,500	7900	4331
Pueraria phaseoloides	C	0.04	14,900	2300	2953
Salvinia molesta	C	3.6	35,200	47,300	30,460
Bassia scoparia	C	3277[1]	589,900	4,570,100	494,581
Andopogon virginicus	M	0.001	40,000	200	197
Cenchrus echinatus	M	63.6	512,900	**324,700**	1,995,000
Eichhornia crassipes	M	2	81,500	31,900	>11,500
Alternanthera philoxeroides	O	1	287,100	**20,000**	800,000
Citharexylum gentryi	O	171	589,900	630,400	141,338
Cleome rutidosperma	O	<10	204,400	93,900	**42,000**
Hypochaeris radicata	O	15	473,400	123,200	9158
Paspalum uvillei	O	1	124,000	20,000	2953
Rubus glaucus	O	5	411,600	59,000	13,904
Spartina angelica	O	75	**337,000**	362,300	>1,000,000
Chondrilla juncea	F	3400	**555,600**	4,684,500	56,000,000

[1] Subsequent checks of infestation records indicate a total net area of 2480ha (F.D. Panetta, unpubl. data).

4.6 Emerging issues

4.6.1 Eradication on mainlands

Much of the technical ability to manage pests and weeds was developed for sustained control strategies on mainlands and later modified to attempt insular eradications. For example, aerial baiting using helicopters equipped with a global positioning system (GPS), which is now common for the eradication of rodents from islands, was developed for sustained control of brushtail possums in mainland New Zealand (Morgan and Hickling 2000). However, as the scale of island successes has increased managers have begun to reconsider the feasibility of eradicating pests and weeds from mainland sites. This is feasible particularly when the pest or weed is just establishing, or when it is widespread but patchily distributed, the boundaries of the distribution are known, and dispersal is limited or manageable.

Founder populations clearly provide the best chance for eradication if they are detected before they establish or spread too widely. In fact even species that eventually become invasive and ubiquitous often struggle to establish (Forsyth and Duncan 2001) and early detection and action may tip the balance in favour of extinction. Among vertebrates, founding beaver populations (*Castor canadensis*) were eradicated from part of mainland France (Rouland 1985), but grey squirrels (*Sciurus carolinesis*) had dispersed too widely in northern Italy before those proposing eradication could overcome the legal impediments to act (Bertolino and Genovesi 2003).

Timely detection and response is even more critical with invertebrates and weeds, and in marine or freshwater habitats. Eradication of marine pests appears almost impossible once they are established. The main constraint in dealing with most aquatic species is that they disperse by prolific production of spores or larvae that are more-or-less immune to attack by humans. For example, the invasive seaweed *Undaria pinnatifida* was eradicated from the hull of a sunken ship in the Chatham Islands, New Zealand before it established (Wotton *et al.* 2004), but has proved impossible to eradicate from harbours once it has established around the coast of mainland New Zealand and Tasmania (Hewitt *et al.* 2005). In a rare example of the eradication of a marine invader, the mussel *Mytilopsis* sp. was eradicated before it escaped from a partially-isolated marina in Darwin, Australia (Bax *et al.* 2002), albeit at a cost of over AU$2 million.

The cost of eradicating even quite small founder populations of invertebrates can be very large. For example AU$175 million was allocated over 6 years in an ongoing attempt to eradicate red imported fire ants (*Solenopsis invicta*) from Queensland, Australia (Davis *et al.* 2004). The best strategy for managing such pests is to be proactive and stop them arriving in the first place. If the risk of this is high and the cost to eradicate incursions is also high, then the cost of the pest's potential impacts would have to be extreme to justify any attempt at eradication.

4.6.2 Does scale count?

There are two general strategies used in eradication that have different effects on the question of increasing scale. The first type is where a single operation is applied with the expectation that all the pests will be killed (e.g. in aerial baiting for rodents where baits are presented across the whole landscape over a short period (Cromarty *et al.* 2002)). The second type is where repeated control events, often using different techniques as densities decline, are applied until no target pests remain (e.g. for weeds especially where seed banks ensure the need for repeated control (Panetta 2007) or for ungulates where the control is by shooting (Cruz *et al.* 2005). There are some intermediate scenarios such as in eradications of rabbits, where aerial baiting has so far almost always failed to kill 100% and the operations have had to detect and rapidly mop-up survivors to achieve eradication (Parkes 2006c).

It is not clear that increasing scale causes any intrinsic increase in risk of failure of aerial baiting of rodents. It may be just a matter of more helicopters and more bait. However, we can speculate that (a) increasing scale increases risks correlated with increasing habitat complexity which means that some pests do not encounter baits; and (b) if any rodent has a minute but positive chance of survival that sheer numbers (correlated with scale) will increase the risk that some will survive despite exposure to baits.

However, scale affects the second strategy more obviously because recruitment and immigration are inevitable during the spaced control events. Recruitment can be managed by more intensive control applied more frequently to meet the first rule of eradication, but immigration can be a major problem as scale increases and the target pest has to be managed in 'bite-sized' units. A comparison between two successful feral pig eradications on large islands, both having access to similar control methods, illustrates this. On 58,465 ha Santiago Island (Galapagos) feral pigs were eradicated without the use of fencing to divide the island, but it took 30 years (Cruz *et al.* 2005). In contrast, on 24,900 ha Santa Cruz Island (California) feral pigs were eradicated within 1 year after the island was fenced into five blocks and the problem of 'back-filling' dispersal into previously cleared blocks thereby resolved (Morrison *et al.* 2007).

Scale is important in terms of the benefits of eradication. Island biogeographic principles (MacArthur and Wilson 1967) would suggest that increasing the size of islands from which pests have been removed increases the biodiversity benefits achieved or made possible, although no one to our knowledge has analysed these benefits across insular eradications.

4.6.3 Delimiting boundaries and detecting survivors and immigrants

One particularly difficult issue on very large islands or continents is how to deal with uncertainty about the distribution of the pest or weed. This is of course not a problem on smaller islands as the sea is assumed to be the boundary in most cases.

Delimitation of distributions is a critical and expensive issue for weeds because, however successful a project is at removing known infestations, undetected plants will spawn further foci of invasion (Panetta and Lawes 2005). For example, in the successful eradication of blackberries (*Rubus megalococcus* and *R. adenotrichos*) from the Galapagos Islands, over half of the costs were expended delimiting the infestations (Buddenhagen 2006). Clearly, increasing efficiencies in delimiting weed distribution will be worth the effort. Development of predictive models based on landscape characteristics (Shafii *et al.* 2003) or these in conjunction with weed dispersal parameters (Pullar *et al.* 2006) are an attempt to do this. Delimitation can also be a problem for some vertebrate eradication campaigns. For example, the attempt to eradicate red foxes from Tasmania is made very difficult by the managers' inability to know where foxes occur over this 6 million hectare island—despite over 1000 reported sightings of unknown reliability (Saunders *et al.* 2006).

The issue of delimitation and detecting survivors or immigrants is a subset of the wider problem of finding rare objects or events, which is part of the field of search theory that evolved out of applications of military science to search-and-rescue (Haley and Stone 1979). Search theory has major potential for the management of pests (Ramsey *et al.* 2009) and weeds (Cacho *et al.* 2006). To be 100% certain that no pest or weed exists in an area one would need to search everywhere in the area with an infallible detector, but as neither condition is usually met managers often have to interpret a string of 'zero found' data (Regan *et al.* 2006). The key parameters required to do this efficiently are a detection probability (the chance that if there is an object in the area searched or within range of the detection device that it will be detected), and some analysis of the meaning of a string of 'zero detected' events.

4.6.4 A particular problem with weeds—seed banks

Seed banks present a particular problem in eradicating weeds. Maximum seed longevity for some species can be decades and so an eradication project must continue until no viable seeds remain. Many plants also produce prolific numbers of seeds with effective dispersal mechanisms. This means that eradication is often compromised when plants recruited from the seed bank during the eradication operation reproduce and recharge the seed bank (Panetta 2007). The need for visits to infestation sites at frequencies set by plant maturation periods, and the costs to search for new plants, appear to be the major determinants setting the upper boundaries for weed eradication feasibility.

Seed banks of some weeds may be depleted rapidly, especially where management involves soil disturbance and prevention of further seed input. However, in some cases there remains a dormant but highly persistent component. Here, frequency distributions of seed longevity are highly skewed and mirror the highly leptokurtic distributions of seeds in space. This can make it rather difficult to set stopping rules in weed eradication projects. Regan *et al.* (2006) have suggested optimal stopping times for weed eradication based on the economic trade-off between the costs of continued monitoring with its intrinsic uncertainties, and the cost of being wrong and declaring success too soon.

4.6.5 Tricky species

Some vertebrates are difficult to eradicate at the scales needed because we do not have effective control tools. For example, species such as musk shrews (*Suncus murinus*) or brown tree snakes (*Boiga iregularis*) that eat only live prey have proved very difficult to eradicate. Success has been achieved only on very small islands (2 ha Ile de la Passe in the Seychelles for the shrew (Seymour *et al.* 2005) and some areas of less than 1ha on Guam for the snake (Rodda *et al.* 2002).

Some species learn more quickly than others to avoid control. Eradication of common mynas (*Acridotheres tristis*) succeeded on 63 ha Aride Island, Seychelles (Lucking and Ayerton 1994), but most other attempts have failed because the effort or tactics required to get the last, smart individuals have not been applied or are unavailable (Parkes 2006b).

Some species also appear to retain higher levels of neophobia in the initial uncontrolled populations than others. For example, one explanation of the failure to achieve eradication of rabbits with aerial poison baiting is that all rabbit populations contain some neophobic individuals (Parkes 2006c).

As already mentioned, aquatic pests are also difficult to eradicate. Fish, crustaceans, and reptiles have been eradicated by rotenone poisoning (Ling 2003), or physical draining in small enclosed ponds (e.g. O'Keefe 2005), but to date we lack the tools to attempt eradication in larger bodies of water or rivers. New developments in concentrating fish to small parts of their habitat where they might be vulnerable to extirpation (e.g. using pheromone lures for sea lampreys (*Petromyzon marinus*) (Li *et al.* 2003)) or in genetic manipulation to alter sex ratios (e.g. the daughterless carp project (Gilligan and Faulks 2005)) hold some hope for control of these pests. Whether the latter approach could be extended to achieve eradication without extra conventional control is highly debatable given the fundamental constraints of natural selection for those that do produce daughters.

Annual weeds are also very difficult eradication targets. Some can reproduce before they are detectable (e.g. the composite bitterweed *Helenium amarum*) (Tomley and Panetta 2002) while others, such as root parasites, are only detectable in some seasons (Panetta and Lawes, 2007). Sustained control may be the only feasible option to manage such species.

4.6.6 Institutional commitment

Eradication can be expensive in the short term, and tends to lead some funding agencies, particularly government ones, to seek to spread the costs across years. This of course increases the risk that eradication will fail, and several authors (e.g. Myers *et al.* 2000) have noted the requirement for institutional commitment to the concept, conditions, and possibility as a prerequisite for success.

Commitment to eradication on islands (where the obligate conditions are most easily met) is increasing for vertebrate pests at least. Since 2001 we are aware of reviews of the problem and priorities for action conducted by Britain and France for their overseas territories (Lorvelec and Pascal 2005; Varnham 2006), by Ecuador for the Galapagos Islands (B. Milstead pers. comm.), for Mediterranean islands (Genovesi 2005; P. Genovesi pers. comm.), for the Pacific (Sherley 2000), for the Aleutian Islands (G. Howald pers. comm.), and for Mexican islands (A. Aguirre pers. comm.). National plans are in preparation for introduced rodents on Australian islands and for all biosecurity issues in Mauritius. To a large extent these national and international plans should reinforce the flow of funds for eradication attempts in the near future. A caution for enthusiasts is that institutional commitment can as easily turn off if a few high-profile attempts fail because of foreseeable flaws in the planning process.

4.6.7 Local elimination

A strategy between eradication and sustained control is where it is technically and logistically possible to remove all the target pests or weeds, but where immigration is certain, more-or-less frequent, but manageable. Three versions of local elimination of mammal pests are being implemented in New Zealand:

- First, pests such as stoats have been removed from several islands in Fiordland. However, the islands are all within swimming distance for stoat populations on the adjacent mainland (King and Murphy 2005) and so immigration is likely and in fact the frequency has been measured for one island cleared of stoats (Parkes and Murphy 2003).
- Second, areas of native vegetation are enclosed with fences that restrict immigration of all alien mammals ('mice to deer' is the claim) and the resident populations of such mammals within the 'mainland island' are eradicated (Parkes and Murphy 2003).
- Third, as we know from islands, some pest species can be eradicated with a single application of a control method. Trials are being conducted for some key pests on parts of New Zealand's main islands to compare the costs of achieving zero density (and managing buffers to slow immigration) with the costs of a normal sustained control strategy with higher residual densities and more frequent control operations (Morgan *et al.* 2006).

4.7 Conclusions

So how much have we progressed in halting the homogenization of the world's biota since 2001? First, the bad news. A glance at journals such as *Biological Invasions* suggests the answer is 'not a lot' for continental mainlands, not much either on mainlands or islands for weeds or invertebrates, and hardly at all for marine species. The answer is also not much for vertebrates used within the pet trade—Australia, for example, has 1200 fish species kept in the pet trade of which 485 may still be legally imported, and 35 of which (so far) have established wild populations (Vertebrate Pest Committee, unpubl. data). However, most countries are attempting to regulate the legal importation of risky new species (e.g. Bomford

2003), and so this risk should at least not get any larger. Illegal importation into the pet trade and smuggling of wildlife and plants remains a concern.

The good news is that on islands managers have aimed higher, although largely for vertebrate pests and mostly for mammals. Rodent and ungulate eradication is becoming routine and is being planned for increasingly larger areas. This is important if eradication is to make a substantial contribution to the conservation of native biota and ecosystems. New Zealand is often regarded as a testing ground for eradication but despite its successes only 0.008% of the total land area of its 710 islands have never had exotic mammals (the main animal threat) and only 0.15% of the area has been rendered free of exotic mammals by eradication (Parkes and Murphy 2003). Eradication is thus a vital pest management strategy, but insufficient by itself. It must be planned along with border biosecurity and sustained control. Prevention of reinvasion of areas that have been cleared of invasive plants or animals is, of course, also vitally important.

Further good news is that there is substantial research to develop new control techniques or to modify older techniques, often developed for sustained control strategies to suit eradication aims. As examples, species-selective toxins or control devices (e.g. Marks 2001) may allow pests to be targeted without the expense of managing non-target animals at risk to the broad spectrum toxins now used. Modern genetic tools allow better understanding of the source-sink dynamics of populations (e.g. Robertson and Gemmell 2004), and can identify and manage reinvasion risks and pathways (Abdelkrim *et al.* 2007). Finally, the wheel has turned full circle and the ability to achieve 100% population reductions learnt on islands are being applied to mainland problems, either to achieve eradication (Parkes and Murphy 2003) or to reduce sustained control intervention frequencies by limiting rates of recovery of pest populations (Morgan *et al.* 2006).

4.8 Acknowledgements

We thank D. Morgan and P. Cowan for comments on drafts of this chapter, and J. Hone and M. Bomford for earlier discussions on the essential conditions for past eradication.

5
Principles of containment and control of invasive species

Tony Grice

5.1 Introduction

Biological invasion can be described as a process involving several stages (Fig. 5.1) though delineation of transitions between stages can be quite arbitrary (Cousens and Mortimer 1995; Groves 1999, 2006; Colautti and MacIsaac 2004). Five stages of invasion can be identified: pre-introduction, introduction, naturalization, expansion, and 'saturation'. The time of initial introduction is followed by naturalization, defined as the stage when a species is able to 'reproduce consistently and sustain populations over many life cycles without direct intervention by humans' (Richardson et al. 2000). The naturalized organism then increases until it reaches the limits of population size and range ('saturation') that will be determined by some combination of biotic and abiotic factors. These may include human intervention.

A strategic approach to the management of an invasive species can be defined as a spatial and temporal distribution of effort that yields the greatest benefit from the available resources (Grice 2000). Four broad strategic goals for countering biological invasions can be identified: prevention, eradication, containment, and control. The first two goals are mutually exclusive but the distinction between the last two is scale dependant. These goals can be aligned with the five generalized invasion stages. Prevention, achieved through risk assessment, quarantine regulations, and biosecurity activities, is the only appropriate goal during the pre-introduction stage (see Chapter 1). Generally, eradication (see Chapter 4) is possible only prior to naturalization, during the earliest part of the expansion stage (Cacho 2004; Mack and Foster 2004; McNeeley *et al.* 2005), or for isolated populations. When eradication of an invasive organism becomes impractical, the strategic options are containment and control. Containment may be the most appropriate strategy for a species during the early expansion stage, whereas a control strategy is likely to be most suitable for an advanced stage with a large and extensive population. This chapter focuses on the principles relating to effective containment and control of invasive species, and the timing and placement of limited resources necessary to maximize the efficiency of their use.

The potential range of an invasive species, as for an indigenous species, is determined by a combination of biotic and abiotic factors. Biotic processes may involve

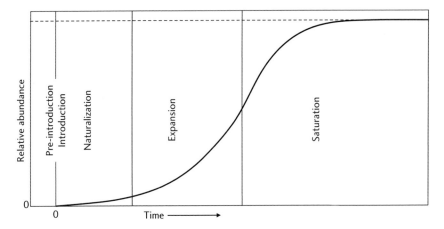

Fig. 5.1 Five phases of invasion: 1. Pre-introduction; 2. Introduction; 3. Naturalization; 4. Expansion; 5. Saturation (Cousens and Mortimer 1995; Groves 1999, 2006; Colautti and MacIsaac 2004).

inter-specific competition, predation, parasitism, and the availability of pollinators and dispersal vectors. Abiotic factors may be climatic or edaphic, or topographic barriers to dispersal. Containment can be defined as a management strategy that seeks to impose additional anthropogenic limits to the distribution of an invasive species so that the realized range is kept to a fraction of the potential range (Kriticos et al. 2006). 'Partial containment', refers to a strategy that involves attempting to slow the rate of spread of an invasive species (Cacho 2004).

I have described containment in terms of restrictions to an invasive species' distribution. A containment strategy can be applied at any management scale, for example, from a whole country down to the level of a single paddock. Whatever the scale, a containment strategy focuses on restricting the species range. Generally, containment strategies applied at the finest scales are more likely to be applicable to sessile organisms such as plants. In practice, the spatial distribution of effort—a key aspect of any strategic approach to the management of an invasive species—may be similar, regardless of whether the objective is containment or control.

Control is defined here as management that attempts to reduce the impact of an invasive species without necessarily restricting its range. Under a control strategy, an invasive species' realized range may approximate its potential range but the goal is to reduce its abundance, and so the impact, to levels below those it would otherwise attain.

5.2 Control and containment—strategies without an end-point

Containment and control strategies, including biological control (see Chapter 6), seldom have an end-point and require activity indefinitely. Containment is a valid strategy only for organisms that are expanding their range and where it may be possible to curb or halt the rate of range expansion. It is conceivable that a 'permanent' barrier to range expansion could be put in place, or that an effective biological control programme alters the demographics of the invasive species so that the likelihood of range expansion is radically reduced. However, there are few, if any, examples of this being effective, and many where it has not. Historical examples of attempts to impose barriers to range expansion such as the construction of thousands of kilometres of fences to prevent the spread of European rabbits (*Oryctolagus cuniculus*) in Australia were notorious failures (Rolls 1969; Noble 1997). The 'end-point' of containment programmes is most likely to be that an invasive species 'is *being* contained' rather than that it 'has *been* contained'.

The effort and resources expended in order to maintain containment and control strategies can indicate success where these decline over time. Rarely, an extremely successful containment strategy may be replaced by an eradication strategy (Mack and Foster 2004).

5.2.1 When to contain, when to control

Once eradication has been rejected as a goal, the options are containment, control, or do nothing. The total area occupied by an invasive species and the complexity of its distribution will influence whether containment is possible. This is because a containment strategy involves putting most effort into managing a species at the periphery of its range. The greater the area the species occupies, the more disjunct its distribution and the more convoluted the range boundary becomes, the longer will be the potential management periphery. Control then becomes a more likely strategic option. Consequently, containment is most likely to be viable during only the early stages of an invasion (Fig. 5.1), consistent with the general benefit of early intervention in pest management (McNeeley *et al.* 2005). Whatever the strategy, the earlier management action is commenced, the more economical and successful it is likely to be.

5.2.2 Feasibility of containment

Three sets of factors determine both invasion potential and the feasibility of containment:

- The characteristics of the invasive species;
- The characteristics of the environment that is being invaded;
- The management regime that is imposed on that environment (Grice 2006).

Together they will determine the resources that would be required in order to contain the species.

Species that are highly fecund, highly mobile and/or readily dispersed, and that have short generation times and broad ecological tolerances, are more likely to rapidly invade suitable environments (Rejmánek *et al.* 2005). 'Propagule pressure', defined as 'the number of individuals released into a region to which they are not native', is proposed as a major determinant of the establishment success of an introduced species (Lockwood *et al.* 2005; Dehnen-Schmutz *et al.* 2007). It is also a key driver of the rate of population growth and range expansion. Possession of a persistent, dormant life-stage (e.g. long-lived seeds) may also increase a species' establishment capacity in environments novel to it. Species that have specific requirements for pollinators, dispersal agents, or other symbiotic organisms are likely to be less invasive where those organisms are absent (Rejmánek et al. 2005).

Species with traits that facilitate rapid invasion are often the most difficult to contain. For highly fecund species, a containment strategy may have to contend with a strong capacity for population growth. Efficient dispersal abilities increase the probability the containment zones will be breached. Short generation times will mean that, following such a breach, there will be only a narrow window of opportunity in which to detect and eliminate new outbreaks before they in turn become source populations. For species with broad ecological tolerances there will be fewer spatial or temporal barriers to range expansion and population increase.

Characteristics that determine how readily individuals and populations of a species can be detected also influence the feasibility of containing them (McNeeley 2005). Species that are cryptic during all, or part, of their life cycle may be more difficult to contain than those that are more readily detected (e.g. Panetta and Lawes 2003; Correll and Marvanek 2006).

Containment of invasive species will be less feasible in some environments and landscapes than in others. In the case of terrestrial weeds, complex topography, dense vegetation, and heterogeneous environments reduce the probability of detecting new populations. Any environmental or landscape characteristics that reduce accessibility increase the costs of containment. Some environments also restrict the management options that are available. For example, the use of many herbicides is not appropriate in wetland or riparian environments because of the risk of those herbicides or their break-down products moving away from the target site (e.g. Krutz et al. 2005).

Many aspects of the management regime imposed on a landscape can influence the feasibility of containing an invading species. In general, it will be easier to contain species invading landscapes that are intensively managed. In such situations, detection and access are less likely to present major problems. Intensively used landscapes tend to support higher human populations that may be in a better position to provide the human, technological, and financial resources required to detect and treat an invasive species that has breached a designated containment line. On the other hand, the higher human populations likely to be found in intensively used landscapes may provide anthropogenic means of dispersing invasive species (e.g. ants), and increase the social challenges to containment (van Schagen et al. 1994; Plowes 2007). Unless there is universal acknowledgement that a particular invasive species is a problem, there will be conflicting views on how important containment is. Community members or organizations that do not view the species as a problem, whether for them or for the community as a whole, may fail to comply with legislation or to cooperate in management actions designed to contain the species. These issues are likely to be especially significant for species that are perceived to have benefits, as well as environmental costs.

Principles of containment and control | 65

5.2.3 Elements of a containment strategy

Any containment programme designed to operate at a fairly large scale is likely to require the following activities on an on-going basis:

- Reliable identification of the target species.
- Effort to detect the invasive species, with that effort being focused on areas outside the area to which the programme is intended to contain it.
- Reporting of occurrences in order to be able to undertake control measures and to document progress of the programme.
- Development of appropriate policies in relation to the invasive species and its management; and the means to implement those policies.
- Regulations relating to the invasive species, its cultivation, transport, and control, and policing of those regulations.
- Provision of suitable resources and direction of those resources to locations/stakeholders in a manner consistent with the specific objectives of the containment strategy; a containment strategy requires decisions about how the costs of containment will be distributed amongst the various stakeholders.
- Education and extension to indicate the need for, approach to, and progress of the containment programme.
- On-ground activity targeted to use the available resources most effectively to eliminate the invasive species from areas outside the containment zone and reduce the risk of dispersal beyond it.
- Research to develop appropriate detection, management, and monitoring techniques.
- Monitoring of progress that can provide feedback to improve the containment programme or inform decisions about changes of strategic direction.
- Coordination of government agencies, community groups, researchers, land managers, and other stakeholders, including exacerbators.

5.2.4 To control or not to control

Where neither eradication nor containment of an invasive species is possible, the only options are to control or to do nothing. The decision about whether to attempt control or not will depend on:

- The importance of the impacts that the invasive species has, or is perceived to have, relative to the likely costs of control.
- The stage to which the invasion has progressed.
- The availability of control measures.

Clearly, if no effective control measures are available, control cannot be attempted. To use ineffective control techniques under such circumstances would be a waste of resources.

The impacts of an invasive species may be ecological, economic, and/or social, and those impacts may be positive or negative. There are many cases in which individual invasive species have positive impacts from the perspectives of one or more interest groups and negative impacts from the perspectives of others (see section 5.4). These issues will influence decisions about how much effort is put into control, where that effort is expended, and how the costs and benefits are distributed amongst stakeholders. Because impacts may be ecological, economic, or social, they cannot be readily quantified in a single currency though some attempts have been made to quantify environmental impacts using economic concepts (e.g. Sinden *et al.* 2003). Decisions about what, if any, action to take will inevitably involve a degree of subjectivity taking various societal values into account.

The relationship between the abundance and impact of an invasive species is important. The best documented cases of this relationship are probably those involving weeds of agriculture or pastoralism (Cousens 1985). The relationship between percentage yield loss (of a crop) and the abundance (density) of a weed has been described by a rectangular hyperbolic curve (Fig. 5.2). Applying this relationship to invasive species in general, impacts increase with higher abundance but at a decreasing rate. At low abundance, an invasive species will have negligible or undetectable impacts, while above some threshold, further increases in abundance result in no greater impacts.

The abundance of an invasive species depends on the carrying capacity of the invaded environment and the stage to which the invasion has progressed. This means that in the early stages of an invasion (Fig. 5.1), impacts are likely to be low. A period of low impact may extend for some time given the time-lags that have been documented to occur during many invasion processes (Mack *et al.* 2000; Grice and Ainsworth 2002; Groves 2006). The costs of control during an early

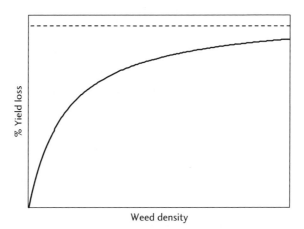

Fig. 5.2 A rectangular hyperbolic curve describes a relationship between percentage yield loss of a crop and the abundance (density) of a weed (Cousens 1985). A similar relationship could be expected to describe the relationships between density and other impacts of invasive species.

stage of low abundance may be relatively low, though detection costs per individual will be relatively high. On the other hand, in the early stages of an invasion, there may be little motivation to take action simply because impacts are low. The perception may be that the species in question does not warrant attention. This highlights the need for a capacity to predict the impacts of an invasive species on the basis of observations and data collected during the early stages of the invasion or from other parts of its introduced range. Such information, therefore, provides some basis for making decisions to control (or eradicate or contain) at a stage of invasion when those options can be pursued cost-effectively.

5.3 Principles of containment and control

5.3.1 Evaluate impacts of invasive species

Priorities for programmes to eradicate, contain, or control invasive species should be based on an evaluation of their impacts. There are few, if any, quantitative data for most invaders but action should not be delayed while waiting for comprehensive measures of their impacts. World-wide there is a reasonable body of knowledge of the processes and impacts whereby invasive species affect ecosystems (Mooney 2005). Extrapolation from data gathered elsewhere, combined with local observations and informed opinion, often enables some evaluation of a species' impact in systems for which there are no or few quantitative data (Grice *et al.* 2004). Convincing evidence of specific impacts is especially useful in the case of contentious invasive species where it may be necessary to use firm evidence to help resolve conflicting interests.

5.3.2 Assemble knowledge of species' biology, ecology, and responses to management

Containment and control of an invasive species must be based on knowledge of its biology and ecology, including the specifics of how the species functions in its invaded range (e.g. Grice 1998; Campbell and Grice 2000). Knowledge of the species' life cycle, basic habitat requirements, means of dispersal, and reproductive capacity are useful. For plants, the time taken to first reproduction is important because it determines the minimum time before propagules newly arrived at a site can produce plants that are themselves sources of seeds. Seed longevity and germination requirements are important in determining how long a site must be monitored after established individuals have been removed. It is also useful to know about a plant's capacity for vegetative reproduction and spread.

Measures to counter incursions of invasive species will always be based on imperfect knowledge of their biology and ecology. Especially in cases of new incursions of high-risk species, countermeasures should not be delayed more than absolutely necessary because an early response can be critical (see section 5.3.6). Additional knowledge can be acquired as containment and control programmes are implemented and used to refine the approaches taken. It is also important to build knowledge of how an invasive species responds to different control measures and the conditions under which they are applied. More effective and efficient measures can be developed through both observations and systematic experimentation where the species is being managed under field conditions.

5.3.3 Map distribution and abundance

Successful containment or control programmes require that their targets are detected. The precision required of detection methods will vary with the circumstances, including the specific objectives of the programme and the control methods. Early in a containment attempt, reliable but general information on distribution and abundance may be adequate to develop and begin implementing a management programme. As knowledge of a species' distribution and abundance improves, the programme can be modified. Detection may use casual sightings but a systematic sampling procedure will yield a more reliable picture of the distribution and abundance of an invasive species on which to base a more refined containment or control programme. There will, of course, be a trade-off between effort spent on detection and effort spent on other aspects of containment and control. Especially in the case of newly discovered, high-impact, rapidly invasive species, an early response to an initial detection can be critical, and such a response should not be delayed by efforts to delineate its distribution in detail.

5.3.4 Set priorities for species and places

As resources available for managing invasive species will always be insufficient to address all problems, the species to be targeted and the locations where effort will be expended must be prioritized. Containment and control are usually devised as strategies to be applied against particular species though management of multiple invasive species may be a general objective in particular locations, for example, in areas of high conservation value (see section 5.3.5). Examples of location-focused priorities include cases of predator control to conserve remnant populations of threatened bird or mammal species (e.g. Risbey *et al.* 2000; Burrows *et al.* 2003; Innes *et al.* 2004) However, even in situations where more than one species is being targeted, each invasive species will often require specific treatment at the operational level. Once a decision has been made that an invasive species is to be targeted for containment or control, a strategy is required that is appropriate to the objective (containment versus control), the species, and the environment (ecological, social and economic) that it is invading.

The containment of an invasive species should generally be directed at the periphery of its distribution. This may or may not be the case where the objective is control. Decisions about where control effort should be directed may be made with reference to local, catchment, regional, or continental scales, and there may be some conflict between decisions at these various scales. The interests of a manager of an individual unit of land will be focused mainly on actions directly relevant to that unit of land even though the same level of effort may be more cost-effective and strategic if it was directed elsewhere. Decisions will be influenced by the measured and perceived impacts, the costs and benefits of control actions, and how those costs and benefits are distributed amongst stakeholders. A number of action scenarios are possible under the influence of these potentially conflicting forces:

- Effort is directed toward places where the invasive species is most abundant. Such a decision could be based on (a) a realization or perception by managers of such locations that the invasive species is having major impacts there; (b) information or perception that by directing effort at areas of high population density, the risk to adjacent areas may be reduced or that population growth by the invasive species will be more severely impaired.
- Effort is directed toward places where the invasive species is least abundant. The decision to take this course of action could be based on a perception or information indicating that the costs of control are relatively low in such circumstances and impacts that would result from a higher density of the invasive species can be avoided or delayed.
- Effort is focused on areas of particularly high value irrespective of the costs. Such a strategy could be taken, for instance, where a particular species, community, ecosystem, or human enterprise is especially susceptible to the invasive species.

5.3.5 Coordinate management of multiple, functionally similar invasive species

Many ecosystems are subject to invasion by multiple alien species that have similar impacts. One example of this is discussed in section 5.4.3 where a number of introduced predators combine to affect the survival and breeding success of endemic New Zealand birds. Often, many species of invasive plants combine to affect the composition and structure of native vegetation (e.g. Lawes *et al.* 2006). Attempts to address problems caused by a single invasive species in an environment in which there are multiple, functionally similar invasive species is likely to be an ineffective use of resources.

5.3.6 Take action early in the invasion process

In the case of containment, the benefits of early intervention will take the form of potential range not occupied. Early containment action has two advantages: one is that the boundary to the species' distribution along which effort has to be expended is shorter, so the effort required is lower; the other is that the ratio of occupied range to unoccupied potential range is lower so that, assuming the containment programme is effective, a greater area will be saved from invasion.

In the case of control, early intervention will generally involve dealing with a lower overall density of the invasive species which may be less expensive. Where it is necessary to kill or capture each individual, the costs per individual will often be higher when the density is low because the same search effort may be required but it is distributed over fewer individuals. Where a broad-scale control method is employed (e.g. the use of fire to control invasive plant species) and it is not necessary to treat each individual separately, the costs per unit area may be the same regardless of the density, though again, the costs per individual treated will be greater at lower densities.

5.3.7 Direct effort where benefit:cost ratio is high

An ability to direct effort to where the benefit:cost ratio is high requires that costs and benefits are sufficiently well known to justify any decisions. It requires some understanding of how the abundance (density) of the particular invasive species relates to the impacts that it has, the risks it poses in terms of further population increase and range expansion, and the costs of management actions against it. Despite control costs generally being lower where densities of the invasive species are lower (though see section 5.3.6), locations with both higher abundances and high value environmental or economic assets may receive greater effort. There may also be situations where areas of high abundance are given low priority because assets are already highly degraded. Effort may target infestations of an invasive species that are located where they present a high risk of giving rise to new infestations or where they are adjacent to high-value assets. Controlling such strategically located populations would yield a high return on investment.

5.3.8 Direct containment effort at the periphery of an expanding distribution

On theoretical grounds, Moody and Mack (1998) showed that control of outlying populations (nascent foci) of an invasive species is important in slowing the rate of invasion. Concentrating effort at an invasion front will likewise probably be the most cost-effective way of containing it, although this may depend on the level of resources available (Taylor and Hastings 2004). Clearly, this requires knowledge of the species' distribution and abundance (see section 5.3.3) and of dispersal mechanisms, timing and distances, and a capacity to kill or remove outlying individuals.

5.3.9 Exploit natural barriers to range expansion

Containment involves, in effect, placing barriers to range expansion by reducing reproductive output, interrupting the dispersal process itself, or curtailing recruitment. For any species a landscape will consist of habitats of varying suitability. Areas of low-suitability or unsuitable habitat represent at least partial barriers to range expansion that can be exploited in containment programmes. Their effectiveness will depend upon the dispersal distances achieved relative to the width of areas of poor habitat. Containment effort may be directed so as to enhance the effectiveness of the barriers. Effective natural barriers could include mountain ranges, rivers and other water bodies, depending on the ecological requirements of the invasive species involved.

5.3.10 Exploit times when invasive species' populations are low

In the annual life cycle or history of an invasive species, some occasions may present particular opportunities for more efficient or more effective control. These may relate to stages of the life cycle that are more prone to particular treatments or to weaknesses imposed by episodic climatic or other circumstances. Applying control measures prior to major reproductive episodes may reduce rates of spread and/or the time during which follow-up treatments may be required. This applies, for example, to invasive plant species that build long-lived soil seed-banks and is especially important in the case of new infestations where seed-banks have not yet accumulated. When rabbit calicivirus was introduced to Australia in 1995, the disease spread rapidly and killed up to 95% of feral rabbits (*Oryctolagus cuniculus*) in some areas. Land managers were encouraged to reinforce the effects of the disease by mechanical treatment of rabbit warrens even though rabbits were in very low numbers.

5.3.11 Acquire continuing commitment

Containment and control programmes should be developed as long-term strategies because there is generally a low likelihood that a particular programme will achieve an end-point where no further action is required. An important implication of this is that there will be a continuing need for resources. Careful consideration should be given to how these resources will be obtained.

Important decisions in the development of containment or control programmes relate to how responsibility for the programme will be attributed to the various stakeholders. This must cover the relative contributions of public and private sectors. Socio-economic factors are crucial here. Decisions about responsibility will be affected by the various stakeholders' perceptions about who is to blame. In Australia, many invasive plant species were originally promoted to the private sector (particularly grazing industries) as beneficial species, only to subsequently be recognized as problematic. Two examples are *Kochia scoparia*, which was introduced as a pasture species for saline soils in Western Australia (Dodd and Moore 1993), and prickly acacia (*Acacia nilotica*), which was promoted as a shade and fodder tree for use in naturally treeless *Astrebla* spp. grasslands in western Queensland (Mackey 1998). The former was subsequently targeted by a largely publicly-resourced eradication campaign (Dodd 2004) while the latter, was declared a Weed of National Significance (Thorp and Lynch 2000) and became the focus of a national management plan that continues to receive substantial funding from Commonwealth and State governments (QNRME 2004a).

There are several other factors that may encourage substantial inputs from the public sector into containment and control programmes. One is a need for very rapid responses in cases where new, high-risk incursions are discovered in the very early stages of invasion. A publicly funded and coordinated programme may facilitate a timely response to such incursions and also help resolve conflicting interests if these arise. Part of this support could be in the form of incentives to those who stand to lose from a containment or control programme. Finally, a need for stronger public support of containment and control programmes may arise when the benefits that are predicted to derive from them are largely 'public-good' benefits.

Effective containment or control programmes will require less effort and fewer resources as they proceed. Changes in the relative contributions of the various stakeholders could be justified or required as a result. However, some level of effort will always be required to ensure an acceptable level of control or containment. Even in an ideal situation where, for example, a biological control has greatly reduced population density and growth rate, there will be a need for a continuing commitment to monitor the status of the invasive species.

5.3.12 Resolve conflicting interests

To make progress against contentious invasive species it is important to resolve conflicting interests. Conflicts can relate to a species being detrimental to one group of stakeholders but beneficial to another; or to an attitude of neutrality by one group in the face of negative consequences for others. Examples include plant species that are commercially exploited or cultivated for ornamental purposes but which are also invasive (Grice 2006). 'Resolution' of these conflicts almost inevitably involves costs to both sides of the debate. Possible scenarios include:

- Prohibition of commercial exploitation of a species that is invasive. This presents a potential opportunity cost to some stakeholders.
- Regulated commercial exploitation of a species; regulation could involve restricting where the species may be cultivated and establishing protocols for that cultivation. This approach is really only applicable to species that are exploited as 'domesticated' populations, rather than to exploitation of feral populations.
- Control of commercially valuable species in locations where they are not exploited. This applies to productive species that are already widely naturalized or that cannot be contained for other reasons. It is effectively the case for widespread, introduced pasture species in Australia, (e.g. *Cenchrus ciliaris*, *Andropogon gayanus*) that have negative consequences for natural ecosystems and effort to control them is generally restricted to threatened conservation reserves. Such a scenario will, rightly or wrongly, tend to put the onus for overcoming impacts on those responsible for lands whose values are threatened, rather than on those who benefit from the commercialization of the invasive species. Various regulatory mechanisms, such as levies to growers, could be used to transfer some of the costs from those who experience the negative consequences of a commercially-valuable, invasive species to those who enjoy the benefits.

5.3.13 Monitor the consequences

The progress and consequences of containment and control programmes must be monitored. This provides feedback so that a programme can be modified, or for that matter abandoned, if it is found to be ineffective. Monitoring could attempt to document changes in the distribution and/or abundance of an invasive species, or changes in components of the system it is invading. In the latter case, monitoring should focus on those elements known to be affected by the invasive species. Invasive species other than those targeted could be monitored to assist in understanding whether control of the target species facilitated population growth or range expansion of others. Well-designed monitoring programmes will contribute to knowledge of the impacts of particular invasive species, and of the value of the containment and control programmes used to help manage them.

5.4 Examples

5.4.1 Containment of rubber vine in northern Australia

Rubber vine (*Cryptostegia grandiflora*) was introduced to Australia from Madagascar as an ornamental shrub/vine in the late 1800s. In the 1940s it was also grown experimentally in plantations as a possible source of commercial rubber. It is now widely naturalized. Its current distribution occupies a large part of tropical northeastern Australia but it is predicted to have a potential range covering 20% of the northern half of the continent. The species is problematic from two perspectives: it forms dense stands that out-compete native species, radically changing the structure and composition of the vegetation and having flow-on effects for other components of invaded communities; it is detrimental to grazing industries that rely on these systems because it replaces plants that are palatable to cattle, is toxic to livestock, interferes with animal husbandry, blocks access to water, and harbours invasive animals such as feral pigs. Rubber vine is especially prevalent in riparian zones and other low-lying parts of the landscape (Tomley 1998).

Rubber vine is universally recognized as a problematic plant and has been declared a 'Weed of National Significance' in Australia's National Weed Strategy (Thorp and Lynch 2000). This status prompted and facilitated the development of a national strategy for the management of rubber vine involving considerable funding support from the Commonwealth Government of Australia, coordination by a department of the Queensland State Government of on-ground action by multiple stakeholders, as well as research and education activities (QNRME 2004b). A key objective of this national strategy has been to contain rubber vine within its current distribution in north Queensland. This aims to prevent the westward spread of rubber vine into extensive areas of potential habitat. Effort has focused on the southern and western areas of the species' Australian distribution. It has required detection of infestations outside the designated containment line and targeting those infestations for control. Previously unknown infestations were located in Western Australia and the Northern Territory, well outside the containment zone and have been the target of control operations. Prior to the development of the national strategy, biological control research was undertaken. This yielded two biological control agents (the rust *Maravalia cryptostegiae* and the moth *Euclasta whalleyi*) (Tomley 1998), both of which are now well established across much of rubber vine's Australian distribution and result in periodic severe defoliation, greatly reduced seed output, and, in at least some areas, significant mortality (Vogler and Lindsay 2002).

This example illustrates many of the principles outlined above. The attempt to contain rubber vine is facilitated by broad recognition (if not quantification) of its impacts and the fact that there are few conflicting interests surrounding it. General knowledge of the species' biology/ecology is good, effective management techniques (e.g. biological control and burning) are available, and the distribution has been documented. It is the target of a nationally-coordinated campaign that has strategic goals. This effort to counter the invasion by rubber vine began many decades after the species first naturalized in Australia so it does not provide an example of early intervention. Also, little overt consideration is given to the consequences of there being many other invasive plant species in the same ecosystems.

5.4.2 Containment of leucaena—a commercially grown fodder shrub in Australia

The South American shrub leucaena (*Leucaena leucocephala*) was deliberately introduced to Australia and widely cultivated for use by the northern Australian cattle industry. It naturalized in many areas and is now also perceived to be an environmental weed (Walton 2003). There are, as a result, conflicting interests surrounding this species. Its status as a weed has not been recognized in legislation though a Queensland State departmental policy statement considers the issue and provides recommendations in relation to management of the species (Queensland Government 2004). Currently, management of leucaena as an invasive species is voluntary. A Leucaena Network has developed a code of practice (Queensland Government 2004) designed to help reduce the risk of new infestations arising from existing plantations. Any other attempts to deal with existing infestations and impacts are dependent upon local motivation. The situation with leucaena contrasts with that of rubber vine, a far less controversial invasive species. The approach to leucaena as an invasive plant is less strategic and there is little government funding. Leucaena also illustrates the challenges of containing commercially important species that are legally cultivated on a very large scale in widely dispersed plantations. The leucaena code of practice does not address problems associated with the many growers who operate outside the voluntary code, nor the issue of naturalized populations that existed before the code of practice was prepared. This means that many infestations remain untreated.

5.4.3 Control of invasive mammalian predators in New Zealand

Invasive mammalian predators are major threats to native species and ecosystems, particularly on islands that lacked similar native predators. In New Zealand, introduced mammalian predators, including cats (*Felis catus*), brush-tailed possums

(*Trichosurus vulpecula*), rats (*Rattus rattus, R. norvegicus*), house mice (*Mus musculus*), stoats (*Mustela erminea*), weasels (*M. nivalis*), and ferrets (*M. furo*) prey on native birds and have been a key factor in driving a large proportion to extinction and dramatically reducing the abundance of those species that survive. One strategy for dealing with this problem has been to maintain threatened species on predator-free off-shore islands. Another approach has involved attempts to create predator-free 'mainland islands'. On the mainland of New Zealand, eradication of invasive predators is generally not feasible even over relatively small areas, even when predator control is combined with predator-proof fencing. This leaves the two options: control or 'do nothing'.

There are many examples of documented benefits for native New Zealand bird species from control of introduced predatory mammals: control of ship rats and brush-tailed possums led to increased fledging success and adult population numbers of New Zealand pigeon (*Hemiphaga novaeseelandiae*) (Innes *et al.* 2004); control of ship rats (*R. rattus*) resulted in significantly higher breeding success of New Zealand pigeon (Clout *et al.* 1995); control of stoats was associated with increased survival of nestlings and densities of adult bellbirds (*Anthornis melanura*) (Kelly *et al.* 2005); successful breeding of kaka (*Nestor meridionalis*) in an area in which stoats were trapped compared with in other areas where there was no predator control (Dilks *et al.* 2003).

Data collected during these control programmes included the number of predators of different species that were trapped, breeding success and abundance of important bird species, aspects of responses by vegetation, and economics of the control operation. These studies enabled an evaluation of impacts and increased knowledge of the ecology of the main elements of the system. Prioritisation, detection, documentation of change, multi-species approaches, and monitoring of both the invasive species and the components of the systems that are affected by them, are incorporated into these approaches. They also illustrate the need for continuing commitment.

5.4.4 Invasive pasture grasses in Australia

Several grasses introduced to Australia exemplify invasive species that have both positive and negative impacts that influence attitudes toward them and management decisions concerning them. Buffel grass (*Cenchrus ciliaris*) was sown in Australian rangelands as a pasture grass for the cattle industry and to reduce erosion in arid and semi-arid pastoral country (Humphreys 1967). The species naturalized widely or expanded its range to suggest it could cover much of the continent (Humphries *et al.* 1991). Buffel grass is an important introduced pasture species for the northern Australian cattle industry (Walker and Weston 1990) but it is now also seen as having deleterious consequences for native species and communities (Griffin 1993; Low 1997). These consequences include reduced native plant species richness (Franks 2002; Clarke *et al.* 2005; Jackson 2005) and fuelling of high intensity, more frequent fires (Clarke *et al.* 2005). The conflicting interests surrounding this invasive species, and its advanced stage of invasion, present challenges to attempts to contain or control it. The most efficient strategy in such cases is to focus control at locations of high conservation value that are not used for cattle production (e.g. conservation reserves).

A different approach could be taken with gamba grass (*Andropogon gayanus*) another species used by the cattle industry in northern Australia. It is currently less widespread, less important to the grazing industry, and probably has even more serious environmental impacts, especially through its effects on fire regimes (Rossiter *et al.* 2003). Containment, or even local eradication, may not only be possible with this grass but it may also be more desirable from an environmental perspective and more socially acceptable. These two examples illustrate how positive and negative economic, environmental, and social factors interplay to determine whether containment or control are the more feasible option, and indeed, whether any management action is likely to be worthwhile.

5.5 Conclusions

In this chapter, 'containment' and 'control' have been defined as management strategies that seek to reduce the impacts of invasive species that cannot be eradicated on a broad scale. 'Containment' is management that specifically attempts to restrict the distribution of an invasive species to a fraction of the range that it would otherwise occupy, while 'control' tries to reduce impacts without necessarily limiting distribution. Sound decisions in relation to these two options will depend on the characteristics of the invasive species, the landscape that it is invading, the management regime imposed on that landscape, and the resources that are available. Both strategies require an indefinite commitment of resources. They also demand some assessment of the impacts of the target species, and knowledge of its biology, ecology, distribution, abundance, and responses to management. This knowledge can be used to set priorities in terms of both species and the spatial distribution of management effort. It is important to consider the multi-species nature of many invasions, and failure to do so may result in an ineffective use of resources. Regardless of the objective (eradication, containment, or control), the earlier in the invasion process action is taken, the more efficient and effective that action is likely to be. Socio-economic factors are critical in deciding what is possible by way of containment or control and what are the most appropriate approaches; it is critical that any conflicting interests are resolved. Finally, it is important that the consequences of the efforts are monitored so that approaches can be modified as the containment or control programme proceeds.

5.6 Acknowledgements

I thank Dane Panetta, Lynise Wearne, and Peter Williams for useful comments on a draft of this chapter.

6
Biological control of invasive species
Sean T. Murphy and Harry C. Evans

6.1 Introduction

Concerns about the impact of invasive species on biodiversity have raised the profile of invasive species to that of an agent of major global threat and change. Thus the discussion on how to prevent and manage invasive species is no longer just an agricultural one. Biological control through the use of host-specific natural enemies, which has essentially been developed and moulded by the agricultural sector for over 100 years, has been seen by many in recent times as a potentially effective and environmentally benign tool, suitable for use against invasive species in natural ecosystems. The biological control strategy includes several techniques. The most common technique that has been used against invasive species is the 'introduction' or 'classical' approach, which involves the utilization of coevolved host-specific natural enemies from the native range of the target species.

Globally, most work on classical biological control has been for the control of arthropod and plant invaders but efforts have been made on invasive mammals, invasive marine organisms, and other species. There have been many successes in biological control and many 'moderate' successes (where only partial control was achieved), but there have also been many more failures. The technology has also been clouded by occasional situations where general natural enemies have been introduced for control of a target, but have also ended up feeding on non-target species. Nonetheless, the practice of classical biological control utilizing host-specific natural enemies has continued and over the years the agricultural sector has, with inputs from the ecological science community, evolved various protocols that try and address safety and minimize ecological risks. Modern biological control is now underpinned with these and a weight of other scientific and technical guidance to maximize success and minimize risk.

Here we review the relevant major points about the science, practice, and economics of classical biological control to illustrate the potential of the tool for use in natural ecosystems; the focus will mainly be on the biological control that has been developed for plant and arthropod invaders as most effort has been against these species. We then discuss some of the methods that have been developed, or are under development, to illustrate how practitioners are trying to promote 'best practice', particularly in relation to safety. We also look at more general regulatory issues and the dilemmas these are posing to practitioners. Finally, we discuss some of the general constraints faced when trying to implement biological control projects. Case studies are presented to illustrate particular points and principles.

6.2 Why classical biological control is an appropriate tool for managing invasive species

One major theory which has been used to explain biological invasions is that when a species is 'released' from its natural enemies, for example, through introduction by some means into a new geographical area, that species becomes invasive because the natural enemies no longer exert control. This theory has been implicit in the ecological literature for many decades but was made explicit through the 'Enemy Release Hypothesis' (ERH) (Keane and Crawley 2002). Whilst this is not the only theory that attempts to explain biological invasions (for a review see Mack *et al.* 2000) many now accept this as at least a core component in generating invasions. There are also elaborations of the ERH, the most cited being that related to plant invasive species. Here it is suggested that a plant released from herbivore pressure is able to evolve more in competitive traits (Blossey and Nötzold 1995).

Numerous studies since the beginnings of exact ecological science in the early 20th century have shown that natural enemies can suppress and/or regulate populations of their host (e.g. see reviews in Cullen and Hassan 1988; Jervis and Kidd 1996). All species in their native range host several natural enemies and it is now well known that these natural enemy communities can sometimes be diverse, especially on hosts such as the majority of plant and arthropod species. These natural enemies can either be specialist or generalist, the long association with a host generating a co-evolutionary relationship. It has been found that specialist natural enemies tend to have a major impact on their host. The species composition of natural enemy communities frequently varies throughout the native range of the host species, particular species being adapted to local conditions within the host's range. But some natural enemies do have a wide geographical range themselves and some studies have shown that, in least in some cases, natural enemies can limit the distribution of their host as well as abundance (Harrison and Wilcox 1995).

Classical biological-control practice is based on this ecological theory and supporting practical studies, but biological-control practitioners were using the notion of the ERH back in the early 1900s. The theory behind biological control is at the core of the argument of why this technology is considered by many (see Hoddle 2002, for example) to be the best way of controlling many invasive species: host-specific natural enemies are argued to be environmentally benign and in the case of classical biological control, the agents are self-replicating and thus able to disperse naturally in natural environments.

6.3 The practice of classical biological control

6.3.1 Early history and development

Classical biological control as we know it today grew as a means of addressing major plant and insect problems affecting crop production. The potential to use natural enemies for control was explored back in the 18th century when several attempts were made across the globe to introduce predators for the control of insect pests At this time and until about 1990, vertebrate predators were included as potential control agents but most introductions involving these were ineffective and had much worse side effects in terms of impacts on non-target species. A more professional approach, factoring in specificity of natural enemies and their likely impact, started in the late 1880s by applied entomologists working in countries such as the USA and Australia who were looking at ways of addressing major species outbreaks affecting whole crop industries (see Debach 1974 and Simmonds *et al.* 1976 for reviews). At these times, invasive species were referred to as 'pests'.

The first documented examples of the international transfer of natural enemies to control invasive species were in the 1870s when the predacious mite *Tyroglyphus phylloxerae* was sent from the USA to France for use against grape phylloxera (*Phylloxera vitufolia*), and the ladybird (*Coccinella undecimpunctata*), was despatched from the UK to New Zealand in an attempt to suppress invasive aphid pests. Neither of these appeared to achieve successful control, although their establishment was confirmed. However, momentum gathered, particularly in the USA, for this invasive species management approach, which resulted in the first great successes in classical biological control in the 1880s—which established it as a major method of pest control for invasive species. One major early success was the project conducted against the scale (*Icerya purchasi*) which was first recorded in the late 1860s in northern California; within two decades, the insect had reached pest status and was threatening the burgeoning citrus industry in southern part of the state, as well as other horticultural crops. Correspondence with Australian entomologists, ascertained that it was not a problem species in Australia, although it was a serious problem in New Zealand. A period of exploration resulted in shipments of a parasitic fly, *Cryptochaetum iceryae,* and a ladybird, *Rodolia* (*Vedalia*) *cardinalis*, from Australia to California. These species were subsequently established in 1888–89, and within a few years of release, all infestations in the State were under control; the ladybird was considered to be the most important agent. The cost of the project has been estimated at around US$5000, with the benefits to the citrus industry of California amounting to millions of dollars annually ever since. Furthermore, it has been calculated that similar successes have been achieved in more than 50 countries, with the Galapagos Islands of Ecuador being the most recent recipient of the rodolia ladybirds.

The importance of fungi as natural enemies of arthropods was realised at a very early stage, encapsulated in the later statements by Petch (1925) 'That such diseases do kill off large numbers of insects periodically and so exercise a considerable natural control is undoubted' and by Steinhaus (1949) 'Entomogenous fungi in nature cause a regular and tremendous mortality of many pests in many parts of the world and do, in fact, constitute an efficient and extremely important natural control factor'. Historically, interest in trying to harness this control potential emerged around the same time (1880–90) as the first attempts to exploit insect natural enemies. However, this was directed at mass producing and applying fungal inoculum from naturally-occurring outbreaks on both indigenous and exotic pests to increase their efficacy, especially using the ubiquitous white and green muscardine fungi, *Beauveria banana* and *Metarhizium anisopliae*, respectively (Samson *et al.* 1988).

As with invasive arthropods, the history of biological control of invasive plants is dominated by entomologists, and, indeed, the use of fungal pathogens as biological control agents is a relatively modern event, not taking off until the 1970s (Evans *et al.* 2001). The highly invasive plant, *Lantana camara*, from South America was one of the first targets and the same scientists who pioneered the biological control of the cottony cushion scale took centre stage. Early success with this was limited, however, due mainly to the complex of intra- and inter-specific hybrids involved. But other targets had less problematic taxonomies, such as the New World prickly pears (*Opuntia* spp.) which became a major problem in Australia in the late 19th century. A well-publicized major success was achieved against these plants in the 1920s through the release of the moth *Cactoblastis cactorum*, which was collected from the Americas. This success stimulated efforts against other invasive plants. Classical biological control has grown since these early times and has now been used extensively across the globe against approximately 550 invasive arthropods and 130 invasive plants in agriculture (see Greathead and Greathead 1992 and Julien and Griffiths 1998 for a review of projects and outcomes).

6.3.2 Biological control projects against invasive species in natural ecosystems

Classical biological control is now being increasingly used for the management of invasive species in natural ecosystems, particularly against invasive plants and insects. But this effort to date has largely been driven by the fact that these plants and insects are also a major problem in agriculture; in some countries, additional analyses have highlighted the importance of the problems caused by invasive species in natural ecosystems. Countries leading the way on this are Australia, New Zealand, South Africa, and the USA but a few other countries across the globe have a low level of activity. Australia is targeting several major invasive plants that affect natural ecosystems: for example, *Acacia nilotica, Ageratina riparia, Cryptostegia grandiflora*, and several others. Success has been achieved against some of these plants. For example, the European plant, St John's wort (*Hypericum perforatum*) has been a problem in parts of south and west Australia since the mid-1880s. The plant is mostly recorded as a problem in pasture (where it is poisonous to livestock) but it also invades native forests where it is a fire hazard in the summer months. A long-standing biological control effort using introduced herbivorous insects has resulted in long-term control in open sites but only partial control in more shaded areas (Briese 1997).

Invasive insect species have also been targeted with classical biological control and this has been a particularly active area of research and activity. The cypress aphid (*Cinara cupressi*), which originates from the Middle East, caused a major threat to native cedars in eastern and southern Africa in the late 1980s/early 1990s but a biological control programme was started because the aphid was considered a major threat to the plantation forestry industry (Day *et al.* 2003). Nonetheless, over the last decade or so, there have been several biological control initiatives or projects set up against invasive insect pests because of the threats these species pose to biodiversity. Examples include the horse-chestnut leaf miner, a North American species that has rapidly spread in Europe causing extensive damage to its host tree (Kenis *et al.* 2005) and an invasive weevil from the central Americas that feeds on native bromeliads in the state of Florida, USA (Frank and Cave 2005). In both these projects, potential biological control agents have been identified and are being considered for further assessment. One of the best examples of how biological control has been used effectively comes from the island of St Helena in the south Atlantic (Box 6.1).

Box 6.1 Saving natural populations of endemic gumwood trees on the island of St Helena in the south Atlantic through biological control

At sometime in the 1970s or 1980s, a scale insect called the Jacaranda bug (*Orthezia insignis*) was found to be attacking a precious endemic tree belonging to the daisy family, the gumwood (*Commidendrum robustrum*), of which there were only about 2000 individual trees left on the island. The Jacaranda bug is native to South America but has long been recorded as a common pest in tropical countries. By the early 1990s the Jacaranda bug had infested many of the gumwood trees and by 1993 it had killed trees; the insect sucks the sap of a tree but it also produces honeydew on which sooty moulds grow and these then smother the tree. Given the polyphagous nature of the bug, there was concern that it would attack other plants on the island.

A biological control agent, a specialist ladybird, *Hyperaspis pantherina*, was already known and had been used for the control of the Jacaranda bug in Hawaii, four African countries, and Peru, where in most cases it had substantial impact on the target. After further study of the taxonomy, life history, and environmental safety of the beetle, the Government of St Helena gave permission for the release of the agent in 1993. Mass rearing was started on the island in May 1993 using the large natural supply of the Jacaranda bug available on the island. In early 1994, 5000 beetles had been reared and released and this soon had the Jacaranda bug under control. Since 1995 there have been no further problems with the bug and restoration projects were started to replant gumwood trees (Fowler 2004).

Classical biological control has also been considered for invasive species in other taxa that have impacts on biodiversity; for example, land snails, amphibians, and mammals. There has also been research effort on biological control for use against invasive sea- and freshwater organisms: comb jellies, mussels, crabs, and fish. Much of the work on biological control has been on researching potential but some projects have been implemented.

One of the most disastrous biological control projects was that conducted against the giant African snail (*Achatina fulica*) on several tropical islands, e.g. the Hawaiian islands in 1955. Several predatory snails and a predatory flatworm were used but major problems arose because the predatory snails, and possibly the flatworm, attacked snails native to the islands and in some cases seem to have driven the species to extinction (Civeyrel and Simberloff 1996). This was one example project that reinforced the principle of the need always to use host-specific natural enemies and the need for ecological risk assessment (see below).

In Australia the potential of using viruses for the control of the cane toad (*Bufo marinus*), a species introduced into several countries for control of white grubs, has been investigated but these have been found to be lethal to native frogs so will not be used. Other options are being investigated. Viruses have also been used for the control of some invasive mammals; the most well publicized example being that of the myxoma virus for the control of the European rabbit (*Oryctolagus cuniculus*) in Australia and in the UK (Fenner and Myers 1978), but more recent work on viruses in Australia has raised public concern about the use of pathogens. More recent research in biological control in Australia and the USA has focussed on fertility control for a range of species.

Active research is being conducted on the potential for the biological control of marine and freshwater invasive species but a central issue has been the difficulty of how to measure the host specify of potential control agents in such complex environments The European green crab (*Carcinus maenas*) has become invasive on the coasts of Australia, Japan, North America, and South Africa. Physical and chemical control measures have been tried but are not effective and thus studies have been underway in Australia and in the USA on biological control. Many pathogens, parasites, and predators are known to attack the crab but the focus of research has been on a rhizocephalan (a barnacle) parasitic castrator from Europe and the mechanisms of host location and compatibility (Kuris *et al.* 2005). Success has been achieved though against the Atlantic comb jelly (*Mnemiopsis leidyi*) which is a serious invader of the Azoz, Black, Caspian, Marmara, and Mediterranean Seas. It was probably spread to these areas via ballast water in trade ships. Several natural enemies of the comb jelly are known but concern has always been about possible non-target effects and how these can be assessed, but in the end biological control has been fortuitous. A predaceous comb jelly, *Beroe ovata*, arrived accidentally in the Black Sea sometime in the late 1990s and seems to have been responsible for the collapse in the Atlantic comb jelly (Anon. 2004).

6.3.3 Success, failures, and the economics of biological control

Although on a global scale classical biological control has been used extensively against a wide range of invasive species, there has been relatively little evaluation of the technical or economic outcomes of the projects; but more effort has been to evaluate projects against plants than insect targets. This situation is a reflection of the fact that most funding for biological control comes from the public sector and this source tends to be limited. But some countries have put reasonable investment into biological control, including follow-up evaluations, and it is of no surprise that biological control projects in these countries tend to have a higher success rate than elsewhere. General reviews of biological control projects against plants and insects have been compiled and these attempt to provide general information about the number of agents used in projects, establishment rates, and the degrees of success in controlling the target (see Greathead and Greathead 1992; Julien and Griffiths 1998).

Complete control of invasive insects through the introduction of one or more agents has been estimated at between 10–15% of total cases (Gurr *et al.* 2000; Hill and Greathead 2000) and for invasive plants between 30–39% of total cases (Julien and White 1997; Syrett *et al.* 2000). But, as mentioned, it is notable that in countries where research effort and evaluation has been more intense, the success rate has been much higher; e.g. 83% in South Africa (McFadyen 1998). In general, these analyses tend to group partial successes in achieving control with failure to achieve any control. But there are many instances where biological control has contributed significantly to the control of a target, even if it has not resulted in the complete control of the target. Of 23 invasive plants targeted in South Africa, six are considered to be under complete control and 13 under 'substantial' control (Hoffmann 1995). Some control projects have developed an explicit integrated approach where biological control forms one of the core components; for example, habitat management and biological control have been successfully used for the management of the invasive plant, *Hakea sericea*, in South Africa (Kludge *et al.* 1986).

The methods used to examine the economics of biological control and the rigours of these analyses have varied widely. Some of the earliest analyses, albeit simple, were done by Paul DeBach in the early 1960s who examined the projects conducted in California, USA. The net savings from 5 projects (US$115 million) versus the cost of the projects (US$4.4 million) were considerable. A more recent in-depth analysis of 27 successful projects including plant and insect targets where data on costs and benefits are available showed that that in all but one case, the projects were highly cost effective with benefit:cost ratios greater than unity (Hill and Greathead 2000). But this particular analysis indicated that benefit:cost ratios are clustered with a few projects providing very high returns (ratios of greater than 100:1). These projects have been ones that have involved plants and insects of major economic importance and that have affected a wide geographical area.

Another example of a more in-depth economic analysis come from Australia where an analysis has been done of the economics of all projects conducted on invasive plants since 1903. In Australia, funds for biological control have come from the public and private sectors. The analysis involved 29 projects and 14 of these returned a net positive benefit. The main findings were that a national investment of AU$4.3 million/year has provided an average return of AU$95.3 million/year (Page and Lacey 2006). These authors highlight the fact that social and environmental benefits have rarely been quantified in the studies and yet it is known that many projects have had a positive impact in one or both. An example of the economic return of one of these Australian projects is given in Box 6.2. It has been written many times that one of the advantages of classical biological control is that benefits accrue year after year, with no need for further investment. The economic benefits certainly show that initial investments have in general been worth making even though the success rate with biological control is very low.

Box 6.2 Economics of the biological control of the rubber vine in Australia

Rubber vine (*Cryptostegia grandiflora*) is native to Madagascar and is a scrambling shrub used as an ornamental. It grows well in semi-arid tropical watercourses and is tolerant of a wide range of soil types. The plant was introduced into Queensland, Australia in the 1860s to cover old coal tips but became naturalized and by 1944 had infested 1200ha. The rubber vine smothers tall trees and pastures and forms impenetrable thickets (Figs. 6.1 and 6.2). By 1973 the plant had become a serious invader of riparian vegetation, floodplains, and natural eucalyptus woodlands. By the early 1970s it was estimated to be spreading at 1–3%/year and by 1989 there were about 120,000ha of dense infestations. It was also estimated that the potential distribution of the plant in northern Australia could be 32,000–160,000km^2. Rubber vine reduces grazing in pasture areas, restricts access to water, and is also toxic to livestock. The cost to the cattle industry in 2001 was estimated to be AU$18.3 million. But rubber vine also has a major impact on native plant communities. In about the late 1980s, rubber vine was reported to threaten: four vulnerable animal species; 13 plant communities; one Ramsar site; 13 important wetlands; and 48 reserves.

On this basis a biological control was undertaken and host-specific agents from the native range of rubber vine introduced. A moth (*Euclasta whalleyi*) was released in 1988–91 and a rust fungus (*Maravalia cryptostegiae*) in 1995–97. Both agents established and spread well, the rust being the more effective agent as the moth is now affected by parasitism. Rubber vine populations had decreased by 25–65% 4 years after the release of the fungus and the agent also prevents recolonization. At the same time, pod numbers were reduced by 85%, leaf cover by 73%, and flower production by 48%. Overall (using 2004–05 AU$ values), the rubber vine biological control project cost AU$3.6 million and this provided a net value by 2004/05 of AU$232.5 million, providing a benefit:cost ratio of 108:1. This did not include the benefits to the environment such as the reduced threat to native plant communities (Page and Lacey 2006).

Biological control of invasive species | 85

Fig. 6.1 Eucalyptus trees covered in rubber-vine (*Cryptostegia grandiflora*). Photo: Harry C. Evans.

Fig. 6.2 Rubber-vine infesting rangeland in central Queensland; note the abundance of pale pink flowers, appearing as white dots. Photo: Allan J. Tomley.

6.4 Modern methods of biological control

The general steps involved in a modern classical biological control project have been discussed in depth by several authors; see for example, Van Driesche and Bellows (1996); these authors also cover taxa other than plants and arthropods Two related topics have dominated in the literature for some time because of their relevance to successful and safe biological control: the characteristics of an efficacious agent and how to assess and manage ecological risks. Although several agents have been introduced in most biological control projects, most successes have come about through the action of just one or two agents. This aspect has recently put further attention on the need for criteria for the selection of agents to avoid introducing 'poor' agents that might increase the risks of non-target effects (Sheppard and Raghu 2005).

6.4.1 The characteristics of efficacious agents

Determining why some biological control projects are successful and yet many more are failures has been an active area of research of population ecology since the 1920s. Ideas and hypotheses have been put forward to try and explain the lack of establishment of agents and also why established agents may or may not control the target (Hopper 1996). The general thrust of work on the latter has been to examine the effects of various demographic parameters of natural enemies and also the population response to the host's distribution on the equilibrium density and population stability of the target (Hochberg 2002). The focus of this work has primarily been on insect natural enemies that attack insect hosts. Numerous mathematical models have been created and these have contributed much to the theoretical basis of population ecology, but none of this has produced a useful predictive framework or guidelines that help with the selection of agents. Some have suggested that such predictions will be almost impossible because of the inherent variation in the systems involved. A basic problem has been that little experimental work or observations have been done to examine the hypotheses and thus to accept or reject these hypotheses, and this was highlighted in a number of reviews written in the 1990s (e.g. Hopper 1996).

An important aspect in the process of the selection agents does dictate that practitioners consider one particular ecological trait in agents: that of complete or, in some circumstances, very narrow host specificity. This is because of the need to minimize the risk to non-target organisms; this is one of the points discussed in the next section.

There are now a number of more recent and important papers that do provide ideas on how to improve biological control projects through careful and well-planned research on the ecology and genetics of natural enemies. Some of these papers have focussed on the topic of criteria for agent selection. For example, in the case of invasive plants several authors have suggested that it is important to identify the stages in the life cycle that herbivores or pathogens might have a major impact and that ideally these studies should be done in the native and invaded ranges (e.g. Raghu *et al.* 2006). These authors also highlight the logistical and other problems of conducting such studies and thus suggest the use of models to simulate plant population dynamics. On another front, the importance of gathering data on the response and/or impact of natural enemies on their host has been emphasized. The use of simulated herbivory has been used to make predictions about insect attack on plants (Wirf 2006) while a wide range of methods have been suggested to assess insect impact on insect hosts (e.g. Jervis and Kidd 1996). These papers have, however, largely been produced by workers from countries where resources for biological control have generally been good, e.g. Australia. Also, few examples exist where data from such studies have clearly helped in the selection of a successful agent. Another issue is that not all these ideas and methods have been universally agreed on such that a best practice can be made available more generally to all.

6.4.2 Issues related to ecological risks

Safety is the single, most important issue for biological control. Unfortunately, however, there is still much misinformation, deliberate or accidental, surrounding this pest management strategy which leads to a climate of apprehension especially directed at the classical biological control approach and the movement and release of exotic natural enemies to control invasive species. There is considerable mileage to be gained by investigative journalists and television-programme makers from highlighting the perceived dangers and actual 'disasters' of the strategy (few, and, all non-specific), whist never reporting on the successes (many; the majority ignored because the invasive species problem no longer exists and by nature, therefore, is not newsworthy). The same questions emerge in any debate on biological control, even in specialist fora: 'what will the agents move on to once the target is eliminated?'; 'will not the pathogen mutate and attack crop plants?' Perhaps some of the terminology used—biotic agents, microbial control, fungal pathogens, exotic or alien natural enemies, and even biological control—needs to be modified or toned down in this increasingly risk-adverse world, perhaps replaced by terms such as 'natural control' (as a general term), 'ecosystem balancing', or 'beneficial organism'?.

To illustrate the situation, the risks associated with the introduction of natural enemies to control invasive species have been seriously questioned in North America (Howath 1991; Simberloff and Stiling 1996a, b) which traditionally has been a very active region in biological control. This has resulted in animated scientific exchanges (Frank 1998; Simberloff and Stiling 1998), as well as to poor public presentations of the issues involved and to bad journalism in general. Specifically, detractors have focused on the example of the European weevil, *Rhinocyllus conicus*, released in North America for the control of invasive thistles species, and which has since established on native thistles, in order to highlight the risks posed by biological control, and the classical approach, in particular (Strong 1997). However, the arguments are somewhat flawed since it was already known, and considered during the initial risk assessment, that the weevil has a broad host range within the thistle family and that was acceptable based on the risk:benefit analysis at that time. Indeed, more recent studies on non-target impacts have shown that the doomsday scenario for this 'cause célèbre' has been greatly exaggerated (Herr 1999), although additional studies have revealed other unanticipated ecological risks when the weevil proved to have indirect impacts on indigenous thistle insects (Louda *et al.* 2003). Given such controversies, this agent would not be released today, based so much on the science *per se* but on the precautionary principle linked to public perception and practitioner ethics. This point is illustrated further with a recent biological control project against giant hogweed in Europe (Box 6.3). Certainly, there is no doubt that plant pathogens with a similarly wide range of hosts as the thistle weevil must never be considered for classical biological control. Specificity has to be at a significantly higher level based on co-evolution.

Box 6.3 Risk analysis and the precautionary principle—the case of the giant hogweed

---

Although the European Union (EU) countries have been the source of almost 400 classical weed biocontrol agents (Julien and Griffiths 1998), biological control of invasive alien plants using exotic natural enemies has yet to be implemented in Europe. In an EU-funded project aimed at developing a sustainable management strategy for the invasive and pernicious giant hogweed (*Heracleum mantegazzianum*), a biological control component was included within an ambitious multidisciplinary approach. The results of surveys in the Caucasus region of Russia revealed a guild of natural enemies associated with this plant which were absent in the western European invasive range. One fungal pathogen in particular, *Phloeospora heraclei*, causing extensive and coalescing leaf lesions showed promise since it impacted heavily on the first-year plants—especially the seedling stage—which are considered to be the most vulnerable stage for biocontrol in a biennial plant species. Host-range screening of the Caucasus fungus confirmed this potential but, later in the centrifugal phylogenetic testing sequence, symptoms were detected on the closely-related parsnip and coriander, although not on the hogweed *H. sphondeylium* despite the fact that there are European records of this pathogen on this host (Seier and Evans 2007). This was perplexing because both in the UK and mainland Europe, there are no reports of *P. hercalei* on the invasive giant hogweed, nor, apart from one isolated record, on either parsnip or coriander even in the horticultural literature listing their diseases. This suggests that there are special forms or pathotypes of the fungus and, moreover, that these results are an artificial extension of host range (see Wapshere 1989; Marohasy 1996).

However, because of the ground-breaking aspect of this project, in relation to classical biological control and Europe, it has been decided to adopt the precautionary principle and not recommend further action, concluding that none of the agents showed high specificity, based on the studies to date, to warrant release as a classical agent (Cock and Seier 2007). What, therefore, are the options? The programme had a finite time frame (2002–05) and the relatively limited budget had to be split between the various disciplines. Closer analysis of the biological control project, and comparison with recent, successful, weed biological control projects, showed that more funding and considerably more time (5–10 years) are required to complete risk assessments of potential agents. In conclusion, therefore, insufficient time and resources were allocated to clarifying the ambiguous and anomalous results of the host-range tests and, consequently, the potential of this pathogen, and other agents, for the classical biological control of giant hogweed.

The term 'pathophobia' was coined to draw attention to the slow progress and lack of funding interest in applied weed pathology, despite the early successes in the 1970s of this still-developing field of invasive species management (Freeman and Charudattan 1985). The use of pathogens in biological control, and, in particular of fungal pathogens of invasive plants, has proven to especially problematic since the risks involved have invariably been judged to be unacceptable, founded not on scientific pest risk assessments but more on emotive, historical narratives on plant disease outbreaks which show what coevolved pathogens can do once they catch-up with their crop hosts in exotic situations However, on the positive side, they also demonstrate just how effective coevolved natural enemies can be for the sustainable management of alien plants which have become invasive in their new habitats.

The introspective capacity for self-analysis and regulation has, fortunately, been a feature of biological control practitioners especially relating to the principles and safe practice of their discipline (Marohasy 1996; McFadyen 1998). Data on weed biological control projects involving insect agents has been examined and it was concluded that, despite the intercontinental movement of over 600 insect species, there are few documented cases of non-target effects: all were considered to be predictable behavioural responses rather than the purported and more evocative 'host shift or jump'. The inherent safety and genetic stability of these coevolved agents was stressed (Marohasy 1996). Similarly, the performances of 26 fungal pathogens used as classical biological control agents for the management of invasive plants has been analysed (Barton 2004). There were no instances of non-target effects and it was even concluded that the central plank of risk assessment—host-range testing—was, in fact, over-rigorous since a number of pathogens which had demonstrated extended host ranges in the greenhouse situation, but, nevertheless, still cleared and released based on risk:benefit analyses, were never recorded from these same species in the field. In the case of the biological control of insects, an analysis showed that there is only data relating to non-target effects in only 1.7% of all documented introductions, and reported impacts are only relatively minor. However, there may have been impacts that were not quantified or reported, particularly in the very early history of biological control (Lynch and Thomas 2000).

Pest risk assessments and management tools have been developed and refined over the past 100 years, almost exclusively by entomologists engaged in invasive plant biological control. A main part of pest risk assessment, host specificity screening in invasive plant classical biological control, is founded on a centrifugal phylogenetic method of host-range testing, proposed in the 1970s (see Wapshere 1975); this replaced an earlier methodology which was more focused on the threat to crop plants in the release area rather than on genetic relatedness. Subsequently, this was adopted for and adapted to the screening of plant pathogens and further modified for insect agents in order to reduce the chances of rejection. Nevertheless, there are currently even more refinements and introspection of the agent-selection procedures in order to meet the more stringent demands of the increasing risk-adverse societies. Hopefully, this should dispel doubts and encourage greater support. Test requirements for fungal pathogens of invasive plants differ considerably from the relatively simplistic choice/no choice tests conducted on potential insect agents, involving additional criteria such as internal analyses of inoculated test plants (Evans 2000), in order to better interpret host–pathogen relationships and the resistance mechanisms deployed.

Host-specificity screening of biological control agents of arthropod invasive species has not been a feature of biological control projects against these targets until quite recently. Indeed, the older, standard biological control texts, which deal primarily with control of arthropod pests, make no mention of safety or risk assessment (e.g. DeBach, 1974). The numerous predators, parasites, and parasitoids moved around the world for classical biological control of invasive species were rarely, if ever, tested for specificity: the assumption being that they were part of the indigenous natural enemy guild of the target, and, therefore, inherently specific and safe. And natural enemies, both arthropods and pathogens of arthropod invasive species, are still sometimes moved between continents with little consideration for safety or quarantine issues. There are also numerous instances where exotic strains of entomopathogens, usually as bio-pesticide products, have been freely exchanged for both laboratory and field-based trials against invasive species. But the situation is changing and there has been much research during the last decade on how to assess the host-specificity of potential insect and fungal agents of arthropods. The topic is more difficult than for agents of invasive plants because the main groups of agents used for arthropods, parasitoids, and predators, have complicated behaviours and ecology. For example, parasitoids respond to two trophic levels, the host and the plants of the host. Nonetheless, protocols for testing have been suggested which recommend criteria for drawing a list of test species (see, for example, reviews in Bigler *et al.* 2006).

6.5 Constraints to the implementation of biological control

Biological control has been used successfully to address some of the major invasive species problems that have threatened countries or sometimes entire continents. In natural ecosystems it is particularly appropriate as other control methods, such as pesticides, may have a larger negative impact than the invasive species. Also, in agriculture, there is a growing demand across the globe for pesticide-free crops and biological control has been highlighted as a major means of providing ecologically safe management. Overall, the technology is particularly useful in developing countries where the needs for cheap and cost-effective management tools are important as resources are limiting. An analysis of five biological control projects in developing countries showed the overwhelming benefits of those projects when compared to the impacts and costs that the target species were causing (Cock 2002). However, it has been estimated that biological control has been used against only about 5% of invasive species problems worldwide (Van Driesche and Ferro 1987) and the situation has not changed. A recent assessment of trends in biological control research and application suggest that there has been little growth despite increased opportunities (Kairo 2005). This 'lack of adoption' is likely to be due to several factors; as we have seen, the concerns about non-target effects have made biological control implementation more difficult in some countries, but factors other than this are important.

In countries with long experience in biological control such as Australia, Canada, New Zealand, South Africa, and the USA, national regulatory frameworks and legislation for biological control are well advanced and many now have the involvement of environment agencies in the assessment of agents process; this has been brought about because of the greater awareness of the risks (see above) that all invasive species (including biological control agents) might pose to natural ecosystems (Sheppard *et al.* 2003). There is much variation between countries in the extent to which the assessment processes meet a full ecological risk:benefit-cost analysis. But despite the benefits this has brought in terms of the broadening the consultation process and criteria on which release decisions are made, it has lead to a greater cost of effort and overall is at risk of reducing biological control initiatives (Sheppard *et al.* 2003). And regulatory restrictions, notably in the USA, have been enforced to the extent that they 'have nearly eliminated classical biological control with exotic pathogens of introduced insect pests' (Lacey *et al.* 2001).

In contrast, in many countries, especially in the developing world, there is no national framework or responsibilities established for those who want to implement biological control. And with invasive species now high on the global agenda, government organizations without experience of biological control have become even more risk adverse. The publication of the FAO Code of Conduct in 1996 for import and release of exotic biological control agents (FAO 1996) was a turning point as it provided important guidance for countries and this Code has been used to develop national frameworks and policy. The Code also contained technical implementation details which were useful but this was not enough for some countries without experience or technical knowledge about biological control (Kairo *et al.* 2003); there is a real need for knowledge transfer, but with the issue of the revised Code (FAO 2005b) there is an intention to produce technical implementation details as a set of documents in support of the Code (Nowell and Maynard 2005).

Lack of adequate funding for biological control also hinders its proper implementation and follow-up in many countries. This problem was alluded to earlier in this chapter. In general, funding is linked to government policy issues referred to above. Most classical biological control is undertaken by public sector organizations and is funded by the public sector (Hill and Greathead 2000) and classical biological control in developing countries is often funded in part by international assistance agencies (Kairo *et al.* 2003). Public and donor funding usually operates on short cycles (3–5 years is typical) but it is recognized that biological control projects take much longer (Cock *et al.* 2000). Thus crucial studies tend to be curtailed and the results of the biological effort are unknown.

6.6 Conclusions

It has been predicted that there will be a massive increase in invasive plant species from those introduced over the past century that have already become naturalized, and from other species that continue to be moved as a consequence of globalization (McFadyen 1998); the same is true of species in other taxa that have been, or are being moved around the globe. This author also argued that 'classical biological control is the only safe, practical and economically feasible method that is sustainable in the long term, and the importation of (beneficial) insects and pathogens must not be prevented by ever-increasing restrictions and demands for pre-release studies'. This has been enforced more recently by the statement that classical biological control is the only method open to resource-poor farmers in the developing world, who, in the absence of control, abandon weed-infested land and clear more forest (Wilson and McFadyen 2000). These authors argued that 'nit-picking' about non-target impacts sends out the wrong messages to countries where there is effort to assess and implement management methods for invasive species. But on the other hand, the increasing concern about ecological risks that classical biological control presents calls for constant vigilance by biological control practitioners to ensure that best practices are followed and high standards are maintained. Methods for all stages of classical biological control that have been tested are now available so there are no reasons why modern projects should not be safe and have a good chance of being successful.

7
Public participation in invasive species management
Souad Boudjelas

> Pest problems cannot be solved without community support, but communities first need educating.
>
> Tim Low

7.1 Introduction

This chapter reviews public participation in invasive alien species (IAS) management. It explores why public participation is important, from ethical, legal, and practical standpoints. How the public engages in IAS management through different modes will be considered and related to the varying drivers of individual initiatives. Examples presented throughout the chapter and more detailed case studies will illustrate the value of public participation. The chapter concludes with a discussion on how to maximize public participation through the process of mainstreaming.

7.2 Why involve the public?

Why involve the public when managing IAS? There are three answers to this question: ethics, legal compliance, and effectiveness.

7.2.1 Ethics

Public participation is a central tenet of any democratic country; it is considered the public's right to be consulted and heard on key issues that directly affect them. Public participation provides the mechanisms for inclusion of the public's values and ideas into the process. Such a consultative culture is also thought to extend good citizenship.

7.2.2 Compliance

As IAS management is a component of environment and resource management, many of the compliance aspects of IAS derive from more general environment legislative instruments. Few countries have implemented significant legislation at the national level to obligate public participation in environment decision making in general or IAS in particular. The compliance burden of IAS management generally rests with national governments adhering to international conventions and frameworks; for example, the 2004 Ballast Water Convention of the IMO (http://www.imo.org). However, The Convention on Access to Information, Public Participation in Decision Making and Access to Justice in Environmental Matters (also known as the Aarhus Convention after the Danish city where the final talks were held), endorsed in 1998, is widely considered the world's foremost international legal instrument promoting public participation. The Aarhus Convention recognizes every person's right to a healthy environment. To support this, it enshrines public involvement in environment decision making. Article 6(4) Aarhus Convention states that '…each party shall provide for early public participation, when all the options are open and effective public participation can take place' (Convention on Access to Information, Public Participation in Decision Making and access to Justice in Environmental Matters United Nations ECE 1998). To date, there are 41 parties to the Convention. Signatories are mainly European and Central Asian countries but the Convention is open to all members of the United Nations.

Few countries have implemented significant legislation concerning IAS management (e.g. New Zealand, Australia, and United States). Widely regarded to be at the forefront of legislation on IAS, New Zealand has also enshrined public participation into the relevant legal tools. The two main Acts dealing specifically with IAS are the Biosecurity Act, 1993 and the Hazardous Substances and New Organisms, (HSNO) Act, 1998.

The objective of the Biosecurity Act, 1993 is to prevent the unintentional introduction of IAS into New Zealand and their spread within the country. The principal mechanism established under the Act is the preparation of regional and national Pest Management Strategies (PMS). In the preparation of a PMS, the Act requires public hearings to be undertaken and public submissions on the proposed PMS to be considered.

The Biosecurity Act was later followed by the HSNO Act; which, in the words of the Act itself, has the purpose '… to protect the environment, and the health and safety of people and communities, by preventing or managing the adverse effects of hazardous substances and new organisms' (Hazardous Substances and New Organisms Act 1996) As part of the HSNO Act the Environmental Risk Management Authority (ERMA) was established. A key function of ERMA is to grant approval to the importation of new organisms. As part of the application approval process the HSNO Act stipulates public consultation as a key step in the consideration of the approval process of applications to introduce non-native species to New Zealand. The outcomes of all decisions are required to be made available to any submitters. ERMA is also responsible for public awareness of the threat of IAS.

In this context, legislative instruments are simply statutory tools to ensure a desired outcome—in this case the participation of the public in matters of IAS management. Making such involvement compulsory is a clear indication of the belief that public participation is a highly effective component of IAS management. The remainder of this section will explore the different benefits of public participation in IAS management.

7.2.3 Effectiveness

We have discussed the ethical and legislative considerations but now turn to the primary reason for public participation in IAS management: it is an extremely effective tool.

7.2.3.1 Locally relevant

Many impacts of IAS are felt directly by communities with close dependencies on the environment. Public consultation of directly impacted communities will result in relevant priority setting. See Box 7.1 for an example of where public consultation resulted in a change of project priorities.

7.2.3.2 Maximize the resource effort

Historically, public participation has been largely confined to the public supplying information and opinions in response to the request for submissions to government-led initiatives. While playing a relatively passive part in this context, the information that the public supplies can be a key informational resource to such initiatives.

Due to its complexity and extent, IAS management is a resource intensive endeavour calling on significant amounts of financial and human resources (including expertise, skills, knowledge, and effort). The public contains a broad and deep resource pool that once suitably motivated can bring considerable effort to bear on a problem. Leveraging this extensive human resource pool to augment government and non-governmental organization (NGO) effort will maximize the resources available to drive a successful solution.

A suitably informed and motivated public can form an effective low cost, extended, passive monitoring and surveillance network. In New Zealand, the effectiveness of such public participation is illustrated by the identification of all three incursions to date of the red imported fire ant (RIFA) having been made by vigilant members of the public alerting MAF Biosecurity New Zealand (M. Sarty 2008, pers. comm.). Early notification greatly increases the likelihood of eradication. In the case of RIFA in New Zealand, such public vigilance has led to successful eradication of two of the incursions, with the third still in progress. Public surveillance has also been instrumental in identifying incursions of moths, e.g. fall webworm, painted apple moth and white-spotted tussock moth, and southern saltmarsh mosquito (M. Sarty 2008, pers. comm.)

Invasive species management

Box 7.1 Case study 1: a Fijian community sets priorities for invasive species management on their island

The Fijian ground frog (*Platymantis vitianus*) has been listed by the World Conservation Union (IUCN) as endangered since 1996 (IUCN 2004). There are only two extant endemic amphibians in Fiji, the larger ground frog and the smaller tree frog (*P. vitiensis*) (Pernetta and Watling, 1978; Morrison, 2003). Fijian ground frogs (FGFs) once lived throughout the country, but now only survive on four mongoose-free islands, including the island of Viwa. The Viwa Island population is considered the smallest and the most vulnerable of the four islands populations.

Viwa Island is a small, 60 ha island situated about 900 m off the east coast of Fijis largest island, Veti Levu. The island's human population is a small community made up of 104 people who are reliant on subsistence living. Large populations of cane toads (*Bufo marinus*) and Pacific rats (*Rattus exulans*) are present on the island and are considered to be the two main threats to the FGF (Denny *et al.* 2005). Over a number of years, researchers at University of South Pacific (USP) have been studying the ecology of the Fijian ground frog on Viwa. In 2004, USP partnered with the Pacific Invasives Initiative (PII) to undertake a feasibility study aimed at protecting the FGF by eradicating the cane toad and Pacific rats.

Fig. 7.1 Members of the community on Viwa Island (Fiji) practice with non-toxic baits in preparation for the eradication of Pacific rats (*R.exulans*). Photo: Rob Chappell.

Public participation in invasive species management | 97

Fig. 7.2 Preparing tracking tunnels and bait stations for surveillance of Pacific rats (*R. exulans*) on Viwa Island, Fiji. Photo: Bill Nagle.

The nature of the eradication project required the full approval and support of the local population.

During an early phase of the feasibility study, the project team consulted extensively with the local community seeking local input and support. Through this consultative process it became apparent that local support was strongest for eradication of the rat, rather than the cane toad. Through the eating of crops and stored foodstuffs, and the known health risks to the local community (e.g. spreading leptospirosis), rats were perceived as the greatest pest. Based on the community's wishes, the project team prioritized the rat eradication over the cane toad eradication. This ensured support from the community for the proposed activities (Denny *et al.* 2005) (Figs. 7.1 and 7.2).

See: Pacific Invasives Initiative website: http://www.issg.org/cii/PII

The world is seeing a rapid increase in community-led initiatives around the environment in general and IAS management in particular. These initiatives grow out of community concerns and effort and are traditional grass roots activities. Some community initiatives leverage government resources over time but remain driven, and owned, by local communities. One example of a community-led initiative is Weedbusters, an awareness programme that alerts people to the damaging effects of weeds and involves them in weed control (http://www.weedbusters.info/). Awareness-raising events around Australia are organized by individuals, schools, community groups, and local and state governments. Activities include weed clean-up efforts, field days and demonstrations, seminars, and displays to assist with weed identification and competitions. Weedbusters started in 1994 as Queensland Weed Awareness Week. In 1995 and 1996 it became Weedbuster Day, with thousands of people participating in events throughout the State. New South Wales also held Weed Awareness Weeks in 1986, 1990, and 1996. In 1997, Weedbuster Week was launched nationally, with encouragement and support from the Australian government, all State and Territory Governments and the Cooperative Research Centre (CRC) for Australian Weed Management. The Weedbusters programme has now spread to New Zealand and South Africa, and other countries have expressed interest in running their own Weedbuster programmes.

Another example of a community-led initiative is the *Mimosa pigra* (mimosa) management programme on Aboriginal lands in the Northern Territory, Australia. This programme is facilitated by the Caring for Country Unit (http://www.nlc.org.au/html/care_menu.html) which was established in 1995 by the Northern Territory Land Council, the principal representative body for aboriginal people in the northern portion of the Northern Territory. The purpose of the unit is to assist aboriginal communities to manage environmental issues such as feral animals and weeds through consultation, participatory planning, building capacity, sourcing of funding, and facilitating training and partnerships. Mimosa, a highly invasive weed, was identified by indigenous communities as a land management priority. This resulted in the development of community-based ranger programmes to control infestations of mimosa on riverine floodplains. How these programmes are implemented is determined by Aboriginal people through intensive consultation and coordination. Over time, these programmes have broadened their activities to address a variety of natural resource management activities including management of fire, feral animals, and other weeds (Ashley *et al.* 2002; Jackson *et al.* 2005; Caring for Country website).

7.2.3.3 Public support

National and local governments and NGOs have a continuing central role to play in driving and funding many IAS management initiatives. In many developed countries there is a natural scepticism of government and government-led initiatives (Petts and Leach, 2000). These challenges are exacerbated by a growing tendency in some cultures to challenge the supremacy of the scientific expert community on which centrally-managed environment strategies are based. These sentiments are based on the growing public desire for a more inclusive decision-making process and the realisation that these are not purely technical decisions, but involve significant value judgements. As public expectations of the availability of information and inclusivity increase, fuelled in part by the ever greater penetration of modern technologies such as the Internet, so will the need to respond to these concerns.

These government-led initiatives and decision-making processes require widespread public support to be successful. In this context, greater public participation has proven to be an effective mechanism for building public support. Hoffmann and O'Connor (2004) identified public acceptance and approval as a key factor for the successful eradication of two invasive ants: African big-headed ant and tropical fire from the Kakadu National Park. Raising awareness activities included: a press release following the discovery of the African big-headed ant to alert the public to the problem and the proposed eradication; public notices in Jabiru Township; notifications in the local paper; creation of an information sheet; as well as information notices for tourists staying at the Cooinda tourist resort. Also, permission for access prior to any inspection or treatment was directly sought from key people within any locality. These activities were crucial in generating support for the project and for preventing re-infestation. In developing countries, particularly those founded on tribal cultures, public support is even more important given land-ownership and access rights (see Box 7.2).

7.2.3.4 Part of the problem; part of the solution

As certain public behaviours are driving the IAS problem it is only logical that public participation needs to be an integral part of the solution. The public must be engaged in order to moderate or desist from such behaviours. In some cases, the public acts in an indirect capacity by creating the demand for IAS introductions, while in other situations it is members of the public that are actively introducing the IAS. For example, many invasive plants have been introduced as ornamentals by botanical gardens, arboreta, and nurseries—driven by public demand for the more familiar or aesthetically pleasing species.

Many examples exist of the public being directly responsible for the introduction and release of exotic species that turn invasive. For example, a significant percentage of the known 185 exotic fish species caught in US open waters are thought to be the result of intentional releases by hobbyist aquarium owners (Fuller 2007). Individuals can also inadvertently facilitate the spread of invasive species. For example, seeds or insects can hitchhike to remote places on camping equipment. To effect widespread changes in those public behaviours and attitudes that are directly and indirectly driving the introduction of exotic and potentially invasive species, public consultation and awareness-raising must form the fundamental component of the solution.

7.3 How to successfully involve the public

The type of public participation in IAS management can be considered in terms of a model of participation continuum—the Participation Model. How best for the participation to occur depends on the degree of participation being sought (Fig. 7.5).

100 | Invasive species management

Box 7.2 Case study 2: consultation leads to community support for the eradication of Pacific rats on Vatu-i-Ra Island, Fiji

Vatu-i-Ra is a small, 2.3 ha island located in the Vatu-i-Ra Channel between Vanua Levu and Viti Levu. It is about 15 km off the north east coast of Viti Levu. The island is owned by the Nagilogilo clan (Yavusa) who live in two villages in Rakiraki province of Viti Levu. The clan is made up of 15 families. The island is listed as a Site of National Significance (SNS) in the Fiji National Biodiversity Strategy and Action Plan, and has also been identified by BirdLife International as one of 14 Important Bird Areas (IBA) in Fiji. The island supports nine species of breeding seabirds including in excess of 20,000 pairs of black noddy (*Anous tenuirostris*) as well as breeding hawkesbill turtles (*Eretmochelys imbricata*) and the endemic pygmy snake-eyed skink (*Cryptoblecephalus eximius*).

A large population of Pacific rats (*R. exulans*) was found present on the island during surveys undertaken by BirdLife International in 2003 and 2004. Also, during 2004, ground-nesting seabird species were found in much smaller numbers than tree-nesting species (30–200 pairs of ground-nesting species compared with about 27,000 pairs of tree-nesting black noddies). Based on evidence from other islands where rats have been implicated in the demise and decline of seabird populations (Atkinson 1985), BirdLife International, concluded that it was

Fig. 7.3 Explaining eradication to community members. Nagilogilo clan members assist the feasibility study team with bird assessment prior to eradicating Pacific rat (*Rattus exulans*) from Vatu-i-Ra (Fiji). Photo: Karen Johns.

Fig. 7.4 Prior to a ground-based eradication of Pacific rat (*R. exulans*), members of the Nagilogilo Clan and the feasibility study team celebrate the erection of a sign, advising visitors of the importance of Vatu-i-Ra Island (Fiji) as a bird-nesting site. Photo: Rob Chappell.

necessary to remove the rats from Vatu-i-Ra Island to protect seabird populations. However, eradicating the rats was contingent on approval from the island owners, the Nagilogilo Clan.

The community consultation process used by BirdLife consisted of a series of meetings with the Nagilogilo Clan to:

- Develop good relationships with the clan members and gain their trust.
- Seek the clan's approval and encourage participation in the project.
- Discuss the clan's aspirations and concerns in relation to the project.
- Raise the clan's awareness about the impacts of rats on seabird populations on Vatu-i-Ra Island.
- Share information about the project.

This process resulted in the Nagilogilo Clan providing their approval and support for the proposed eradication and BirdLife committing to assist them in their efforts to position Vatu-i-Ra Island as an eco-destination thereby, creating a revenue stream for the Clan (Johns *et al.* 2006) (Figs. 7.3 and 7.4).

In 2006, the eradication project was successfully implemented by BirdLife International with support from Nagilogilo Clan, the Pacific Invasives Initiative (PII), and the New Zealand Department of Conservation. Vatu-i-Ra Island was declared rat-free in 2008.

See: Pacific Invasives Initiative website: http://www.issg.org/cii/PII

Invasive species management

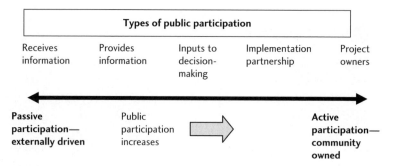

Fig. 7.5 A Participation Model.

There are different types of public participation in IAS management:

1) **Passive participation:** the public's involvement is limited to a passive receiving of information. People involvement in this case consists of being told what is going to happen or what has already happened. This one-way type communication is a feature of externally-driven projects. These are typically government or NGO-led projects. Campaigns can take many forms depending on location, budget, and purpose. Typically formats include: public meetings; poster campaigns; media (TV/radio) campaigns; direct mailing; targeted meetings, e.g. schools, tribal leaders, and other governance groups. Also, increasingly in developed countries, the Internet is being used as an information channel to the public. In the extreme case of the externally-driven management model the public is informed of government decisions on policy or implementation plans late in the lifecycle of the project with no ability to contribute. This mode of participation is increasingly considered unsatisfactory because it does not allow for inclusion of the public's views and aspirations in the planning of the project. Often this leads to a lack of support from the public for planned activities.

2) **Participation by giving information:** the public participate by providing information to project implementers. Often, implementers are not under any obligation to act on the views received from the public. As with passive participation, if the public's views and aspirations are excluded—as is it often the case—any proposed IAS management activities are likely to lack public support. This can result in behaviours that will adversely impact the outcomes of a project. For example, when visiting an island from which rats have been eradicated, members of the public may not feel compelled to take precautions to reduce the risk of re-introduction.

3) **Joint decision making:** the public constitutes a significant pool of expertise, skills, and perspectives on IAS issues and can effectively contribute to the decision-making process. This mode of participation is channelled through public consultative processes both informally and as part of formal processes such as public hearing and submissions, e.g. HSNO

Act in New Zealand. As public input is included in the decision-making process this drives the need for public participation earlier in the management life cycle.
4) **Participation in implementation:** this type of participation is concerned with the involvement of the public in project implementation. As well as a source of information, the public can act as a significant resource effort for actual project implementation and monitoring (Box 7.3).
5) **Leading implementation:** at the opposite end of the continuum to passive participation lie fully publicly owned and driven IAS management initiatives. Increasingly, local groups are forming, whether completely independently or facilitated by external bodies, to address the impacts of IAS (Box 7.4).

Where public attitude itself lies on the Participation Model is a function of their knowledge and interest in the issue of IAS. To mainstream the issue of IAS a shift to higher levels of participation must occur. By doing so, a positive feedback reaction will occur, i.e. greater understanding and commitment will lead to greater call for participation and awareness.

Governments are instrumental in the shift to participation through public education and awareness campaigns. While such campaigns, themselves, lie to the passive end of the model in that the purpose is to inform, the actual objective is to move the public into a more participatory mode. Well-designed public awareness and education campaigns will bring about significant beneficiary changes in public behaviour. Sometimes, these involve the cessation of harmful behaviours, e.g. introducing exotic species, to encouraging knowledge and motivation towards beneficial behaviours, e.g. passive public monitoring.

Box 7.3 Case study 3: mobilizing volunteers to control invasive plants in wetland and riparian areas

Invasive plants are difficult to eradicated or control once they have become established. In most cases, invasive plants are prolific seed producers and any attempt to eradicate such species will require many years of monitoring and removal of new seedlings. This cannot be achieved without consistent commitment and funding.

Some states in the US are using trained volunteer groups as an inexpensive and high-quality source of labour to control various invasive plants in wetland and riparian areas. Cited benefits of engaging volunteers include, the potential for:
- Expanding the control area because of increased man-power
- Extending limited funding
- Increasing the number of people who can help with long-term monitoring
- Increasing the number of people who can help with raising awareness of impacts of invasive plants on the environment

> Involving volunteers in control projects requires good planning and coordination. Three involvement categories have been identified:
>
> 1) One-time events;
>
> 2) Regular working volunteer days;
>
> 3) Independent volunteers that have been trained and certified to monitor and remove invasive plants in a designated site.
>
> See: Alliance for the Chesapeake Bay (2003). Citizen's guide to the control of invasive plants in wetland and riparian areas. http://www.alliancechesbay.org/pubs/projects/deliverables-251-1-2005.pdf

Box 7.4 Case study 4: community groups and schools in Australia rally against a weed of national significance, bridal creeper

A native of South Africa, bridal creeper ((*Asparagus asparagoides*), is now one of the worst weeds in Australia. It invades and smothers native vegetation and forms dense mats under the soil surface that can prevent seedlings of native plants from establishing. It also smothers planted seedlings in forestry areas and in citrus orchards. It is both very difficult and costly to control using herbicides and physical removal. Hence, biological control was identified as a more effective solution.

Following identification and testing for host specificity, three biocontrol agents (the leafhopper *Zygina* spp., the rust fungus *Puccinia myrsiphylli*, and the leaf beetle *Criocers* spp.) were released between 1999–2002. A national release programme was established in 2002 by the Commonwealth Scientific and Industrial Research Organisation (CSIRO) with funding from the Natural Heritage Trust. The aim of the programme was to facilitate and accelerate the redistribution of the leafhopper and rust fungus across the entire range of bridal creeper infestations. Community groups, landholders, and school students—often with the involvement of Weedbusters (Fig. 7.6 and 7.7)—have been the key to the successful implementation of this programme.

Community groups and landholders around Australia have been engaged in the programme and taught the basic skills to rear, release, and monitor the agents. Over 2000 release sites have been established by these groups. Many schools across the country have also been involved in the programme through rearing and releasing the agents at local infested sites under the supervision of experts. Through their engagement in the programme, students not only learn about weed control but also gain knowledge and awareness of wider environmental issues, new skills, a sense of stewardship, and involvement with community organizations.

Fig. 7.6 'Woody Weed' (a costume character that is widely used to publicize weed control) at the 'Bush Friendly' garden, Floriade Canberra, in 2006. Photo: Jenny Conolly.

Fig. 7.7 The Weedbusters 'Grow Me Instead' garden at Floriade Canberra, in 2008. Photo: Jenny Conolly.

> Workshops and Internet-based resources have been used to raise awareness and generate public support for the programme.
>
> See: Australian Weeds Strategy—A national strategy for weed management in Australia. Natural Resource Management Ministerial Council (2006). Australian Government Department of the Environment and Water Resources, Canberra ACT. http://www.weeds.org.au/docs/Australian_Weeds_Strategy.pdf. The bridal creeper website: http://www.ento.csiro.au/weeds/bridalcreeper

7.4 Conclusions

This chapter has explored why the public needs to be fully involved in IAS management and a Model for Public Participation was proposed. Whilst public participation and adoption of the most appropriate mode is essential for maximizing the success of IAS management and protecting biodiversity and community livelihoods, public participation is not without its own challenges. Key ones include:

- Scepticism: a perception that the public involvement is tokenism will result in poor uptake by the public.
- Lack of resources: as time demands on the public at large become greater the availability of time and resources to actually respond to the opportunity to participate will be challenged.
- Apathy: while in many countries there is a growing environmental awareness, there remains a significant amount of apathy to overcome to release the combined energies of the whole public.
- Animal rights concerns: unfortunately one of the key tools of IAS management is eradication of species. Every effort must be made to ensure the ethical considerations are fully addressed and the public concern with animal rights and humane behaviour during control operations are assuaged.

As the Public Participation Model has demonstrated, the issue of IAS management will be successfully addressed by an approach that combines an appropriate suite of initiatives across the spectrum of public involvement from passive activities to those totally community-owned. The foundation for enabling the IAS management community to address the problem is mainstreaming the issue of IAS.

To borrow the concept of 'Crossing the Chasm' from the sphere of technology adoption (Moore 1999), IAS can be considered to still be in the early adopter phase of its life cycle.

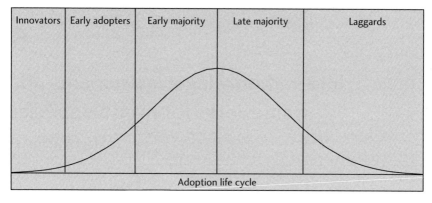

Fig. 7.8 Technology Adoption Life Cycle (from Moore 1999).

The uptake of new ideas differs across the demographic of the public. In this model, innovators equate largely to the likes of IAS and environmental professionals. The public are spread throughout the other sectors depending on their awareness and willingness to adopt change in support of the IAS issue. At any given time, different countries will lie on different phases of the life cycle. The unique nature of New Zealand's isolation and hence its vulnerability to IAS threat has resulted in New Zealand being relatively mature in this model with it lying further to the right than most other countries.

To mainstream the IAS issue, and leverage the full resources of a country, education campaigns need to drive public awareness of the IAS to the early and late majority of the model. Once this has been achieved, the combined impacts of large-scale positive behaviour change (i.e. a reduction in behaviour exacerbating IAS) and increased large scale public participation in IAS management activities will result in unprecedented reductions in the impacts of IAS.

8
International legal instruments and frameworks for invasive species
Maj De Poorter

8.1 Introduction

The natural biogeographical barriers of oceans, mountains, rivers, and deserts provided the isolation in which species and ecosystems evolved. Increasingly, these barriers have lost their effectiveness as economic globalization has resulted in an exponential increase in the movement of organisms from one part of the world to another (Carlton 1999; McNeely *et al.* 2001). Increasing volumes of trade, travel, and tourism have led to more species than ever before being moved around the world, on land, in the air and sea. For instance, a billion tons of ballast water is contained in ships per year and *daily* at least 10,000 species are being transported around the world in it (Carlton 1999). Invasive alien species (IAS), as defined by the Convention on Biological Diversity (CBD), are those alien species whose introduction and/or spread threaten biological diversity (CBD 2002).

Similar IAS problems are repeatedly faced in different parts of the world. For example, water hyacinth (*Eichhornia crassipes*) is a problem in many tropical freshwater bodies worldwide, including waterways in Florida (USA), Kafue (Zambia), Lake Victoria (Kenya), and Bhopal (India). Sharing information and expertise internationally on the ecology, impacts, and management of such IAS is hence important (De Poorter and Browne 2005). In addition, knowledge and information on a species' past invasiveness elsewhere is crucial to prevention, early detection, and rapid response, as these are key factors in identifying risks of invasiveness for species newly introduced, and species not yet present. (Wittenberg and Cock 2001; De Poorter and Clout 2005; De Poorter *et al.* 2005). International programmes can assist with this.

IAS have the ability to spread across administrative or political boundaries and they do not respect national borders. Species introduced into one nation can often easily spread to neighbouring nations, either without further human agents (if there are no biogeographical barriers) or by secondary introductions (e.g. unintentionally via transport or trade). In order to be effective, management must be able to cut across political boundaries, because unilateral action by countries will not be sufficient to prevent or minimize new biological invasions. International action is required, at global as well as regional level (Clout and De Poorter 2005).

IAS are found in all taxonomic groups: (e.g. Lowe *et al.* 2000; UNEP 2001; and see http://www.issg.org/database). They are associated with many pathways, and they have invaded and affected terrestrial, freshwater, and marine habitats in virtually every ecosystem type in almost all regions of the world (UNEP 2001, 2005a,b; Matthews and Brand 2004; Matthews 2004, 2005). Management to address IAS must be able to deal with any IAS taxa, any pathway for introduction, and any 'receiving' habitat or ecosystem type, in any area. In addition, different aspects of management are not isolated from each other (e.g. Cromarty *et al.* 2002; Chapters 7 and 15, this volume). For instance, an IAS eradication plan's success may depend on research, human dimensions such as public awareness and acceptance or attitude (Genovesi and Bertolini 2001; Cromarty *et al.* 2002; Chapter 7, this volume), and on political and financial support as much as on the technical feasibility of the methodology proposed. In other words, practical management is best formulated within an overall strategy.

Legal and institutional arrangements are crucial to support and underpin practical management. Without them, it might not be possible to address IAS effectively; for instance when existing wildlife or pollution laws result in impediments to manage IAS, such as prohibitions to hunt them, wildlife laws actually protecting them, blanket restrictions on biocide use against them etc. (see Shine *et al.* 2000 for more examples and discussion). This is illustrated by the Indian Wildlife Protection Act, 1972, for example. Many of the species that are alien and invasive in the Andaman Islands cannot be removed because they are native on the Indian mainland and are hence protected by the national legislation—these include chital (*Axis axis*) and elephant (*Elephas maximus*) which were introduced to the islands and are damaging their biodiversity (Sivakumar 2003). There can also be significant impediments to management due to lack of institutional mandate; for example, in a survey undertaken by IUCN's Global Marine Programme and the IUCN/SSC Invasive Species Specialist Group in 2005, most respondents were aware of the threat that would be posed by marine invasives if they arrived in their Marine Protected Areas (MPA), but over half of the respondents reported that existing MPA regulations did not have provisions for eradication, even if they were to find such new incursions. In other words, they did not have a mandate. While some respondents elaborated that they would be able to seek approval to take action on a case-by-case basis, others simply stated that they would be able to do 'nothing'. A lack of mandate to deal with IAS can be a significant impediment to conservation managers, and the creation of such an institutional mandate is one of the key roles of legal instruments, which is often undervalued (Shine *et al* 2000). Addressing IAS has to be done within a strategic framework that integrates: an overall strategy or vision; institutional arrangements; and legal aspects with the practical day-to-day implementation. These four major components have areas of overlap with each other; they influence each other and support each other (Wittenberg and Cock 2001; De Poorter 2006).

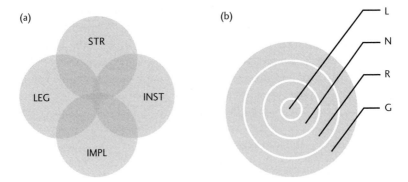

Fig. 8.1 (a) Addressing IAS has to be done within a strategic framework that integrates: an overall strategy or vision (STR), institutional arrangements (INST), and legal / regulatory aspects (LEG) with the practical day-to-day implementation (IMPL). After De Poorter (2006). (b) IAS management requires action internationally at global (G) and regional (R) levels as well as at national (N) and local/sub-national (L) levels.

The above points illustrate the importance of legal instruments (including the institutional mandates they underpin or create) and strategies, and why addressing alien invasive species needs to happen at the international level (globally and regionally), as well as nationally and locally (Fig. 8.1).

8.2 Scope and types of international instruments

A starting point for the legal and institutional approach to addressing biological invasions is that it is not possible to simply regulate against 'harmful' or 'invasive' alien species, since the impacts of 'new' alien species on the environment cannot be known with certainty. Law must therefore, to some degree, cover *all* alien species, and include coverage of how to predict and identify invasiveness and how to set appropriate limits on activities and species that might create invasive problems now or in the future (Young 2006).

Another important factor to keep in mind is that most activities leading to the introduction of potential IAS have legitimate economic and social objectives (primary production, trade, travel, transport, etc.). Legal instruments will be required to strike a balance between legitimate socioeconomic activities and appropriate safeguards for the environment, communities, and public health (Shine *et al.* 2005). Most internationally agreed legal instruments have developed independently from each other, in different sectors, in different times and in the context of different international mandates, so the terminology used in them reflects this. Definitions vary significantly between different international instruments and this presents quite a challenge to implementation at the national level (see, for example, Shine *et al.* 2000; Miller and Gunderson 2004). For example, 'introduction' (of a species) is defined very differently in the following two cases: it refers to the 'movement by human agency, indirect or direct, of an alien species outside its natural range (past or present)' in the context of the Convention on Biological Diversity and to 'the entry of a pest resulting in its establishment' in the context of the International Plant Protection Convention (IPPC). The concept of 'preventing new introductions' hence means quite different things in these two conventions

A mosaic of over 50 internationally agreed legal instruments, binding as well as non-binding, now deal with some aspects of IAS. Many focus on a particular dimension of the IAS issue, with regards to a particular protection objective. Binding instruments, such as treaties and conventions, are agreements of a mandatory nature between States: they must be observed and their obligations performed in good faith. They often require lengthy negotiation processes, and rarely contain detailed rules. It is not unusual for their text to include qualifiers such as 'as possible', 'as practical', 'as appropriate'. While this may be seen as detracting from their mandatory nature, it is in fact a reflection of reality. Countries have widely diverging capacities, diverse values, and different priorities—without such qualifiers it would become virtually impossible to reach international agreements to which a significant number of countries would agree to become signatories. Young (2006) states this succinctly: 'unless it is grounded in practical motivations, capabilities and situational realities, the most beautifully crafted legislative instrument will be meaningless or ineffective'.

Non-binding instruments are adopted by inter-governmental forums in the form of resolutions, recommendations, action programmes, action plans, plans of work, technical guidelines, codes of conduct, etc. Their non-binding nature usually means that they can contain more detail (e.g. technical guidelines), can be negotiated relatively quickly, and can be reviewed, amended, or updated more easily, to keep pace with changes in knowledge or new developments in technology. In addition, non-state actors, including non-governmental organizations (such as IUCN), may develop guidelines and other advisory material (e.g. IUCN 2000; McNeely *et al.* 2001; Hewitt *et al.* 2006).

International instruments of relevance to IAS have been developed in diverse sectoral context and this is reflected in the diversity of international institutional arrangements and processes that may apply to them. The following sections look at some global and regional instruments in more detail. These include instruments for biodiversity or environmental conservation (including specific reference to aquatic ecosystems and fisheries); instruments that relate to sanitary and phytosanitary measures, and instruments relating to international transport. Any measures against (invasive) alien species that result in restrictions or prohibitions in international trade need to be consistent with applicable rules and disciplines adopted within the World Trade Organization (WTO) framework, and hence some aspects of the multilateral trading system are also included. Also covered are the international framework for some approaches to IAS management, examples of international strategies and plans, relevance of international sustainable development initiatives, and international programmes.

8.3 Invasive species and global instruments for conservation of biological diversity

8.3.1 The Convention on Biological Diversity (CBD)

At the 1992 Earth Summit in Rio de Janeiro, Brazil, world leaders agreed on a comprehensive strategy to meet human needs, while ensuring a healthy and viable world for future generations. One of the key agreements adopted was the CBD. The Convention establishes three main goals: the conservation of biological diversity, the sustainable use of its components, and the fair and equitable sharing of the benefits from the use of genetic resources.

With regards to IAS, the CBD is the only global instrument to provide a comprehensive basis for measures to protect all components of biodiversity against those alien species that are invasive. Article 8(h) of the text of the convention requires Parties, as possible and as appropriate, 'to prevent the introduction of, or control or eradicate, those alien species which threaten ecosystems, habitats or species'. Recent Conferences of the Parties to the CBD, and meetings of its subsidiary body, have focused on how to help Parties to the convention with implementation of Article 8(h). The 6th Conference of the Parties (2002) adopted Decision VI/23 on *Alien Species that Threaten Ecosystems, Habitats and Species,* to which are annexed *Guiding Principles for the Prevention, Introduction and Mitigation of Impacts of Alien Species that threaten Ecosystems, Habitats or Species.* The text of the Decision is very long—34 items plus the guiding principles as an annex. It encourages capacity building and regional approaches, encourages cooperation with other relevant international instruments of relevance for IAS, and encourages funders and donors to provide resources for IAS management. It urges Parties, other Governments, and relevant organizations to promote and implement the voluntary Guiding Principles which are contained in its annex. The 15 Guiding Principles outline recommended approaches to: the precautionary approach; a three-stage hierarchical approach (prevention, eradication, control); ecosystem approach; the role of States; research and monitoring; education and public awareness; border control and quarantine measures; exchange of information; cooperation (including capacity-building); intentional introduction; unintentional introductions; mitigation of impacts; eradication; containment; and control. The full text of this Decision can be found in CBD 2002, and any Decisions of the Conference of the Parties that relate to IAS can also be accessed via the CBD Invasive Alien Species portal (http://www.cbd.int/invasive/cop-decisions.shtml). Unfortunately, there are procedural issues relating to the adoption of this decision, (see CBD 2002), which leads some to argue that the Decision is not valid. Others, such as the IPPC, the Council of Europe, and the European Union, on the other hand have already used it as the basis for developing IAS policy and measures in their own spheres of jurisdiction.

The 8th Conference of Parties (2006) adopted Decision VIII/27 which focused on further consideration of gaps and inconsistencies in the international regulatory framework on IAS and which notes that actions to address IAS need to be taken at regional as well as other levels, and the importance of consistency among actions and efforts at the various levels. It encourages the development of regional guidance to address particular gaps in the international regulatory framework; and procedures and/or controls to ensure that cross-border impacts of potential IAS are considered as part of national and regional decision-making processes.

IAS have been formally designated as a 'cross cutting' issue, to be taken into account within the CBD's programme of work. Geographically and evolutionarily isolated ecosystems, including islands, were identified as needing special attention because of their vulnerability to biological invasion. Not surprisingly, the IAS issue features extensively in the work programme on islands biodiversity (Decision VIII/1). However, IAS provisions also appear to various degrees in the CBD work programmes on marine and coastal biological diversity (Decision VII/5); biological diversity of inland water ecosystems (Decision VII/4); dry and sub-humid lands (Decision VII/2 and VIII/2); mountain biological diversity (Decision VII/27); forest biological diversity (Decision VI/22); and agricultural biological diversity (Decision VI/5). In addition, IAS have been addressed in CBD cross-cutting programmes, such as the Global Taxonomy Initiative (Decision VIII/3) and Protected Areas (Decision VII/28) and they also feature in the 2010 target indicators: 'Trends in invasive alien species' is one of the Provisional Indicators for Assessing Progress towards the 2010 Biodiversity Target[1]. IAS are also formally addressed through the Global Strategy on Plant Conservation (COP Decision VI/9 and VII/10). Further details can be found by accessing http://www.cbd.int/programmes/cross-cutting/alien/default.shtml.

8.3.2 The Convention on Wetlands (Ramsar)

The Convention on Wetlands, signed in Ramsar, Iran, in 1971, is an intergovernmental treaty which provides the framework for national action and international cooperation for the conservation and wise use of wetlands and their resources. IAS in coastal and inland wetlands were addressed by the Conference of the Parties to the Ramsar Convention on Wetlands in November 2002. Resolution VIII/18 (Invasive Species and Wetlands) urges Ramsar Parties:

- To address the problems posed by invasive species in wetland ecosystems in a decisive and holistic manner.
- To undertake risk assessments of alien species which may pose a threat to the ecological character of wetlands.
- To identify the presence of IAS in Ramsar sites and other wetlands in their territory, the threats they pose to the ecological character of these wetlands,

---

[1] 2010 Biodiversity target: to achieve by 2010 a significant reduction of the current rate of biodiversity loss at the global, regional, and national level as a contribution to poverty alleviation and to the benefit of all life on Earth

including the risk of invasions by such species not yet present within each site, and the actions underway or planned for their prevention, eradication or control.
- To cooperate fully in the prevention, early warning in trans-boundary wetlands, eradication, and control of invasive species.

8.3.3 The Convention on International Trade in Endangered Species (CITES)

CITES works by subjecting international trade in specimens of selected species to certain controls. All import, export, re-export, and introduction from the sea of species covered by the Convention has to be authorized through a licensing system. The species covered by CITES are listed in three Appendices, according to the degree of protection they need (for additional information see http://www.cites.org).

The 13th meeting of the CITES Conference of Parties in 2004 addressed trade in IAS. Resolution 13.10 on trade in alien invasive species recommends that the Parties of CITES should consider the problems of invasive species when developing national legislation and regulations that deal with trade in live animals or plants. It is recommended that the exporting Party should consult with the Management Authority of a proposed country of import, when possible and when applicable, when considering exports of potentially invasive species, to determine whether there are domestic measures regulating such imports. At the EU (regional) level of implementation, the EU has used The Wildlife Trade Regulations, which are the basis for implementation of the CITES in the European Community, to list four invasive alien animal species as prohibited for import into EC territory: the red-eared slider (*Trachemys scripta elegans*), the American bullfrog (*Rana catesbeiana*), the painted turtle (*Chrysemys picta*), and the American ruddy duck (*Oxyura jamaicensis*).

8.3.4 Convention on the Conservation of Migratory Species of Wild Animals (CMS)

Article III (4c) of the Convention on Migratory Species (Bonn Convention), which relates to endangered migratory species, states that 'parties that are Range States of a migratory species listed in Appendix I shall endeavour to the extent feasible and appropriate, to prevent, reduce or control factors that are endangering or are likely to further endanger the species, including strictly controlling the introduction of, or controlling or eliminating, already introduced exotic species'.

This provision has been elaborated in the Agreement on the Conservation of African Eurasian Migratory Waterbirds, concluded pursuant to the CMS. This prohibits the intentional introduction of non-native waterbird species into the environment and requires that appropriate measures are taken to prevent the unintentional release of such species if such introduction or release would prejudice the conservation status of wild fauna and flora; when non-native waterbird species have already been introduced, appropriate measures must be taken to prevent them from becoming a potential threat to indigenous species.

8.3.5 The UN Convention on the Law of the Sea (UNCLOS)

The UNCLOS (New York, 1982) requires parties to take all measures necessary to prevent, reduce, and control pollution of the marine environment resulting from the intentional or accidental introduction of species, alien or new, to a particular part of the marine environment which may cause significant and harmful changes thereto (Art. 196). (For more details, see Shine *et al.* 2005.)

8.3.6 The Code of Conduct for Responsible Fisheries

Through the Food and Agriculture Organisation (FAO), the voluntary Code of Conduct for Responsible Fisheries was adopted in 1995. The Code provides guidelines for the responsible introduction, production, and management of fish species under managed conditions. It urges States to adopt measures to prevent or minimize harmful effects of introducing non-native species or genetically altered stocks used for aquaculture. (For more details see Shine *et al.*2005; Hewitt *et al.* 2006.)

8.4 Invasive species and regional instruments for conservation of biological diversity

8.4.1 The International Council for the Exploration of the Sea (ICES) Code of Practice

The ICES, through its working group on Introductions and Transfers of Marine Organisms (and in cooperation with other ICES working groups and with the European Inland Fisheries Advisory Committee of FAO), has addressed IAS through a series of successive Codes, representing a risk management framework. This is a response to three specific challenges relating to the global translocation of species to new regions:

1) The ecological and environmental impacts of introduced and transferred species, especially those that may escape the confines of cultivation and become established in the receiving environment.
2) The potential genetic impact of introduced and transferred species, relative to the mixing of farmed and wild stocks (as well as to the release of genetically modified organisms).
3) The inadvertent coincidental movement of harmful organisms associated with the target (host) species. (ICES 2005).

The most up-to-date version of the ICES Code of Practice sets forth recommended procedures and practices to diminish the risks of detrimental effects from the intentional introduction and transfer of marine (including brackish) organisms, including the risks associated with ornamental trade and bait organisms, research and import of life species for immediate consumption, and also includes biocontrol agents. The ICES views the Code of Practice as a guide to recommendations and procedures. As with all Codes, the current Code has evolved with experience and with changing technological developments (ICES 2005).

While initially designed for the ICES Member Countries, concerned with the North Atlantic and adjacent seas, all countries around the globe have been encouraged by ICES Members and other international instruments (e.g. CBD) to implement the Code of Practice. The Code has become a recognized instrument and has been applied in the evaluation process for several species introductions, both in ICES Member Countries and outside the ICES area (Hewitt *et al.* 2006).

8.4.2 United Nations Environment Programme (UNEP) Regional Seas Programme

Many agreements and action plans that were developed under the UNEP Regional Seas Programme include provisions on alien species. Binding requirements are laid down by four protected area protocols, concluded for the Mediterranean, the wider Caribbean area, the southeast Pacific, and eastern African region (Shine *et al.* 2005). In the case of the Mediterranean, an action plan has also been adopted concerning introductions of species and invasive species. Its main objectives are to promote the development of coordinated measures and efforts throughout the Mediterranean region in order to prevent, control, and monitor the effects of species introduction (UNEP-MAP-RAC/SPA 2005).

8.4.3 Other agreements

Details on other regional agreements that include provisions on (invasive) alien species, such as (but not limited to) the following, can be found in Shine *et al.* (2000).

- The African Convention on the Conservation of Nature and Natural Resources.
- The Agreement for the Preparation of a Tri-partite Environmental Management Programme for Lake Victoria.
- The Convention for the Establishment of the Lake Victoria Fisheries Organization.
- The Environmental Protocol under the Antarctic Treaty.
- The Convention on the Conservation of European Wildlife and Natural Habitats (Bern Convention).
- The Convention on Fishing in the Danube.
- The Convention of Conservation of Nature in the South Pacific.
- The Association of South East Asian Nations (ASEAN) Convention on the Conservation of Nature and Natural Resources.

8.5 Invasive species and instruments relating to phytosanitary and sanitary measures

8.5.1 The International Plant Protection Convention (IPPC)

The International Plant Protection Convention (IPPC) provides a framework for international cooperation to prevent the introduction of pests of plants and plant products and to promote appropriate measures for their control (see the International Phytosanitary Portal (IPP) at http://www.ippc.int for further information and for the text of the International Standards for Phytosanitary Measures (ISPMs).

The IPPC defines 'pests' as 'any species, strain or biotype, animal life or any pathogenic agent injurious or potentially injurious to plants or plant products' e.g. fungi, bacteria, phytoplasmas, viruses, and invasive plants. Official IPPC definitions can be found in the International Standard for Phytosanitary Terms (ISPM) # 5 Glossary of Phytosanitary Terms, which is revised annually. Until the 1990s, the IPPC mainly focused on phytosanitary certification with an almost exclusively agricultural focus. However, since 1999 the IPPC has clarified its role with regards to IAS that are plant pests (see, for example, Tanaka and Larson 2007). This includes a revision to clarify how environmental impacts are included under the term 'economic harm' (ISPM 5) revisions relating to bio control (ISPM 3), and to Pest Risk Analysis (including ISPM 2 and ISPM 11). Pest Risk Analysis (PRA) underlies Import Health Standards or other import restrictions across country borders. Of particular relevance is the revised standard for Pest Risk Analysis for Quarantine Pests (ISPM 11 Rev1). The Revised ISPM 11 spells out clearly that such analysis may include:

- Invasiveness of the commodity itself (e.g. the garden plant that is proposed for import can be a potential weed);
- Secondary effects of plant pests on other taxa;
- Effect on plants via effect on other taxa;
- Effects on native plants (i.e. not just cultivated ones).

It is expected that in future more countries will increasingly apply their established phytosanitary systems to protect the environment and biological diversity from the risks posed by plant pests. (For further discussion, see, for example, Unger 2003; Tanaka and Larson 2007; and for a discussion of challenges, see De Poorter and Clout 2005.)

Implementation of the IPPC is supported by nine regional plant protection organizations established to strengthen the capacity of countries in a region to address phytosanitary issues (Shine *et al.* 2005). As an example, the European and Mediterranean Plant Protection Organization (EPPO) has reflected the IPPC revisions in its new working programme of an IAS (Schrader 2004).

8.5.2 Other regulations

For a discussion on the *FAO Code of Conduct for the Import and release of exotic biological agent*s and the relevance to IAS of Animal Health and the Office

International des Epizooties (OIE), and Human Health and Human Health regulations see Shine *et al.* 2000, 2005.

8.6 Invasive species and instruments relating to transport operations

8.6.1 International Maritime Organization (IMO)

The Convention establishing the IMO was adopted in Geneva in 1948 and the IMO first met in 1959. The IMO's main task has been to develop and maintain a comprehensive regulatory framework for shipping and its remit today includes safety, environmental concerns, legal matters, technical cooperation, maritime security, and the efficiency of shipping.

The member countries of IMO developed guidelines for the control and management of ships' ballast water, to minimize the transfer of harmful aquatic organisms and pathogens. These guidelines were adopted by the IMO Assembly in 1997. Management and control measures recommended by the guidelines include:

- Minimizing the uptake of organisms during ballasting.
- Cleaning ballast tanks and removing mud and sediment that accumulates in these tanks on a regular basis.
- Avoiding unnecessary discharge of ballast.
- Undertaking ballast water management procedures, including:
    - Exchanging ballast water at sea. Implementation of this measure is subject to ships' safety limits.
    - Non-release or minimal release of ballast water.
    - Discharge to onshore reception and treatment facilities.

In recognition of the limitations of the guidelines, the current lack of a totally effective solution, and the serious threats still posed by invasive marine species, IMO member countries also agreed to develop a mandatory international legal regime to regulate and control ballast water, and the *International Convention for the Control and Management of Ships' Ballast Water and Sediments* was adopted in February 2004. It is not yet in force, but will be so 12 months after ratification by 30 States, representing 35% of world merchant shipping tonnage. In it, Parties undertake to:

- Give full and complete effect to the provisions of the Convention and the Annex in order to prevent, minimize, and ultimately eliminate the transfer of harmful aquatic organisms and pathogens through the control and management of ships' ballast water and sediments.
- Ensure that ports and terminals where cleaning or repair of ballast tanks occurs have adequate reception facilities for the reception of sediments.
- Promote and facilitate scientific and technical research on ballast water management; and monitor the effects of ballast water management in waters under their jurisdiction.

Ships are required to be surveyed and certified; to have on board and implement a Ballast Water Management Plan; and to have a Ballast Water Record Book. All ships using ballast water exchange should, whenever possible, conduct ballast water exchange at least 200 nautical miles from the nearest land and in water at least 200 metres in depth. In cases where the ship is unable to conduct ballast water exchange as above, this should be as far from the nearest land as possible and, in all cases, at least 50 nautical miles from the nearest land. Details and the full text of the new Convention can be found at the IMO website (http://globallast.imo.org/index.asp?page=mepc.htm&menu=true).

There are no internationally-agreed prevention measures for hull-fouling as an IAS vector, although CBD Decision VI/23 called on the IMO to develop mechanisms to minimize this as a matter of urgency.

8.6.2 The International Civil Aviation Organization (ICAO)

The ICAO recognizes that civil air transportation represents a potential pathway for IAS introduction. Contracting States have been urged to take mutually supportive efforts to reduce the risk of introducing potential IAS via this pathway to areas outside their natural range. (For more details, see Shine *et al*. 2003, 2005.)

8.7 Relationship with multilateral trading systems

Non-native species are introduced through trade intentionally (imported products) or unintentionally (e.g. as by-products, parasites, and pathogens of traded products, hitchhikers, and stowaways in vessels, vehicles, or containers that deliver products or services). National measures to minimize unwanted introductions—quarantine and border controls on live species, commodities, packaging, and other vectors—therefore have a direct interface with the multilateral trading system and need to be consistent with applicable rules and disciplines adopted within the WTO framework. The WTO Agreement on the Application of Sanitary and Phytosanitary Measures (WTO-SPS Agreement, 1995) provides:

- That a WTO member may adopt national measures to protect human, animal, or plant health/life from risks arising from the entry, establishment, or spread of pests, diseases, or disease-causing organisms and to 'prevent or limit other damage' within its territory from these causes.
- For the use of international standards as a basis for national protection measures that affect trade. The aim is ensure that national measures have a scientific basis and are not used as unjustified barriers to international trade. The Agreement recognizes standards set by three organizations: IPPC (pests of plants and plant health); OIE (pests and diseases of animals and zoonoses); and Codex Alimentarius Commission (food safety and human health).
- Four key principles (reflected in the revised 1997 IPPC Agreement) that include consistency in the application of appropriate levels of protection, least

trade restrictive alternatives, acceptance of equivalent but different SPS measures, and transparency through advance notification of measures.

Consistent with these principles:

- Countries may take action when necessary to protect plant/animal health by preventing introduction or carrying out eradication/containment.
- Such action should be based on the appropriate level of protection for that country.
- Pest risk analysis is to be used in the development of measures.
- Countries should base national measures on international standards where available. Where no international standard exists or a higher protection level is sought, the State concerned must justify a national measure through scientifically-based risk assessment.
- Emergency (or provisional) measures are permissible without such analysis, when situations require urgent action, or there is insufficient information on which to base action. However, such measures must be reviewed for their scientific justification and modified as appropriate.

(For further discussion see Shine *et al.* 2000, 2005; Werksman 2004; Cooney and Dickson 2005.)

8.8 International instruments and approaches relevant to invasive species

Existing international instruments underpin the use of some of the key approaches and principles that are of critical importance in dealing with IAS. These include:

Ecosystem approach

IAS management must be approached in a wider ecosystem context (see Zavaleta *et al.* 2001) and the use of the ecosystem approach is one of the CBD *Guiding Principles for the Prevention, Introduction and Mitigation of Impacts of Alien Species that threaten Ecosystems, Habitats or Species*. The ecosystem approach in general has been defined and elaborated in 12 broad principles in a series of CBD Decisions (Decisions V/6 and V/8). In the context of IAS, the most important ones are: centralized management to the lowest level; consideration of effects of management activities on adjacent and other ecosystems; conservation of ecosystem functioning and structure; recognition of lag effects in ecosystem processes; involvement of all relevant sectors and scientific disciplines. (For further discussion, see Shine *et al.* 2000.)

Precautionary principle/approach

Precaution relates to decision making in situations of scientific uncertainty. Principle 15 of the Rio Declaration holds that 'lack of full scientific certainty shall not be used as a reason for postponing cost effective measures to prevent environmental degradation'. In the CBD preamble it is stated as 'Noting that where there is a threat of significant reduction or loss of biological diversity, lack of full scientific certainty should not be used as a reason to postpone measures to avoid or minimize a threat of significant reduction or loss of biodiversity'.

IAS are a form of biological pollution. Impacts on biodiversity are wide ranging, complex, insidious, and often irreversible (e.g. De Poorter and Clout 2005). The precautionary principle is therefore a cornerstone of the IUCN approach which states that in the context of alien species, 'unless there is a reasonable likelihood that an introduction will be harmless, it should be treated as likely to be harmful' (IUCN 2000). In the invasives context, the precautionary approach/principle has been widely incorporated into international guidance on invasives, including the CBD guiding principles, the CBD Jakarta mandate on marine and coastal biodiversity, the European IAS strategy, and the African Eurasian Waterbird agreement. However, in the context of trade measures, precaution remains controversial (Cooney 2004; Cooney and Dickson 2005).

8.9 Relation between invasive species and sustainable development programmes

In 1994 The Barbados Programme of Action for the Sustainable Development of Small Island Developing States (SIDS), a non-binding instrument states that 'the introduction of certain non-indigenous species' as one of the four most significant causes of the loss of biodiversity in SIDS (Section IX para 41) and specifically identifies the need to 'support strategies to protect Small Island Developing States from the introduction of non-indigenous species' (Section IX para 45, C. (vi)).

An International Meeting for the 10-year Review of the Barbados Programme of Action took place in 2005 in Mauritius and adopted the Mauritius Strategy for further implementation of the Barbados Programme of Action. The Mauritius Strategy reiterates the recommendation to control major pathways for potential IAS in Small Island Developing States.

The World Summit on Sustainable Development (WSSD) took place in 2002, in Johannesburg, South Africa. The WSSD's goal, according to UN General Assembly (UNGA) Resolution 55/199, was to hold a 10-year review of the 1992 UN Conference on Environment and Development (UNCED) at the Summit level to reinvigorate global commitment to sustainable development. The Plan of Implementation of the WSSD (Paragraph 44(i)) calls for countries to 'Strengthen national, regional and international efforts to control IAS, which are one of the main causes of biodiversity loss, and encourage the development of an effective work programme on invasive alien species at all levels'. (See http://www.un.org/esa/sustdev/documents/WSSD_POI_PD/English/POIChapter4.htm.)

8.10 Regional strategies and plans

8.10.1 South Pacific Regional Environment Programme (SPREP): invasive species strategy for the Pacific Island region

An Invasive Species Programme was initiated by the SPREP. One of the objectives of this programme was to develop a strategy for invasive species for use by all countries and relevant agencies in the region. The aim of the Regional Strategy is to promote the efforts of Pacific Island countries in protecting and maintaining the rich and fragile natural heritage of the Pacific Islands from the impacts of invasive species through cooperative efforts (Sherley 2000).

8.10.2 European Strategy (Council of Europe)

The Bern Convention (1979) provides regional frameworks for implementing the Convention on Biological Diversity. A European Strategy on IAS was developed and recently adopted (December 2003) to promote a comprehensive and cross-sectoral approach to all aspects of IAS, with a focus on trans-boundary cooperation within Europe (Genovesi and Shine 2004). It includes:

- Building awareness and support;
- Collecting, managing, and sharing information;
- Strengthening national policy, legal and institutional frameworks;
- Regional cooperation and responsibility;
- Prevention;
- Early detection and rapid response;
- Management of impacts;
- Restoration of native biodiversity.

8.10.3 European Union

In the European Union, the cross-cutting issue of IAS is addressed at the community level through a range of legal instruments that apply to prevention of unwanted introductions of organisms harmful to plants or plant products, animal and fish diseases, and species that may threaten the wild fauna and flora (Demeter 2006). As an important milestone in the currently ongoing review of implementation of the European Commission (EC) Biodiversity Strategy, a broad consultative process culminating in a conference in Malahide, Ireland, reconfirmed Community-level actions on IAS as a priority issue. The Directorate General Environment in the EC has started up an inter-service consultative group to plan action on IAS. A study has been contracted which will provide important advice for the development of an EU policy on the issue (Demeter 2006).

8.10.4 Pacific Invasives Initiative

The Cooperative Initiative on Invasive Alien Species on Islands (Cooperative Islands Initiative or CII) was launched in 2002 following calls from countries with islands for more effective efforts to manage invasive species. The CII is a global initiative. With requests from Pacific countries for more coordinated and cooperative approaches to addressing invasive species threats the multi-partner 'Pacific Invasives Initiative' (http://www.issg.org/cii/PII/index.html) or PII became the first programme of the CII to be funded. The goal of the PII is to conserve island biodiversity and enhance the sustainability of livelihoods of men, women, and youth in the Pacific. Activities are focused on raising awareness of invasive species issues, building capacity in the region to manage invasives, and facilitating cooperative approaches to achieve and sustain desired outcomes.

8.10.5 Pacific Ant Prevention Programme (PAPP)

The PAPP is a regional multi-agency initiative endorsed by Secretariat of the Pacific Community (SPC) member countries and territories. The main objectives of the PAPP are to increase awareness on the potential threats posed by invasive ants in Pacific Islands, develop and put in place invasive ant emergency response systems, management methods, and develop national capacities to deal with new incursions.

To ensure the sustainability of activities on invasive ants—including surveillance and awareness—SPC have taken responsibility for managing of the PAPP. However, full implementation of PAPP is multi-sectoral, involving many partners including the Samoa-based regional Secretariat of the Pacific Regional Environment Programme, the Pacific Invasives Initiative, Biosecurity New Zealand, and the US Department of Agriculture.

8.11 International programmes and organizations

8.11.1 The Global Invasive Species Programme (GISP)

The GISP was founded in 1997 as a partnership programme, by the Scientific Committee on Problems of the Environment, IUCN - The World Conservation Union and CAB International. The GISP mission is to conserve biodiversity and sustain human livelihoods by minimizing the spread and impact of IAS (http://www.gisp.org/index.asp). The main focus of GISP is to promote global cooperation in invasive species prevention and management; it aims to prevent the international spread of IAS, minimize the impact of established IAS on natural ecosystems and human livelihoods, and create a supportive environment for improved IAS management. It has a mandate under the CBD, and a specific interest in fostering cross-sectoral collaboration between relevant international instruments and organizations. This will be done through cooperation amongst its Member Organizations, as well as a wide range of partners across the globe.

The First Phase of GISP produced many outputs, including: *A Guide to Designing Legal and Institutional Frameworks on Alien Invasive Species* (Shine *et al.* 2000), *A Toolkit for Best Prevention and Management Practices* (Wittenberg and Cock 2001), a *Global Strategy on Invasive Alien Species* (McNeely et al. 2001), and the initial development of the Global Invasive Species Database (see section 8.11.4).

8.11.2 The International Union for Conservation of Nature and Natural Resources (IUCN)

IUCN (http://www.IUCN.org) was founded in 1948. It brings together 83 States, 110 government agencies, more than 800 non-governmental organizations (NGOs), and some 10,000 scientists and experts from 181 countries in a unique worldwide partnership. Its mission is to influence, encourage, and assist societies throughout the world to conserve the integrity and diversity of nature and to ensure that any use of natural resources is equitable and ecologically sustainable. IUCN has been involved with the issue of IAS for many years and is one of the founding partners of GISP. IUCN Regional Programmes address IAS Eastern and Southern Africa, Asia, and in Meso and South America, while IUCN at a global level integrates IAS into all parts of its programmes and themes (including Marine, Protected Areas, Ecosystem Management, Forests, Wetlands, Biodiversity Policy) and the Environmental Law Centre. The IUCN *Guidelines for the Prevention of Biodiversity Loss caused by Alien Invasive species* were adopted by IUCN Council in February 2000 and are available in English, French, and Spanish (IUCN 2000).

8.11.3 The Invasive Species Specialist Group (ISSG)

The Invasive Species Specialist Group (ISSG) of the IUCN Species Survival Commission (http://www.issg.org/) is a network of over 170 expert volunteers from more than 40 countries who are knowledgeable about invasive species problems. The mission of ISSG is 'to reduce threats to natural ecosystems and the native species they contain, by increasing awareness of alien invasions and of ways to prevent, control or eradicate them'. The ISSG's role is global and includes providing technical and policy advice to IUCN, to IUCN members, and, more generally, to organizations and individuals requiring advice in their fight against biological invasions. In addition, ISSG produces technical publications, publishes the *Aliens* newsletter, and manages the Aliens-L Listserver and the Global Invasive Species Database (section 8.11.4).

8.11.4 The Global Invasive Species Database (GISD) and Global Invasive Species Information Network (GISIN)

The GISD is managed by ISSG and is a free, online public resource of authoritative information about IAS (http://www.issg.org/database/). The GISD aims to increase public awareness about IAS and to facilitate effective prevention and management activities by disseminating specialists' knowledge and experience globally to a broad audience. It focuses on those IAS that threaten biodiversity and covers all taxonomic groups from micro-organisms to animals and plants. GISD profiles include information about the negative impacts of invasive species, causes and vectors of introduction and spread, prevention strategies, lessons learned during management projects, and the contact details of experts who can help. All of this information is either supplied by or reviewed by expert contributors from around the world.

A related project is the development of a Global Register of Invasive Species. The development of a GISIN was proposed at the sixth meeting of the Conference of the Parties to the Convention on Biological Diversity. It is currently coordinated by an interim steering committee. GISIN will provide a platform through which IAS information from participating databases can be accessed (see http://www.gisinetwork.org/ for more information).

8.12 Conclusions

A lack of legal provisions and/or mandate to deal with IAS can be a significant impediment to conservation managers. In order to be effective, IAS management must be able to cut across political boundaries, and the addressing of IAS and the threats that they pose to the environment must happen at international (global and regional) levels as well as at national and local levels. A mosaic of binding and non-binding international instruments addresses IAS and can be used to underpin practical management. The remaining challenges include: increasing cooperative action and approaches at regional level; gaining greater acceptance of the application of precaution in the multilateral trading system; and financing and developing sufficient capacity at national level, in all countries worldwide, to implement effective IAS management.

9
Management of invasive terrestrial plants
Jodie S. Holt

9.1 Introduction

Management of terrestrial plants is based on principles common to management of all organisms, regardless of the type or identity. In practice, however, plants possess features that necessitate unique approaches and tools for management. As sessile organisms that are rooted to the ground, plants possess adaptations, or evolved traits, that result in survival and reproduction in particular environments. Tremendous genetic variation occurs among terrestrial plants and in addition, most plants possess plasticity, or the ability to change growth and form in response to environmental factors. For these reasons management techniques may not apply uniformly across all environments or work equally well on all parts of the same plant.

9.2 Classification of weeds and invasive plants

Management of weeds and invasive plants generally requires some level of identification of the plants in question. Invasive plants are most often vascular seed plants and most are also flowering plants (angiosperms). Some notable exceptions occur (e.g. some ferns, which are seedless, and conifers, seed plants that have no flowers, are considered weeds). Angiosperms are further divided into dicots and monocots (the classes Dicotyledones and Monocotyledones). In situations where broad-scale or non-selective methods of weed control are used, distinction among grasses and sedges (monocot) and broadleaf (dicot) plants may be sufficient for identification purposes. In other situations, successful management depends on knowing the specific identity of the undesirable plant and having some understanding of its characteristics and life history, including mode of reproduction (Radosevich *et al*. 2007).

In addition to taxonomical classification, weeds and invasive plants are classified by their length of life cycle (annual, completing the life cycle in 1 year or less; biennial, living longer than 1 but less than 2 years; or perennial, living longer than 2 years), season of growth (winter or summer), and time and method of reproduction. Perennial plants may resprout every year from below-ground reproductive structures that store carbohydrates over the winter (Fig. 9.1). This seasonality in

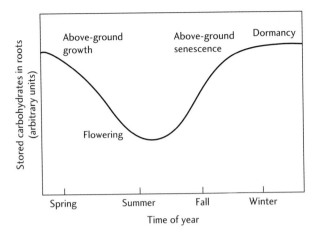

Fig. 9.1 Seasonal progression of stored carbohydrates in roots and below-ground shoots of perennial plants. Reproduced with permission from Holt J.S. and Radosevich S.R. (1989). Plants. In California Weed Conference (ed.) *Principles of Weed Control in California*, p. 20. Thomson Publications, Fresno, CA.

carbohydrate storage and in direction of carbohydrate translocation can determine whether a control method will be successful. For example, shoot removal from a perennial plant in spring using physical or chemical means is unlikely to kill the plant since the majority of carbohydrates are being mobilized upwards to produce new shoots at that time (Fig. 9.1).

Plants are also classified in terms of undesirability. Legally, a noxious weed is any plant designated by a Federal, State, or County government as injurious to public health, agriculture, recreation, wildlife, or property (Sheley *et al.* 1999; Radosevich *et al.* 2007). Many states, provinces and countries maintain at least one official list of such weeds so that their introduction can be prevented or restricted. Noxious weeds usually create a particularly undesirable condition in crops, forest plantations, grazed rangeland, or pastures. Therefore, both the identity and legal status of a particular species should be considered when developing a management programme.

9.3 Plant characteristics important in management

Plants are characterized by growth and development from regions of cell division (meristems) located at the apices of shoots and roots, which result in a repetitive modular structure in the plant body. A unique feature of the plant body is roots and, in many cases, reproductive structures (such as rhizomes, tubers, etc.) below ground, which possess vegetative buds that can develop into new plants. Even shoots can develop roots adventitiously in response to various environmental cues and thereby propagate the plant vegetatively. Above- and below-ground modules are relatively autonomous such that herbivory or physical damage may harm the plant but not kill it. For example, in most plants removal of vegetative branches will lead to the production of new tissues rather than death. In addition, fragmentation of an individual into independent clones (ramets) may arise through physical means, including many control techniques. Cloning is an important characteristic leading to the persistence and dispersal of many perennial weeds (Silvertown and Charlesworth 2001; Radosevich *et al.* 2007).

Flowering plants reproduce sexually by means of seeds, which are usually dispersed physically away from the maternal plant and may remain dormant for very long time periods in the soil. The soil seed bank is a reservoir of seeds in which both deposits and withdrawals are made (Fig. 9.2) (Harper 1977; Allessio Leck *et al.* 1989). Deposits occur by seed rain from seed production and dispersal both on-site and from off-site locations, whereas withdrawals occur by germination, predation, senescence, and death. Storage of seeds results from the vertical distribution of seeds through the soil profile, although most weed seeds occur at shallow depths (Allessio Leck *et al.* 1989; Baskin and Baskin 1998; Benech Arnold *et al.* 2000). As noted above, the soil may also contain a reservoir of vegetative buds that can revegetate a site following weed removal. The dynamics of seed and bud banks should be considered when developing a plant management programme (Altieri and Liebman 1988; Baskin and Baskin 1998; Swanton and Booth 2004) (Fig. 9.3).

9.4 Management of terrestrial invasive plants

The objectives of invasive plant management in wildland ecosystems are similar to those of weed management in agroecosystems, although the specific considerations in evaluating whether to control invasive plants differ from those pertaining to agricultural weed control (Holt 2004). Economic cost:benefit analysis is more difficult in ecosystems where no harvestable commodity is produced and the value of the land or services it provides are intangible (Pimentel *et al.* 1997, 2000). The principles for setting priorities for management of invasive plants are based on the risk of invasion and the value of the land (Hobbs and Humphries 1995; Hiebert 1997). Assuming some form of management is indicated, many of the same approaches and tools for weed control in agroecosystems can be used in wildlands, as described below (Holt 2004; Smith *et al.* 2006).

9.4.1 Principles of prevention, eradication, containment, and control

Management of weeds and invasive plants is a general strategy that encompasses the approaches of prevention, eradication, and control (Table 9.1) (Ross

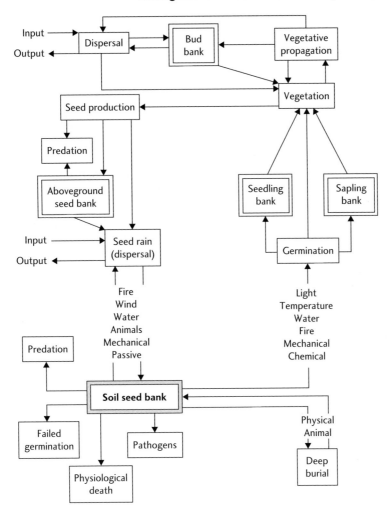

Fig. 9.2 General model of the soil seed bank and bud bank showing environmental factors determining inputs and outputs. Reproduced with permission from Simpson R.L., Allessio Leck M., and Parker V.T. (1989). Seed banks: General concepts and methodological issues. In Allessio Leck M., Parker V.T., and Simpson R.L. (eds.) *Ecology of Soil Seed Banks*, pp. 3–8. Academic Press, San Diego, CA.

and Lembi 1999; Radosevich *et al.* 2007). Prevention involves procedures that inhibit the establishment of specific plants in areas that are not already inhabited by them. These practices restrict the introduction, propagation, and spread of weeds or invasive plants on a local or regional level. Preventive measures

Fig. 9.3 (a) When developing a plant management programme, the dynamics of seed banks should be considered, as illustrated by the knee-deep pappus and seeds (b) of artichoke thistle (*Cynara cardunculus*) in California. Photo: Jodie Holt.

for invasive plants include cultural practices, such as using weed-free feed for stock animals that graze on rangeland or travel through forests and parkland. Prevention also includes the use of quarantines, weed and seed laws, and, more recently, weed risk assessment and inventories to screen and restrict potential invaders from introduction (Reichard 1997; Groves *et al.* 2001; Wittenberg and Cock 2005). Preventive measures for invasive plants are discussed in more detail below.

Eradication is the total elimination of a plant species from a field, specific area, or entire region. Eradication also requires the complete suppression or removal of all seeds and vegetative parts of a species from a defined area, which is particularly difficult when a seed or bud bank is present. The goal of eradication is rarely, if ever, achieved without monumental effort and cost. Because of the difficulty and high costs associated with these practices, eradication is usually attempted only in small defined areas, areas with high-value crops or land use, or areas of particularly sensitive species or habitat (Zamora *et al.* 1989; Rejmánek and Pitcairn 2002; Wittenberg and Cock 2005).

In contrast to eradication, control practices reduce or suppress plants in a defined area but do not necessarily result in the elimination of any particular species. The goal of weed or invasive plant control is usually reduction of

Table 9.1 A list of approaches and some examples of techniques used in the management of weeds and invasive plants.

Approach	Techniques
Prevention	Using weed-free feed for stock animals; adhering to weed laws and quarantines; conducting risk assessment; screening plants for suitability for introduction; conducting surveys and monitoring; using clean machinery; using clean soil and gravel; inspecting nursery stock for weed seeds; avoiding planting invasive species or cultivars.
Eradication	Complete removal by any means.
Control:	
Physical	Hand pulling; using manual implements; using fire; using machines to remove plants; mowing; chaining; using mulches; solarizing the soil in defined areas.
Cultural	Prevention (see above); manipulating competition by native or other desirable species.
Biological	Introduction of natural enemies; augmentation of natural enemies; habitat management to enhance predators and parasitoids; using grazing.
Chemical	Using herbicides.
Integrated weed management	Using a combination of methods from techniques listed above.

impact to an economically acceptable level in agroecosystems or an ecologically acceptable level in wildland ecosystems (Fig. 9.4) (Radosevich et al. 2007). Similar to control, containment is often a goal of management of invasive plants, where the infestation is held to a defined geographic area and not allowed to spread (Wittenberg and Cock 2005). Weed control, therefore, is a matter of degree that depends upon the goals of the people involved, effectiveness of the tool or tactic used, and the abundance and competitiveness of the species present. Just as in agroecosystems where weed control occurs in the context of crop production, control of invasive plants requires careful consideration of the native plant community in which the invaders occur. In many cases, effective management will also require restoration of desired species concurrent with control efforts.

There are four general methods of controlling unwanted plants: physical, cultural, biological, and chemical, which are described below and summarized in Table 9.1 (Ross and Lembi 1999). In view of the complexity and diversity of most wildland ecosystems, the best approach in many cases is an integrated one that uses a combination of the most appropriate methods and tools in the context of the system in question.

132 | Invasive species management

Fig. 9.4 The goal of weed or invasive plant control is usually reduction of impact to an economically acceptable level in agroecosystems or an ecologically acceptable level in wildland ecosystems. *Arundo donax* (giant reed) is illustrative of a highly invasive plant that displaces most of native riparian vegetation. This weed dries in winter, breaks off and washes downstream in floods, burns readily, creating a considerable hazard. A few years ago a huge mat of giant reed washed downstream and took out Van Buren Bridge in Riverside, California, USA, costing the city US$5 million to replace it with this one. Photo: Michael Rauterkus.

9.4.2 Physical methods of invasive plant control

Physical methods of control include any technique that uproots, cuts, buries, smothers, or burns vegetation. Depending on the habitat and any restrictions pertaining to it, these methods can employ manual or machine power. In general, many more techniques of physical control can be used in agroecosystems where the soil is bare prior to crop planting than in wildlands where native vegetation might be present. Physical methods that are suitable for use in wildland ecosystems consist of hand pulling or using manual cutting or hoeing implements; using fire in the form of controlled burning; using machines for mowing, shredding, chaining, or dredging; and in some cases, using mulches or solarization (Ross and Lembi 1999; Wittenberg and Cock 2005; Radosevich *et al.* 2007).

9.4.2.1 Hand pulling and using manual implements

Hand pulling and hoeing are the oldest and most primitive forms of weed control, and in general, have fewer negative impacts on the environment than other forms of control (e.g. mechanical or chemical methods). A variety of hand implements, from primitive to very sophisticated, have been developed for the removal of weedy vegetation. Hand pulling and using manual implements are most effective on annual or simple perennial plants that are not able to sprout from roots or other vegetative organs. However, these methods are very labour intensive, and over wide areas may be prohibitively expensive or unfeasible.

Even though manual methods of controlling unwanted vegetation have declined in agroecosystems in favour of mechanized approaches, hand pulling and using manual implements are still practised in certain high-value crops or habitats, or when other types of plant suppression are not possible. For example, hand tools are often used in wildland situations where selective herbicides are not available or registered for use, where the area to be weeded is too small for most equipment, or where the habitat contains species or other elements that are sensitive to disturbance and may even be protected by law (Wittenberg and Cock 2005). These methods are also widely used in areas where volunteer crews are available for weed control but may not have the equipment, funds, or training to use mechanized tools. Hand methods can also be effective and economical for removing the few individual plants that escape other control measures. These few individual plants, if left unattended, have the potential to expand and replenish the soil seed bank. Similarly, the earliest stages of colonization by an invasive plant can sometimes be eradicated by hand pulling in areas where financial or environmental constraints prohibit use of other methods (Wittenberg and Cock 2005).

9.4.2.2 Fire

Fire is another tool that has been available to humans for centuries for the manipulation of vegetation. It is still used extensively to remove crop residues and other unwanted vegetation in agriculture and to prepare forest lands for regeneration after clearcut logging (Radosevich *et al.* 2007). Fire is also sometimes used to remove weeds and other residue from along roadsides, canal banks, ditches, and vacant areas. In addition, controlled burning has been used to manage fuel breaks in shrublands that are prone to wildfire. Broadcast burning is also an accepted and effective method to periodically increase rangeland productivity by stimulating growth of certain fire-adapted grass species (DiTomaso and Johnson 2006; DiTomaso *et al.* 2006).

Fire can be an effective weed management tool in certain wildland ecosystems when applied carefully and selectively (Muyt 2001). Applying fire for management of unwanted vegetation can destroy mature plants, eliminate seedlings, exhaust weed seed banks, improve access to a site for follow-up treatments, remove dried weed material, and stimulate germination, growth, and spread of desirable vegetation. However, fire also favours some weeds and invasive plants and can have a detrimental effect on wildland ecosystems (Brown and Kapler-Smith 2000; Muyt 2001; Rice 2004). Therefore, it is critical to assess whether fire is an appropriate invasive plant management technique before conducting any controlled burning (Wittenberg and Cock 2005; DiTomaso and Johnson 2006; DiTomaso *et al.* 2006).

9.4.2.3 Using machines for invasive plant control

Weeds and invasive plants can be controlled mechanically, which suppresses plants by breaking, cutting, or tearing them from the soil, thus exposing tissues to desiccation, and by smothering them with soil. Where it is possible, repeated mechanical operations may deplete weed seeds or vegetative propagules in the soil, providing that escaped or surviving plants are not allowed to reproduce. Repeated mechanical operations may also be used to exhaust the carbohydrate reserves of perennial weeds, thus suppressing them by reducing regrowth (Ross and Lembi 1999; Radosevich *et al.* 2007). Seeds near the soil surface are usually not injured by mechanical operations, while seedlings are quite vulnerable to such disturbance. Annual, biennial, and simple perennial weeds are most susceptible to mechanical control, since shoots are typically separated from roots or the entire plant is uprooted. However, for creeping and woody perennial plants, typically only the shoots are killed by mechanical disturbance and resprouting often occurs (Fig. 9.5). Best results from these operations occur when the soil is dry and the weather is hot since uprooted plants will desiccate and die following soil disturbance. Even vegetative reproductive organs (rhizomes, stolons, etc.) of perennial plants can be destroyed if they are uprooted and left on the soil surface under hot dry conditions.

Due to the size and constraints of using equipment, mechanical control methods are typically performed on a larger scale than manual control, which can damage native vegetation, disturb the soil, and leave debris that must be removed. Thus, mechanical methods of plant removal are generally site- and species-specific and may even be prohibited where weeds occur in plant communities containing native species (Whisenant 1999).

Mowing is also used to control weeds and invasive plants by cutting or shredding their foliage. In some situations, mowing is used to decrease the amount of plant biomass that is present but not kill the plants, for example, along power line rights-of-way, roadsides, ditch banks, abandoned cropland, or vacant lots. Mowing of herbaceous vegetation should occur before the plants set seed in order to avoid weed seed dispersal. Since mowed plants, even annuals, may regenerate new shoots, frequent mowing is required to prevent seed production (Ross and Lembi 1999; DiTomaso 2000; Huhta *et al.* 2001).

Mowing can suppress some perennial weeds through carbohydrate starvation, as described above. Similar to mechanically removing or disturbing plant shoots, frequent mowing also stimulates new shoot development, which eventually depletes the plants' carbohydrate reserves if performed frequently enough. Mowing every few weeks for at least one to two growing seasons is usually necessary to suppress herbaceous perennial vegetation in this way. However, the plants being controlled by mowing cannot be allowed to replenish their underground supply of carbohydrates for this system of weed control to be effective (DiTomaso 2000; Huhta *et al.* 2001). Such frequent, repeated mowing operations can damage desirable species as well as select for prostrate weed phenotypes with multiple shoots

(a)

(b)

Fig. 9.5 Many perennial plants have below-ground reproductive structures such as these rhizomes of *Arundo donax* (a) that store carbohydrates over the winter. These resprout every year, even from very small pieces (b) making them difficult to control. Photos: Joseph Decruyenaere.

that are relatively prostrate. Therefore, this method of mechanical control is not feasible in many situations.

In some wildland situations, chaining is performed to remove unwanted woody vegetation. This practice uses a heavy chain that is dragged between two tractors and in some cases a metal blade is welded across each link of the chain (Radosevich *et al.* 2007). As the chain is dragged between the tractors, shrub stems are crushed and some plants are uprooted. This procedure is used to prepare shrublands or chaparral for rangeland improvement. A similar practice is used to remove submerged and immersed aquatic weeds from canals and rivers. Chaining is very destructive to all plants in its path so is only appropriate for use where an entire site requires renovation or restoration.

9.4.2.4 Mulching and solarization

A relatively non-destructive method of killing weeds and invasive plants in some situations is the use of mulches and solarization. The purpose of mulches used for weed control is to exclude light from germinating plants, which inhibits photosynthesis and causes the plants to die (Radosevich *et al.* 2007). Some of the more commonly used mulches include straw, bark, manure, grass clippings, sawdust, rice hulls or other crop residues, paper, and plastic. Artificial mulches made of woven plastics are also available—these exclude particular wavelengths of light but allow water, nutrients, and air to penetrate into the soil. Mulches are most effective for controlling small annual weeds but larger plants and some perennials can also be suppressed by using mulches. Mulching is commonly used in agriculture, home gardens, and landscapes, but can be used in some wildland areas as well. Organic mulches that have not been composted or degraded have been shown to reduce available soil nitrogen due to nitrogen immobilization by soil microflora, which use the mulch as a food source (Zink and Allen 1998). Mulching with organic matter in this way has been used to suppress nitrophilous exotic species and promote establishment of native species during restoration efforts (Zink and Allen 1998; Reever and Seastedt 1999).

Soil solarization is performed by covering moist tilled soil with clear plastic—which permits light to pass through—to kill imbibed weed seeds (Horowitz *et al.* 1983). The plastic sheets are left covering the soil surface for about 4 weeks. Long periods of high intensity solar radiation that elevate temperature in moist soil under the plastic are required for best results. Weed seeds are killed either by the prolonged period of high temperature or by other factors. For example, high temperature stimulates germination of some seeds but seedlings cannot survive under the plastic (Horowitz *et al.* 1983). Solarization was initially developed for control of soil microflora and microfauna, which can also be injured or killed by this practice (Horowitz 1983). Solarization has also been investigated for use in restoration of wildlands, such as grasslands and abandoned farmland, to reduce abundance of resident exotic weeds prior to planting native species (Bainbridge 1990, 2007; Moyes *et al.* 2005).

9.4.3 Cultural methods of invasive plant control

In agricultural or natural resource production systems, cultural methods of weed suppression often occur during the normal process of land preparation and crop production (Radosevich *et al.* 2007). Some of these practices are also relevant for the control of invasive plants in wildlands, such as prevention and using competition against unwanted plants.

9.4.3.1 Prevention

Surveys and monitoring are the first step in prevention of invasive plants in wildland ecosystems or natural resource production systems other than agriculture (Dewey and Anderson 2004). The prevention or quarantine of a weed problem is usually easier and less costly than control or eradication attempts that follow introduction because plants are most persistent and difficult to control after they become established (Ross and Lembi 1999). If invasive plants are allowed to develop a reservoir of seeds or buds, they usually will be present in that location for many years, even decades. Several measures can be used to prevent the introduction of weeds into non-inhabited areas. These include the following: cleaning mechanical implements before moving to non-weed-infested areas; avoiding transportation and use of soil or gravel from weed-infested areas; inspecting nursery stock or transplants for seeds and vegetative propagules of weeds; avoiding planting exotic or invasive plants around homesites; removing weeds from near irrigation ditches, fence rows, rights-of-way, and other non-crop land; preventing reproduction of weeds; using seed screens to filter irrigation water; and restricting livestock movement into non-weed-infested areas (Radosevich *et al.* 2007).

In many countries there are legal means of preventing potential weed problems at the national, regional, or state level. Weed laws are one example of a possible way of restricting introductions of unwanted plants and other pests. For example, the Federal Noxious Weed Act, enacted in the USA in 1975, prohibits entry of weeds by providing crop inspection for weed seeds at ports of entry. This law also allows establishment of quarantines to isolate and prevent the dissemination of noxious weeds within a defined area or region, and provides for the control or eradication of weeds that are new or restricted in distribution (Radosevich *et al.* 2007). Other more local weed laws can mandate that property owners or public agencies must maintain a programme of weed prevention or control on their lands. The success of such laws will of course depend upon the level of funding available, the knowledge of the authorities about weed control measures, and the cooperation of public and private land owners in establishing weed suppression programmes. Although these legal restrictions on weeds were developed for agriculture, they can provide models for enactment of similar regulations for introductions of horticultural species that have potential to escape into wildlands (Reichard 1997).

9.4.3.2 *Competition*

In agriculture, cultural practices that shift the balance of competition toward the crop will usually disfavour weed occurrence and improve crop yields. Similarly, in some habitats native or other desirable plants can be used to outcompete invasive plants (Luken 1997; Hoshovsky and Randall 2000). Plant canopy cover of desirable species can be manipulated to suppress weeds by rendering the site less suitable for seed germination or sprouting of vegetative buds, or by eliminating the conditions that are optimal for growth and development of particular species. For example, following weed removal, grassland or rangeland areas can be reseeded with native species in order to provide a dense native canopy cover that suppresses reestablishment of invaders (Hoshovsky and Randall 2000). In other areas native communities may reestablish without additional planting and resume dominance of a site as long as weeds are not reintroduced. This approach usually requires long-term monitoring as well as management of disturbances that might reintroduce invasive species or shift competitive balance away from natives.

9.4.4 Biological control

Biological weed control is the use of living organisms to lower the population level or competitive ability of a plant species so it is no longer an economic problem (DeLoach 1997; Cruttwell McFadyen 2000; Coombs *et al.* 2004). Biological control can have longer lasting effects on weed populations than other forms of control; however, a longer time frame, often years, is required for biological control agents to become established and exert their effect. The object of biological control is not to eradicate the weed; ideally, some of the weed population should always be present to maintain a population of the natural enemy. However, such control can be permanent once the weed and natural enemy populations are in equilibrium (Julien and Griffiths 1998).

Biological control does not necessarily kill plants directly; rather their competitive ability and fecundity are typically reduced. Thus, biological control should not be attempted when the eradication or rapid removal of a weed is the goal. Control agents must be host specific for the weed that they affect to prevent damage to crops or native plants. This specificity is an advantage when control is directed at only one weed species but a disadvantage when several plant species must be suppressed in the same area. Closely related native species in the vicinity of the target weed may also be sensitive to the control agent and should be tested for host specificity (Wittenberg and Cock 2005). Because of the potential for problems when organisms, even presumed beneficial ones, are deliberately introduced into new areas, biological control is tightly regulated in some countries. Where no protocols are in place, the International Plant Protection Convention's Code of Conduct for the Introduction of Exotic Biological Control Agents provides such guidance (IPPC 1996; Wittenberg and Cock 2005).

There are several approaches to biological control, including the introduction of natural enemies from the original range of the target species to the new invaded range; augmentation of natural enemies during pest outbreaks; or habitat management to enhance native populations of predators and parasitoids (Wittenberg and Cock 2005). Grazing is perhaps the oldest and most common form of biological weed control. It can be accomplished using a wide array of animals that eat vegetation, including large ruminants and ungulates, birds, insects, and fish. However, grazing can be an agent of weed propagule dissemination and can suppress native plant populations as well as the weed. The use, timing, and rotation of grazing animals for invasive plant suppression should, therefore, be done with care to minimize their negative impacts on native vegetation (Radosevich *et al.* 2007).

9.4.5 Chemical control

Chemicals, like other methods of weed control, have been used for centuries to suppress or remove weeds in agroecosystems; their use for control of invasive plants in wildlands is relatively recent by comparison. Herbicides are organic, synthetic chemicals used to kill or suppress unwanted vegetation. There are many advantages to the use of herbicides as tools for weed control in agricultural and wildland ecosystems, including control of weeds where other methods are difficult or impractical; reduction in the number of mechanical operations needed; reduction in the amount of human effort and cost expended for hand and mechanical weeding; and greater flexibility in the choice of management systems (Ross and Lembi 1999; Radosevich et al. 2007). There are potential problems associated with herbicide use, however, including injury to non-target plants; residues in soil or water; toxicity to other non-target organisms; and concerns for human health and safety. The increased legal and regulatory requirements for herbicide application and worker safety are other concerns associated with the use of herbicides. In many cases, these problems or disadvantages can be overcome by proper selection, storage, handling, transportation, and application of the chemical (Vencill 2002).

If herbicides are found to be necessary or indicated, specific sites for application should be identified and the application should be restricted to specific plants or stands. The criteria for such site selection would be a quantifiable risk assessment of weed impacts to desirable species and potential herbicide impacts to non-target plants, animals, and their habitats. Adequate buffer zones should be allowed to minimize herbicide drift, runoff, or leaching to riparian areas, waterways, or areas of human habitation. Indigenous and recreational food sources in the vicinity of the area to be treated should be avoided or alternatively, established residue tolerances should not be exceeded. Herbicides are only one type of tool available for weed or invasive plant control and not all vegetation management problems can be controlled effectively by chemical means.

9.4.6 Integrated weed management

Over the past several decades, concerns about the environmental and health hazards of pesticides, soil erosion and degradation, and pest adaptation to control methods, such as population shifts and pesticide resistance, have led to the development of the concepts of integrated pest management (IPM) and integrated weed management (IWM) as alternatives to sole reliance on pesticides (Kogan 1986, 1998). In contrast to using single weed control tools at the field scale, IWM uses multiple control tools at the field, farm, landscape, or regional scale within the context of a management system (Buhler et al. 2000; Liebman 2001). IWM also considers other organisms besides weeds, such as insects and other plants, to be integral parts of the system that are not necessarily detrimental. These broader spatial scales and higher complexity represent the levels at which impacts of invasive species often occur. Perhaps because invasive plants often occur in proximity to native species and/or in sensitive or endangered habitats, integrated programmes in which chemical and mechanical control tools and tactics are used in conjunction with other approaches for vegetation management are particularly promising for the management of invasive species (Hobbs and Humphries 1995; Sakai *et al.*

2001). A comprehensive integrated programme for management of invasive species would include many of the approaches discussed above as well as active restoration and ongoing monitoring to sustain a healthy and productive ecosystem (Hulme 2006; Smith *et al.* 2006). Such a programme must be based on knowledge of the biology of the organisms, the characteristics of the invaded habitat or ecosystem, and the potential for human impacts to drive change in the system. Armed with proper information, a programme of risk assessment, prevention, eradication, and integrated management, combined with education, is the best means of combating invasive plants.

10
Management of invasive aquatic plants

Julie A. Coetzee and Martin P. Hill

10.1 Introduction

Aquatic ecosystems throughout the world are continually threatened by the presence of invasive aquatic plants, both floating and submerged (Table 10.1), which cost governments vast sums of money every year to control. These invasive plants are predominantly anthropogenically spread, and their presence is usually a symptom of the enrichment of waters through pollution, as a result of increasing urbanization, industry, and agriculture. These plants have significant ecological impacts on the environment, and associated cascading socioeconomic effects.

Dense impenetrable infestations restrict access to water, negatively impacting fisheries and related commercial activities, the effectiveness of irrigation canals, navigation and transport, hydroelectric programmes and tourism (Navarro and Phiri 2000). Poverty-stricken rural communities whose livelihoods depend on access to clean freshwater waterways are arguably the most negatively impacted communities. Ecologically, increased biomass and dense canopy production of aquatic plant infestations affect water quality, especially dissolved oxygen, which significantly reduces benthic and littoral diversity (Masifwa *et al.* 2001; Toft *et al.* 2003; Midgley *et al.* 2006), and infestations are also associated with increases in the populations of vectors of human and animal diseases, such as bilharzia, malaria, elephantiasis, encephalitis, and cholera (Pancho and Soerjani 1978; Creagh 1991/1992; Harley *et al.* 1996). For these reasons, invasive aquatic plant infestations need to be controlled to mitigate their negative impacts on ecosystems, livelihoods, and economies.

10.2 Plant characteristics important in management

Invasive aquatic plants display variable growth forms, and may be free-floating (e.g. water hyacinth, *Eichhornia crassipes*), attached and emergent (e.g. alligator weed, *Alternanthera philoxeroides*), or submerged (e.g. spiked water milfoil, *Myriophyllum spicatum*). The degree to which these species rely on sexual or vegetative modes of reproduction, or both, is also highly variable. Because of this variability in growth form, mode of reproduction, and the impacts that they have on systems, there is no uniform method for the control of invasive aquatic species.

Table 10.1 Major aquatic weeds that have invaded water bodies around the world.

Species	Common names	Family	Mode of reproduction	Region of origin	Effective methods of control
Free floating					
Azolla filiculoides	Pacific Azolla, red water fern	Azollaceae	Spores, plant fragments	South America	Biological
Eichhornia crassipes	Water hyacinth	Pontederiaceae	Seed, vegetative budding	South America	Chemical, biological
Pistia stratiotes	Water lettuce	Araceae	Seed, vegetative budding	South America	Chemical, biological
Salvinia molesta	Giant salvinia, floating fern, Kariba weed	Salviniaceae	Fragmentation (sterile hybrid)	South America	Chemical, biological
Stratiotes aloides	Water soldier	Hydrocharitaceae	Seeds, stolons	Europe	Mechanical
Floating attached					
Alternanthera philoxeroides	Alligator weed	Amaranthaceae	Seeds, stem fragments	South America	Chemical, biological
Myriophyllum aquaticum	Parrots feather, thread of life	Haloragaceae	Stem fragments	South America	Biological
Nymphaea mexicana	Yellow waterlily	Nymphaeaceae	Stolons	North America	Mechanical
Trapa natans	Water chestnut, bull nut	Trapaceae	Seeds	Eurasia	Mechanical, chemical

Emergent					
Hydrocotyle ranunculoides	Pennywort	Apiaceae	Seeds, stem fragments	North America	Manual, mechanical, chemical
Submerged					
Cabomba caroliniana	Fanwort	Cabombaceae	Seed, stem fragments	Temperate and tropical America	Biological control under investigation
Egeria densa	Dense waterweed	Hydrocharitaceae	Stem fragments	South America	Mechanical, hydrological manipulation
Elodea canadensis	Canadian waterweed	Hydrocharitaceae	Stem fragments	North America	Mechanical
Hydrilla verticillata	Hydrilla	Hydrocharitaceae	Seed, stem fragments, reproductive turions and tubers	Asia, Australia, Europe, central Africa	Mechanical, chemical, biological
Lagarosiphon major	Lagarosiphon, curly water thyme, curly waterweed	Hydrocharitaceae	Stem fragments	Southern Africa	Mechanical
Myriophyllum spicatum	Spiked watermilfoil, Eurasion watermilfoil	Haloragaceae	Seeds, stem fragments	Europe, Asia, North Africa	Chemical, hydrological manipulation

The primary factor influencing the abundance and composition of aquatic plant assemblages is nutrient availability, particularly nitrogen and phosphorus as these are usually the most limiting nutrients in aquatic ecosystems affecting the growth of aquatic plants. When these nutrients are readily available, growth of invasive aquatic species can increase unchecked to the detriment of native aquatic flora. For example, infestations of the floating aquatic macrophyte water hyacinth, arguably the world's worst aquatic weed, are usually the symptom of eutrophication. Under high nitrate and phosphate conditions, water hyacinth doubles its biomass every 11–18 days (Edwards and Musil 1975), and under ideal conditions, red water fern (*Azolla filiculoides*) can double its biomass every 5–7 days (Lumpkin and Plunckett 1982). Floating macrophytes absorb nutrients directly from the water through their root systems, but because of the normally greater abundance of nutrients in sediments compared to the water of most aquatic systems, sediments provide a potentially large source of nutrient supply to rooted aquatic macrophytes (Barko and Smart 1981). Submerged invasive species such as hydrilla (*Hydrilla verticillata*), spiked water milfoil, and lagarosiphon (*Lagarosiphon major*), have all been shown to increase in biomass under elevated sediment nutrient conditions, forming dense canopies that shade out native flora (Agami and Waisel 1985; Barko *et al.* 1988; Rattray 1995; Van *et al.* 1999).

The majority of invasive aquatic plants rely predominantly on vegetative modes of reproduction, while sexual reproduction through seed production plays less of a role in their spread. Most submerged plants are able to increase in population size through stem fragmentation, which is particularly problematic in water bodies utilized by recreational boaters and fishermen because the spread of submerged invasive weeds is enhanced by, and directly related to, recreational boating activities (Johnstone *et al.* 1985; Buchan and Padilla 1999; Johnson *et al.* 2001; Muirhead and MacIsaac 2005; Leung *et al.* 2006). These invasive species are largely introduced to new waters from reproductive fragments attached to boats, their motors, and trailers. New infestations develop from stem fragments that root in the substrate, commonly beginning near boat ramps. Once established, boat traffic continues to break up the infestations, spreading the plants throughout the water body (Langeland 1996). Furthermore, mechanical removal of these plants is difficult because any stem fragments that remain after clearing are capable of regenerating.

Similarly, floating aquatic macrophytes such as water hyacinth, water lettuce (*Pistia stratiotes*) and salvinia (*Salvinia molesta*) are capable of reproducing via vegetative budding and fragmentation. These species produce daughter plants from buds that break off, resulting in new plants capable of regeneration. These plants are spread by boats, animals, and wind and water currents. Red water fern reproduces both sexually via spores, which are both cold tolerant and drought resistant, and vegetatively via detached plant fragments that are easily transported from one water body to another on the feet or feathers of waterfowl (Hill 1999). An example of an invasive aquatic plant that relies solely on seed production for its spread is the floating, attached water chestnut (*Trapa natans*), indigenous to Eurasia and

North Africa, but becoming increasingly problematic in North America. It produces large, sinking nuts, with barbed spines that cling to moving objects, including the plumage of geese, mammal fur, human clothing, nets, wooden boats, construction equipment, and other vehicles (Hummel and Kiviat 2004).

10.3 Modes of introduction and spread

Most invasive aquatic plants have been introduced to areas outside of their native range via the horticultural and aquarium trades, and continue to spread in this fashion. Water hyacinth is indigenous to South America but has been spread throughout the world because its flowers are attractive to gardeners. Similarly, the beautiful yellow water lily (*Nymphaea mexicana*), native to Florida, has been introduced to new habitats for ornamental purposes (Capperino and Schneider 1985). The majority of submerged invasive plants, such as hydrilla in the USA and lagarosiphon in New Zealand, were introduced via the aquarium trade as ornamental aquarium plants (Schmitz *et al.* 1991), and continue to be sold despite their status as declared invaders in many countries, particularly via the internet (Kay and Hoyle 2001). Recently, in 2006, hydrilla was discovered from one water body in South Africa, and it is likely that it too was introduced via the aquarium trade. Genetic analysis of South African hydrilla revealed that it is most closely related to hydrilla from Malaysia and Indonesia (Madeira *et al.* 2007), and reportedly, the majority of aquarium plants imported into South Africa originate from Singapore and Malaysia (N. Stallard, pers. comm.). There are fears that it will spread to other water bodies via recreational boaters and fishermen who frequent the dam, as this is the main mode of spread of the weed in the USA (Langeland 1996).

Other invasive aquatic species, such as water chestnut, have been introduced for agricultural purposes, making their eradication and control difficult due to conflicting interests. An interesting case of conflict of interests is that of alligator weed in Australia. Here it is placed in the top 20 weeds of National Significance because of the negative ecological and economic impacts it has on the environment (Thorp and Lynch 2000). It was accidentally introduced into New South Wales, Australia, in ship ballast in 1946 (Hockley 1974) and has subsequently spread throughout the country, producing dense mats that disrupt the ecology of riparian areas, stream banks, and water bodies (Sainty *et al.* 1998). The main mode of spread is via boat transport between water bodies and the transportation of contaminated turf, soil, and sand from infested to uninfested locations (Julien and Bourne 1998). However, between 1995–2000, it was discovered that alligator weed was being cultivated in gardens as a leafy vegetable by the Sri Lankan community all over Australia, but predominantly in Victoria (Gunasekera and Bonila 2001), because it had been mistaken for a popular leafy vegetable, sessile joy weed (*Alternanthera sessilis*). Following these discoveries, it became clear that alligator weed had been widely distributed by hand and post to many parts of Australia, which prompted the start of an awareness campaign highlighting the danger of the weed, in an attempt to curb the spread (Gunasekera and Bonila 2001). An alternative to alligator weed had to be found to ensure public participation in the eradication programme, and the native lesser joy weed (*A. denticulata*) was considered suitable and distributed to Sri Lankan families (Gunasekera 1999).

10.4 Management of aquatic invasive plants

There are a number of options available for the control of aquatic invasives, offering varied success, and in most cases, an integrated management approach, combining more than one method, is essential for acceptable control. However, deciding on the acceptable level of control must be the first step in an integrated management approach and will vary greatly between the species of plant being controlled and the function of the water body.

10.4.1 Utilization

Because of the sheer biomass of aquatic plant infestations, utilization of these infestations is often encouraged, particularly in poorer rural areas where local communities are perceived to benefit from the use of such infestations. In most instances, however, utilization is not sufficient to control invasive aquatic plants, and may even promote their spread, as was the case with alligator weed in Australia. In addition, most aquatic plants have a very high water content; for example, water hyacinth is nearly 95% water (Harley 1990), and to gain 1t of dry material, 9t of fresh material have to be collected (Julien *et al.* 1996), making the cost of processing water hyacinth commercially unviable (Julien *et al.* 1999).

10.4.2 Manual/mechanical control

Manual and mechanical control methods, involving removal by hand or specialized machines, are generally effective only for small infestations, as they are often labour intensive, and ineffectual against large, dense infestations (Fig. 10.1). These methods are only temporarily effective and require repeated follow-up treatments because often removal of the infestation is not complete, so any plant fragments or buds that remain are capable of regenerating into new infestations. Furthermore, seeds remaining in the hydrosoil germinate as a result of light penetration following removal.

A plant similar in structure to hydrilla, and similar in the effects that it has on freshwater systems in the USA, is spiked water milfoil, a submerged aquatic, native to Eurasia. Control of this plant is extremely difficult once it has established, and although herbicidal control is the preferred method of control using 2,4-dichlorophenoxyacetic acid (2,4-D) or fluridone, infestations in oligotrophic small systems have been effectively controlled physically by means of hand harvesting, suction harvesting, and benthic barriers. However, typical of these methods,

Management of invasive aquatic plants | 147

Fig. 10.1 Manual removal of water hyacinth (*Eichhornia crassipes*) from the Vaal River, South Africa, by teams employed by Working for Water of the Department of Water Affairs and Forestry, a government initiative tasked to reduce the impacts of alien vegetation in South Africa.

control is usually short term as milfoil eradication is often not achieved unless a maintenance programme of harvesting every 2–3 years is conducted (Boylen *et al.* 1996).

In New Zealand, submerged aquatic weeds, particularly lagarosiphon, elodea (*Elodea canadensis*), and egeria (*Egeria densa*) have recently invaded lakes and rivers, threatening hydropower stations (Howard-Williams 1993; Wells *et al.* 1997). Management of these infestations in hydro lakes relies on annual to biannual mechanical harvesting because the plants continually grow back from remaining stems and roots (Howard-Williams *et al.* 1996).

Floating aquatic macrophytes are just as difficult to control manually and mechanically. In the early 1980s, a manual/mechanical removal programme was initiated in an attempt to control water hyacinth on Lake Chivero in Zimbabwe (Chikwenhere and Phiri, 1999). Although almost 500t of water hyacinth were removed by the manual removal team, comprising 500 workers working 8 hours per day, the rapid regeneration of the weed slowed down the effort and proved expensive, with no obvious impact 6 months later. A decision was then taken to move to heavy machinery, using a bulldozer, boat, conveyor, and dump trucks, and while almost 2ha of plants were cleared daily, the amount of water hyacinth in the lake was not effectively reduced (Chikwenhere and Phiri, 1999).

10.4.3 Herbicidal control

Herbicidal control of aquatic plants is usually most successful against small infestations accessible by land, air, or boat, but is relatively expensive, although it has the advantage of being quick and temporarily effective. Like manual/mechanical control, though, new infestations develop from untreated plants, and reproductive structures that remain in the water column or hydrosoil, and so repeated applications are often necessary.

Hydrilla is a major economic and environmental weed in the USA, with control measures amounting to millions of dollars annually (Milon *et al.* 1986; Center *et al.* 1997). The most widespread method of control is chemical, using the herbicide fluridone, which selectively and economically controls hydrilla, particularly in large Florida lakes (Fox *et al.* 1996; Langeland 1996). At low concentrations, fluridone has offered selective treatment for large areas of hydrilla at a relatively low cost, but recent research has revealed that several populations of hydrilla, particularly in large Central Florida lakes, have become resistant to these low concentrations of fluridone (Michel *et al.* 2004; Dayan and Netherland 2005). Hydrilla can still be controlled at higher, sustained doses, but these high doses impact non-target native aquatic macrophytes, and the cost of control becomes much greater at higher concentrations. Despite its apparent success, herbicidal control provides only short-term relief and, subsequently, must be regularly and frequently reapplied. Moreover, the recent occurrence of herbicide resistance in the field of invasive aquatic plant management is concerning because management of these species often relies heavily on a single chemical tool due to efficacy, cost, and environmental considerations (Dayan and Netherland 2005).

Control of water hyacinth in many parts of the world still relies on herbicides. Water hyacinth is very susceptible to herbicides such as 2,4-D, diquat and paraquat, and glyphosate (Gopal 1987). Successful control can be obtained using these chemicals in small, single-purpose water systems such as irrigation canals and dams (Wright and Purcell 1995). In South Africa, the terbutryn herbicide Clarosan 500FW was used to control a severe water hyacinth infestation on the Hartebeespoort Dam in the late 1970s (Ashton *et al.* 1979). As with hydrilla, herbicidal control against water hyacinth is usually temporary, requiring regular reapplication. Furthermore, the use of chemical sprays contaminates sites used for drinking water, for washing, and for fishing, thereby threatening human health (Julien *et al.* 1999).

10.4.4 Biological control

The aim of biological control is to reduce weed populations to manageable levels through a balance between populations of the host plant and its natural enemies which are host specific, depending entirely upon their host plant for survival (DeBach, 1974). Biocontrol is the preferred method of control for large infestations of aquatic plants because it is both environmentally sustainable and economically effective. In many instances, complete control of major aquatic weeds can be achieved through biological control alone, although the overall effectiveness is sometimes compromised by environmental factors such as floods and frost, and alternative management practices, such as nutrient enrichment and herbicidal control (Fig. 10.2).

(a)

(b)

Fig. 10.2 A 7km-long infestation of water hyacinth (*Eichhornia crassipes*) on the Vaal River, South Africa, a high altitude site that experiences cold winters (a). Biological control agents, namely *Neochetina eichhorniae*, *N. bruchi*, and *Niphograpta albiguttalis*, were released onto this system in the early 1990s. It has taken more than 15 years for these agents to have a significant impact on this infestation which is now significantly reduced, and is characteristic of biocontrol of tropical weeds in temperate regions (b).

Aquatic weeds have made good candidates for biological control in comparison to the success achieved against terrestrial weeds (McFadyen 1998). Hoffmann (1995) divided biological control success into three broad categories: complete control where no other control measures are needed to reduce the weed population to acceptable levels; substantial control where other methods such as herbicide application or mechanical control are still required but less frequently; and negligible control where other forms of intervention are required to reduce the weed populations to acceptable levels. In the biological control of aquatic weeds worldwide, there are a number of species, including water lettuce (Fig. 10.3), salvinia, and red water fern that have been brought under complete control through the introduction of host-specific natural enemies to a point where they need no longer pose a threat to aquatic ecosystems (Hill 2003). Furthermore, these weeds have been brought under control with the introduction of a single agent on each of them and the time required to achieve control has been relatively short (less than 2 years).

Water hyacinth has been brought under complete control through the introduction of a suite of agents in some areas of the world, most notably Lake Victoria where the introduction of the two weevils, *Neochetina eichhorniae* and *N. bruchi* reduced the weed infestation from 20,000ha to 2000ha in a period of 5 years (Moorhouse *et al.* 2001). However, in other regions the biological control against water hyacinth has been less successful and this has been ascribed to highly eutrophic waters that allow luxuriant growth of the plant, cooler climates that slow the build-up of the biological control agent populations, and inappropriate application of other control methods such as herbicide application that may directly affect the agents or cause a catastrophic reduction in the weed population thereby decimating the agent population (Julien 2001). Under these conditions water hyacinth infestations regenerate from seedling recruitment or from unsprayed plants which flourish in the absence of the agents.

Biological control of alligator weed is considered the first aquatic weed biocontrol success story (Buckingham 1996). The introduction of the flea beetle (*Agasicles hygrophila*) has been highly successful in controlling this weed where it grows in an aquatic environment, but less successful where it grows on the banks of rivers, lakes, and impoundments (Buckingham 2002). Agents have also been released against parrot's feather (*Myriophyllum spicatum*) in South Africa (Cilliers 1999) and hydrilla in the USA (Balciunas *et al.* 2002) with varying levels of success. Additional agents are being considered for the weeds where control is not considered complete.

Chinese grass carp have also been used in a number of countries around the world for the control of submerged aquatic plants. In many cases these are indigenous plants that have become problematic due to eutrophication (Schoonbee 1991), but hydrilla has been controlled in some impoundments in the USA by grass carp (Langeland 1996), and they were considered for control of lagarosiphon in New Zealand, but studies showed that their feeding would not be specific enough, threatening indigenous New Zealand flora (Howard-Williams and Davies 1988).

Fig. 10.3 An infestation of water lettuce (*Pistia stratiotes*) at Cape Receife, South Africa (a). The weevil, *Neohydronomus affinis* was released here in 2005, and has succeeded in reducing the infestation to a few isolated plants that occur on the waters edge (b).

10.4.5 Integrated control

The long-term management of aquatic weeds requires an integrated management approach utilizing all appropriate control methods with special emphasis on the need to reduce the inflow of nitrate and phosphate pollutions into the aquatic environment (Hill 2003).

10.5 Prevention, early detection, and rapid response

There is a large suite of invasive aquatic plant species creating severe problems in many water systems throughout the world. Not every species has invaded and established on all continents, and some of these species could pose an even greater threat to aquatic environments than those weeds already present. Increased global trade and travel have created many new routes for the intentional and accidental spread of these invaders, significantly increasing the threat of both new and recurring invasions (Ielmini and Ramos 2003). The best defence against the threat imposed by invasive aquatic weeds is prevention, followed by the second line of defence, early detection and rapid response (EDRR). New Zealand, Australia, and many states in the USA have developed EDRR programmes, particularly preventing the introduction of aquatic invasives (e.g. Champion and Clayton 2000, 2001). These could be adopted by other countries around the world, such as Botswana, which is the only country in southern Africa where water hyacinth has not invaded (Navarro and Phiri 2000). An EDRR system could ensure that it remains free of this scourge.

11

Management of invasive invertebrates: lessons from the management of an invasive alien ant

Peter T. Green and Dennis J. O'Dowd

> Declare the past, diagnose the present, foretell the future; practice these acts...make a habit of two things—to help, or at least to do no harm.
> Hippocrates *Epidemics,* Book I, Section XI

11.1 Introduction

In the literature on management of invasive species, heuristic models that outline and integrate management processes across the entire spectrum of invasion, from pre-border risk assessment and surveillance to operational control and monitoring of established populations, are now commonplace (e.g. Wittenberg and Cock 2001; Wotton and Hewitt 2004; Hulme 2006; Lodge *et al.* 2006). Generic frameworks for rapid response to a biological invader typically follow a sequence from detection, to assessment, then action, and monitoring (Fig. 11.1). These models are triggered by detection and diagnosis of an invasive species, followed closely by informing stakeholders and initial assessment of the situation. This includes delimitation of the range and density of the invader, establishment of operational authority, defining and evaluating operational options, seeking initial funding, and commencing interim management. Following the initial assessment a decision is taken on the method of operational control; concerted action is taken to suppress the invader so as to mitigate its impacts. This is typically linked to a monitoring programme to assess the efficacy of control, and, if needed, planning for follow-up action. Unfortunately, few published studies put the flesh on the bones of these idealized generic frameworks to reveal the complex realities of coordinating and implementing on-the-ground control for specific invasive species (but see Anderson 2005, Coutts and Forrest 2007) to test the ideal against the real.

Some of the most intractable and serious of biological invaders are invertebrates (e.g. woolly adelgid and gypsy moth, Lovett *et al.* 2006; social wasps, Beggs 2001; earthworms, Holdsworth *et al.* 2007; land snails, Cowie 2005; and ants, Holway *et al.* 2002). Although invertebrates comprise most described biodiversity in both

154 | Invasive species management

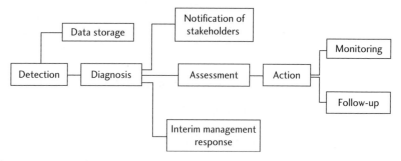

Fig. 11.1 A generic response framework for control of an invasive species. Detection and diagnosis sets off initial assessment of the status of the invasive species, including delimitation, evaluating operational options, funding arrangements, and interim management. Once operational control methods are decided upon, action is taken to suppress the invader and linked to a monitoring programme.

terrestrial and aquatic environments, and an even greater fraction of estimated biodiversity (Hawksworth and Kalin-Arroyo 1995), it is curious that they make up only 26% of the listed 100 of the world's worst invaders (17 terrestrial, nine aquatic; Lowe et al. 2000) and just 20% (10% for aquatic and terrestrial invertebrates, respectively) of the 485 species listed on the Global Invasive Species Database (GISD; ww.issg.org/database). This suggests that invertebrates, in both terrestrial and aquatic habitats, are either under-represented among invasive species or that their variety and importance are under-emphasized by invasion biologists, managers, and policymakers. We think that the latter is much more likely. The increasing recognition of ants as key biological invaders of natural environments reinforces this point.

Hundreds of ant species have been moved by humans across biogeographic barriers (McGlynn 1999; Wilson 2005) so it is not surprising that ants figure prominently among lists of invasive invertebrates. For example, five species (*Anoplolepis gracilipes, Linepithema humile, Pheidole megacephala, Solenopsis invicta,* and *Wasmannia auropuncata*) are listed as among 100 of the world's worst invaders (Lowe et al. 2000) and ants comprise 3% of all invasive species listed on the GISD. This prominence is probably recognition that key features of ants, including sociality, may lead to a greater capacity to dominate as invaders (Passera 1994; Moller 1996). This subset of invasive ants share overlapping attributes that increase the probability of transport, survival, establishment, and spread (Holway et al. 2002; Tsutsui and Suarez 2003; Suarez et al. 2005), and high impact (Davidson 1998). Many of these invasive ants form expansive supercolonies with high, sustained densities of worker ants that extend from hectares to many square kilometres (Tsutsui and Suarez 2003). This key attribute leads to major impacts on natural ecosystems (Holway et al. 2002). This may be especially so on islands, where native species richness and functional redundancy are low, and propagule pressure can be high (Denslow 2003, Daehler 2006). Indeed, native ant species are uncommon on most oceanic islands (Wilson and Taylor 1967) and even some large archipelagos, like New Zealand (Valentine and Walker 1991).

The perceived vulnerability of island ecosystems to invasion and impact by invasive alien species has led some natural resource managers to consider the protection and restoration of insular environments as impossible (Reaser *et al.* 2007). This pessimism is probably related to both the attributes of islands and island species (e.g. limited ranges), and the particular operational difficulties of managing islands including isolation, the ongoing lack of sufficient resources, lack of operational capacity, and high rates of staff turnover, all of which lead to loss of morale and institutional memory. However, numerous successes in invasive species management on islands belie these obstacles, making for renewed optimism (Simberloff 2002; Veitch and Clout 2002).

We have four straightforward aims in this chapter. First, since documented case histories of control programmes for invasive invertebrates are few, especially in natural areas and on islands, we describe the evolution of the control campaign against the yellow crazy ant *Anoplolepis gracilipes* in rainforest on Christmas Island (Indian Ocean). Second, we crystallize the key ingredients that led to the climax of the operational programme in an aerial baiting operation. Third, because every control programme operates under unique circumstances, we illustrate the complexities of the actual response against generic integrated response frameworks. Fourth, and perhaps most importantly, we evaluate the campaign to produce a list of issues and lessons that apply not only to ongoing efforts to suppress this invasive ant on Christmas Island, but that might also resonate with, and inform efforts to, manage other intractable invasive invertebrates.

11.2 History

11.2.1 The yellow crazy ant as a pantropical invader

The yellow crazy ant (*Anoplolepis gracilipes*, hereafter YCA; Fig. 11.2a), is one of the world's 100 worst invaders (Lowe *et al.* 2000). Its area of origin is obscure, but is typically cited as Africa where all other congeneric species are found (Wilson and Taylor 1967). This generalist consumer has invaded many oceanic islands across the tropics, and continents including Australia and North America (Lowe *et al.* 2000). Propagule pressure (estimated by interception rates at Australian and New Zealand ports) and vector diversity are both high, and source regions are diverse (Commonwealth of Australia 2006; Ward *et al.* 2006).

As in other important invasive ant species, kinship and intraspecific aggression in the yellow crazy ant are negatively correlated, suggesting that relatedness facilitates supercolony formation (Drescher *et al.* 2007). Extensive, polygynous supercolonies can form where worker ants are sustained at high densities. Given its numerical abundance, rapid recruitment to food resources, and aggressive

Fig. 11.2 Key players and elements in the management programme. (a) YCA tending the mooncake scale *Tachardina aurantiaca*. (b) The Bell 47 Soloy helicopter with underslung bait hopper. (c) Presto®01 pelletized bait containing fipronil as the active ingredient at 0.01%. This bait was dispersed from the hopper above the rainforest canopy at 4kg per ha. (d) Dedicated field crew loading Presto®01 into the hopper for aerial broadcasting.

behaviour, the YCA appears to break the resource discovery—resource dominance trade-off (Davidson 1998) to disrupt ant and invertebrate communities on both islands and continents (e.g. Haines and Haines 1978; Hill *et al.* 2003; Sarty *et al.* 2007; Bos *et al.* 2008). Impacts may extend to vertebrates, including nesting seabirds (Feare 1999) and land birds (Davis *et al.* 2008).

11.3 YCA invasion of Christmas Island

Christmas Island (10°30′S 105°40′E) is an elevated, oceanic limestone island, 360km south of Java in the north-eastern Indian Ocean. The 134 km2 island rises sharply to a central plateau (maximum elevation 361m) in a series of cliffs and terraces running parallel to the coast. Average rainfall is ca. 2000 mm, most of which falls between November–May. The island is covered by structurally simple, broadleaved rainforest. Of the 51 tree species present, ca. 25 are common canopy trees. The island is an Australian external territory of outstanding national and international conservation significance. BirdLife International has listed the island as an Endemic Bird Area and there are two Ramsar Wetland Sites of International Importance. Over 75% of the rainforest that originally cloaked the island still remains, making Christmas Island one of the best-preserved insular tropical ecosystems anywhere in the world. Rainforest dynamics are dominated by the activities of abundant land crabs, including the red crab (*Gecarcoidea natalis*). This native omnivore regulates seedling recruitment and litter breakdown across the island rainforest (e.g. Green *et al.* 1997, 1999, 2008).

The YCA has been present on Christmas Island since at least the 1930s, but supercolony formation is a relatively recent phenomenon, and has occurred mostly since the mid-1990s. The two supercolonies that triggered the emergency response were detected incidentally in 1997, during long-term research on the effect of the native redland crab on seedling recruitment (Green *et al.* 1997, 2008). By 2002, YCA supercolonies had formed across 3000 ha of rainforest—about 30% of all island forest. Most of these supercolonies had formed in the Christmas Island National Park, which comprises 63% of the island.

On the forest floor, *Anoplolepis* has extirpated millions of red crabs from large tracts of rainforest (O'Dowd *et al.* 2003), which has resulted in the formation of distinctive forest states across the landscape, with altered resource levels and habitat structure (O'Dowd *et al.* 2003). In the forest canopy, the ant forms new associations with herbivorous, honeydew-secreting Hemiptera (Fig. 11.2a; O'Dowd *et al.* 1999; Abbott and Green 2007) that result in reciprocal increases in their population sizes. The combined direct and indirect effects of the YCA and several species of scale insects have been rapid and multidirectional, affecting forest structure and composition, species of special conservation value (O'Dowd *et al.* 1999, 2003), ecosystem processes (Davis *et al.* 2008), and secondary invasions (O'Dowd and Green 2009). This demanded a coherent, coordinated response among scientists, managers, and policymakers.

11.3.1 The interim response

Although the identity of the ant species was unknown when expansive supercolonies were detected in 1997, it was immediately obvious that this ant affected the red land crab, seedling recruitment, scale insect populations, and litter decomposition (Fig. 11.3). Even in the absence of identification, the urgency of a management response was clear. Parks Australia North Christmas Island (PANCI), the responsible management authority, was notified immediately. This sense of urgency was heightened by the subsequent authoritative identification of voucher specimens, followed by a literature search (O'Dowd *et al.* 1999) that revealed some aspects of its biology and impacts, especially in the Seychelles (e.g. Haines and

Year	Event	Date	Detail
1995	Research, begun in 1986, continues on red crab	19 December 2001	Steering Committee (SC) recommends options paper for the control programme.
1996	Research on red crab extends across island	8 February 2002	SC endorses aerial baiting using a helicopter.
1997	YCA supercolonies detected; mass crab deaths; Parks Australia notified; impact assessment begins	15 February	SC endorses rapid deployment of helicopter.
		15–24 March	Pilots visit island; indicate that aerial baiting is feasible.
1998	YCA impact assessment continues; scoping supercolony distribution begins	26 April	SC reviews draft Environmental Assessment for aerial baiting.
		7 May	Director of National Parks approves purchase of 13 t of Presto ant bait.
		14 May	Island-wide supercolony boundary delimitation begins.
1999	YCA impact report; media coverage; formation of steering committee	20 May	SC endorses Environmental Assessment for aerial baiting
		30 May	Tender for aerial baiting operation awarded to McDermott Aviation.
		28 June	Contract for non-target assessment of aerial baiting on canopy arthropods and vertebrates
		27 June	Trial plots set up for aerial baiting operation.
2000	PA-Monash University Linkage Grant; bait testing begins; IWS is designed and approved	12 July	Referral for aerial baiting submitted to Approvals and Legislation Division
		1 August	Teleconference with Approvals and Legislation Division on referral.
2001	1st IWS conducted over 3 months; CIGIS integrated into programme; evaluation of ground baiting	5 August	Effort begins to minimize impacts of aerial baiting on robber crabs
		12–27 August	YCA activity assessed at 44 waypoints in supercolonies before aerial baiting
2002	**Build-up, conduct, and evaluation of the aerial campaign**	14 August	Approvals and Legislation Division determines that aerial baiting is not a Controlled Action and that the operation could proceed
		2 September	Pilot/engineer arrives on the island
		3 September	Ship transporting the bait and helicopter arrives
		5 September	Helicopter re-assembled for flight testing, trial of coordinate systems, and potential impacts of aircraft on seabirds; Presto®01 arrives too moist for broadcasting. Drying of 13 t of bait begins.
2003	2nd IWS conducted, surveillance and follow-up control; linage grant ends	6 September	Assessment of non-target impacts of aerial baiting on canopy vertebrates and arthropods begins.
		8 September	Trial begins to assess efficacy of aerial bait delivery for YCA control.
2004	Surveillance and follow-up control	13 September	SC endorses island-wide aerial baiting of all supercolonies.
		14–21 September	All targeted supercolonies treated by helicopter.
2005	3rd IWS conducted; resurgent supercolonies; YCA listed as key threatening process under EPBC Act	26 September	Non-target impact assessment of aerial baiting on Robber crabs.
2006	Parks Australia submits 10-year plan and funding request	10–21 October	YCA activity assessed at 44 waypoints after aerial baiting.
2007	Long-term funding guaranteed by Australian government	28 October–23 December	Areas excluded from the aerial baiting campaign, including research plots, are baited on foot.

Fig. 11.3 (Con't.)

Haines 1978). The events of 1997 crystallized observations of an ant 'infestation' made much earlier, in 1989. At this time the ant was identified as *A. gracilipes*, but the discovery of what in hindsight was a small YCA supercolony occurred prior to most of the subsequent research on red land crabs, and the supercolony and its potential significance were almost forgotten until the discoveries in 1997.

The task of engaging key stakeholders began. Local PANCI managers were immediately convinced of the threat. However, persuading the administrative officers in Darwin (Parks Australia North) and Canberra (Environment Australia) was more challenging, because unlike the local staff, they had no first-hand experience of YCA supercolonies and their impacts, especially on red crabs. Furthermore, the notion that a single invasive ant species could extirpate tens of millions of the dominant red crab was met in at least one instance with open scepticism. Others felt that the ant invasion could be transient—an irruption soon followed by collapse and recovery (cf. Simberloff and Gibbons 2004). Rapid, quantified assessment of impacts at several sites on the island (O'Dowd *et al.* 1999) helped tip the scales. Public interest generated through the media also helped to maintain focus on the problem.

Initial scoping of the distribution of the supercolonies was ad hoc, involving infrequent but epic treks through remote tracts of the rugged island. This illustrated that supercolonies were widespread, but reinforced the need for a systematic island-wide survey (IWS) of the invasion (O'Dowd *et al.* 1999). The IWS was based on a grid of 1024 waypoints spread across the island (including rainforest, built environment, and areas cleared for phosphate mining) on a grid of 364 m intervals. This interval coincided with an existing network of overgrown 'drill-lines' bulldozed across much of the island plateau in the 1960s for phosphate exploration. Drill-lines were crucial because they provided ready access for field crews conducting the survey. Survey also depended on the existing Christmas Island GIS system. Each waypoint was offset into undisturbed forest and field crews used hand-held GPS units to locate them. At each waypoint, a 50 m transect was set out on which YCA activity was recorded using card counts (Abbott 2005); red crab burrow density was used as an indicator of impact at each waypoint. Supercolonies were defined operationally (rather than biologically) as those areas where there were sufficient ants to cause death of the red crab. Data were displayed in ARCView to show the spatial distribution of ant supercolonies in relation to crab burrow densities.

The initial IWS took 3 months to complete and revealed three key findings. First, YCA were widespread and occurred at 47.6% (359/754) of waypoints in undisturbed rainforest (Fig. 11.4a). Second, supercolonies were found at 24.0%

Fig. 11.3 Timeline from 1995–2007 showing key events leading to the aerial control campaign of the yellow crazy ant in 2002 and its aftermath. Key dates and events in the rapid build-up, conduct, and evaluation of the aerial campaign are telescoped to the right.

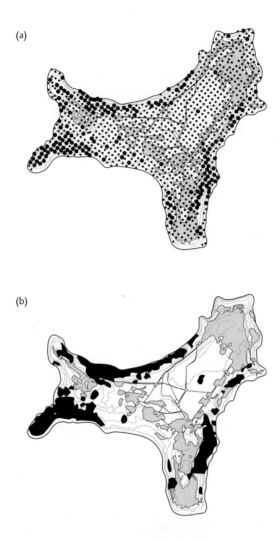

Fig. 11.4 The island-wide survey for supercolonies of the yellow crazy ant *Anoplolepis gracilipes* prior to the aerial campaign of 2002. (a) Supercolony occurrence at 1024 waypoints spaced at 364 m intervals across the island in 2001. Large dots indicate waypoints with supercolonies of the YCA. The grid was rotated 27° from north to align with the existing network of drill lines. Actual survey transects at each waypoint were offset 25 m away from the drill line. (b) Final distribution of supercolonies following detailed boundary delimitation by ground crews in 2002. Some supercolonies areas discovered during the IWS were baited by ground crews before the aerial operation, and do not appear in (b). In both maps, grey areas indicate clearings without forest cover, mostly abandoned phosphate mining areas. Contours are indicated at 50 m intervals.

(181/754) of waypoints, equating to ~25 km² of island forest. Third, burrow densities at waypoints indicated that YCA had killed c.15 million red crabs (c.25% of the total island population).

Simultaneously with development of the IWS, two other key elements for the response were initiated. First, PANCI commenced efforts to source and evaluate an effective ant bait. Initial trials with commercially available bait formulations were unsuccessful. Eventually, Presto®01, a fishmeal bait with an active constituent of fipronil, was identified as effective (Fig. 11.2c). At the time this bait was unregistered in Australia, but use was permitted on Christmas Island under an emergency permit issued by the National Registration Authority. Ground crews treated a total of 371ha of YCA supercolonies between 2000–2001, achieving a knockdown of $>95\%$ of ant workers within days.

Second, a small steering committee comprising volunteer scientists, managers, and policymakers was formed, independent from the management authority. The key functions of the committee were to provide advice and support for the operational programme, strategic direction, and, on occasion, advice to the Director of National Parks. Committee members met regularly by teleconference, and routinely produced discussion papers to present, evaluate, and recommend options for the programme, establish timelines for actions, and review and monitor progress.

11.3.2 The aerial control campaign

The genesis of the aerial control campaign lay in the compelling findings of the IWS (Orchard *et al.* 2002); two-thirds of supercolonies were in areas too rugged for field crews to operate safely and effectively, and in any case, ground control was impractical given the sheer pace and scope of the invasion relative to the number of personnel available. In late 2001, the steering committee canvassed the idea of an aerial approach to YCA control, based largely on the aerial campaigns in New Zealand to control invasive rats (e.g. Towns and Broome 2003) and the newly initiated campaign to eradicate the red imported fire ant in Brisbane (Vanderwoude *et al.* 2003).

Given the novelty of an aerial approach, the Steering Committee recommended in February 2002 that a trial be conducted to determine if YCA supercolonies could be effectively controlled from the air. Two timelines were considered: (1) where both the trial and island-wide control programme were conducted during a single 3-week period in September 2002, whereas (2) involved a staged approach with a trial in 2002 followed by control in 2003. The first plan was considered feasible within the 3-week window because the ant bait is fast acting so the efficacy of aerial broadcasting of bait would be evident within days. Three factors weighed in the recommendation:

1) The urgency of controlling all supercolonies as soon as possible.
2) The high cost involved in transporting a helicopter to the island twice.
3) The high purchase and transport costs of stockpiling tonnes of ant bait if the first plan was adopted but the trial failed.

The first plan was endorsed, thus imposing an extremely tight timeline—in effect to plan, test, and execute the operation within just 7 months (Fig. 11.3). There were seven components to the overall planning and implementation of the aerial baiting campaign.

11.3.2.1 Legislative approval

Approval to aerially broadcast toxic ant bait in the Christmas Island National Park was sought under the Environment Protection and Biodiversity Conservation Act (1999). First, a Referral was submitted that outlined the extent of the invasion, its documented and suspected impacts, and that aerial application of toxic bait was the only effective method of control. A full environmental impact assessment of the potential benefit and risks was submitted (Green *et al.* 2002) and reviewed by Environment Australia. The key conclusion was that the probable consequences of not acting were far worse than the potential non-target impacts. The aerial baiting operation was endorsed by Environment Australia (see http://www.environment.gov.au/cgi-bin/epbc/epbc_ap.pl?name=referral_detail&proposal_id=722), but the pace was such that official approval was given while the helicopter and bait were en route to Christmas Island.

11.3.2.2 The helicopter

It was necessary to source a civilian contractor with an appropriate aircraft, suitable delivery technology, and experience to fulfil the operational requirements of the baiting programme. In March 2002, pilots from two helicopter companies were bought to Christmas Island to reconnoitre potential loading sites and local flying conditions, especially the hazards posed by seabirds. The successful tenderer had previous experience of aerial baiting as part of the eradication programme for the red imported fire ant in Brisbane. They used a Bell 47 Soloy helicopter, which was partially dismantled for shipment to the island, and reassembled at the Christmas Island airport (Fig. 11.2b).

The accurate treatment of YCA supercolonies presented special challenges for the pilot, because boundaries were irregular, many supercolonies were relatively small, and most lay within continuous forest and were not identifiable from above the rainforest canopy. The pilot developed a new system of precision navigation to deal with these challenges, involving the use of two independent GPS units—a highly accurate (sub-metre) Trimble differential GPS unit to stay on track as each run was flown and a hand-held Garmin GPS unit to delineate supercolony boundaries at the start and end of each run. Great skill was required as the pilot was flying the aircraft at around 100 km/hour, while simultaneously keeping track of the aircraft's instrument panel, the two GPS units, and seabirds.

Quality control was an important consideration when choosing the successful tender. An on-board differential GPS recorded the exact routes flown while the bait stream was switched on. Once downloaded to the CIGIS, and buffered to a width of 12 m (the width over which bait was dispersed on each run), these records allowed a detailed assessment of how well target areas had been treated, in terms of both accuracy around the perimeter, and coverage across the treatment area.

11.3.2.3 Dispersion of Presto®01 ant bait

Ant bait was dispersed from a bucket suspended beneath the helicopter (Fig. 11.2b). This bucket was an inverted cone into which about 90 kg of bait was loaded (Fig. 11.2d). A key factor in bait dispersal was its moisture content—the pelletized bait had to be dry enough to flow without blockage through a 25 mm diameter hole at the bottom of the bucket, onto a petrol-driven, rotating spreader. Despite prior testing on the mainland to establish the ideal moisture content, the Presto®01 shipped to the island was too moist and would not flow. So, all 13,000 kg of Presto®01 had to be dried by spreading it thinly with garden rakes on sheets of black plastic in full sun. This was an incredibly labour-intensive exercise, required 250 person hours to complete, and risked photodegradation of fipronil. Drying barely kept pace with the demands of flight operations.

11.3.2.4 Mapping supercolonies

The accurate mapping of YCA supercolonies was crucial to the success of the control programme. The map was based largely on the results from the 2001 IWS, followed by boundary delimitation during the four months leading up to aerial operations in September 2002 (Fig. 11.4b). All boundaries were mapped in the field with hand-held GPS units, and maps were generated using the CIGIS. Field crews used several cues to determine the boundaries of supercolonies, including subjective assessments of crazy ant abundance, both on the ground and as 'trunk traffic' on trees, and the presence or absence of dead crabs. In areas where supercolony boundaries did not correspond with a physical feature of the landscape (e.g. a cliff, forest edge), three field workers walked abreast 10–20 m apart along the length of the boundary, with the two outer people keeping the middle person accurately positioned on the boundary—the outer person continually confirmed the absence of ants, while the inner person continually confirmed their presence. The middle person held the GPS unit and coordinates were taken every 20–50 m. Some boundaries were easily identifiable by observers on the ground, but often there was a wide 'transition zone' between heavily ant-infested forest and intact forest (Abbott 2006). These boundaries proved too finely resolved to be practicable for aerial operations. Accordingly, boundaries were rounded on the CIGIS, but this process never pared off sections of supercolonies. This increased the total treatment area increased by 167 ha from 2378 ha to 2545 ha.

Sections of several supercolonies were excluded from the aerial control programme. These included all freshwater streams and soak areas, including the two Ramsar Wetlands of International Importance. Fipronil is reported to have strong negative effects on freshwater fauna, so exclusion zones of 100 m were imposed around these areas. Five supercolony research plots were excluded from the aerial baiting programme but later treated by ground baiting. The total area excluded from the baiting programme was 76.2 ha.

11.3.2.5 Trial of aerial baiting

The aims of the trial of aerial broadcast were threefold:

1) To assess the efficacy of Presto®01 delivery by helicopter for controlling supercolonies against a target of 99% knockdown.
2) To identify the lowest effective rate of Presto®01 for supercolony control.
3) To assess the degree of bait penetration through the canopy to ground level.

The efficacy of aerial bait delivery for supercolony control was assessed using a before-after-control-impact design. The CIGIS was used to delineate 6 plots (each 9–53 ha) in supercolonies, two for each application rate plus two untreated control plots. In each plot YCA activity was monitored on cards (Abbott 2005) placed at 44 stations along four parallel transects (each 150 m long and 40 m apart), three times before and eight times after aerial baiting.

Some aerial broadcast bait is likely to be intercepted by the forest canopy, reducing the amount reaching the forest floor. Even though the YCA forages extensively in the forest canopy (O'Dowd *et al.* 2003, Abbott and Green 2007) and would be likely to collect suspended bait, decreased bait reaching the forest floor could compromise control. Penetration of bait dispersed through the canopy was estimated with 30 one m^2 catch bags along the transects in each plot. Large plastic bags were stapled to a circular wire hoop held up by wooden stakes. Catch bags were placed in the plots 2–3 hours before aerial baiting and emptied 3–4 hours after treatment. The catch bags were emptied again 3–4 days later in case more pellets fell from the canopy, and the catch dried and weighed.

Bait application rates averaged 5.9 kg/ha and 4.4 kg/ha on the high and low application plots, respectively, very close to the target rates of 6–4 kg/ha. Eighty percent of catch bags intercepted >90% of bait within 3–4 hours of application. The mean rate of bait penetration was 59.2% across all plots. Although a considerable fraction of bait was intercepted by the canopy, YCA activity declined dramatically following aerial baiting. Aerially-dispersed Presto®01 had a significant negative impact on crazy ant activity within days of treatment, at both rates of application. Ant activity on the control plots was high during the week preceding treatment and remained so during the week after treatment. Conversely, on the baited plots, ant activity in the week after aerial treatment declined by an average of 91% of pre-treatment levels, and was essentially nil after 12 days, regardless of application rate and sufficient to achieve >99% knockdown. Five days after the trial, the Steering Committee endorsed the full treatment of all remaining supercolonies on Christmas Island, at a nominal rate of 4 kg/ha.

11.3.2.6 Measuring the success of the island-wide operation

Success of aerial operations was assessed in terms of both the coverage and accuracy of the baiting operations, and the impact on YCA activity. GPS downloads from the helicopter indicated almost blanket coverage of all target supercolonies, more than 2500 ha in just 8 days. The few mistakes included 3 ha that were baited in

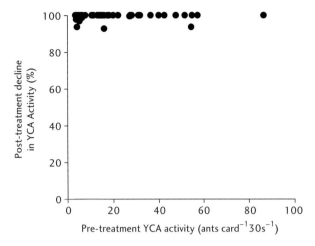

Fig. 11.5 The knock-down effect of Presto®01 ant bait on the activity of the yellow crazy ant at 44 waypoints, 1 month before, 1 month after aerial broadcast of Presto 01. X axis is the number of ants running across a 100 cm² card in 30 s, averaged across 11 cards placed on the soil surface along a transect at each waypoint. Note that over the entire range of ant activity, knockdown was consistent, i.e. density independent.

error, mostly as the result of occasional overshoots on baiting runs. A total of 9 ha were missed, including a 6 ha in the north-west of the island which could not be baited because of the high density of flying seabirds.

Pre- and post-baiting surveys of ant activity were conducted at 44 waypoints in supercolonies from the IWS to measure efficacy of control. At each waypoint, card counts of YCA were used as per the IWS and the aerial baiting trial. Ant activity at each site was assessed once 1 month before and 1 month after waypoints were treated. YCA activity declined precipitously following aerial treatment (Fig. 11.5). Ant activity fell from 21.6 ± 18.7 ants/30s (SD) prior to treatment, to 0.13 ± 0.50 ants/30s after treatment, an average decline of 99.4%. The decline was >97% at 40 of 44 waypoints, and 100% at 77% of waypoints. This met the target of 99% knockdown set by the steering committee prior to the aerial baiting operation.

11.3.2.7 Non-target impacts

Aerial broadcasting of tonnes of toxic bait in the National Park had the potential to cause serious impacts to non-target species of special conservation significance. These included endemic reptiles, several endemic land birds, an endemic seabird, and an unknown number of endemic invertebrates.

Although fipronil is toxic to crabs, the risk posed to the red crab population by the aerial baiting campaign was considered minimal because resident red crabs were already annihilated within supercolonies. Furthermore, live crabs near the boundaries of supercolonies were unlikely to encounter bait because they rarely emerge from their burrows in the dry season, when aerial baiting operations were conducted. Robber or Coconut Crabs (*Birgus latro*) were of more concern because they forage more widely and are more active under dry conditions. Moreover, the robber crabs on Christmas Island are genetically distinctive from robber crabs elsewhere (Lavery *et al.* 1996), and, until the YCA invasion were the largest and least impacted population anywhere in the world (Schiller 1988). Although the YCA extirpates robber crab populations in supercolonies, baiting operations posed a significant threat because *Birgus* have a well-developed olfactory system, move large distances for food, are attracted to the fishmeal matrix in Presto®01, and are extremely susceptible to fipronil.

Considerable effort was invested in minimizing non-target impacts on this charismatic species. Two kinds of attractive food lures were used to entice robber crabs from baited areas during helicopter operations. First, 44 senescent trees of the endemic palm *Arenga listeri* were felled several weeks prior to aerial operations, several hundred meters apart and up to 300m away from baited supercolony boundaries: the pith of this monocarpic palm attracts robber crabs in large numbers from hundreds of metres. Second, where possible, targeted supercolonies were also ringed with depots of poultry food pellets mixed with shrimp paste. This lure had been used during initial ground-based control. Using the CIGIS, over 250 sites were selected around supercolony perimeters. One or two days before areas were aerially treated with ant bait, 15 kg of the lure was dropped from the helicopter at intervals of 150–250 m around the perimeter and at distances between 50–200 m outside supercolony boundaries.

The effect of these lures on mitigating robber crab mortality was assessed after the aerial baiting operation. Mortality was estimated at 5.3% for sites close to edges of baited supercolonies, but because all supercolonies were ringed with food lures, there were no supercolonies without lures to act as controls. However, estimates made during initial ground baiting indicated mortality rates of 15% in the absence of food lures (D. Slip, pers. comm.). In spite of this considerable effort, a comprehensive assessment of food lures to reduce *Birgus* mortality in association with baiting has since shown them to be ineffective (Thomas 2005).

An analysis of non-target impacts on litter invertebrates during ground control operations using the identical bait formulation (but at a higher rate of application) found no evidence for off-target impacts (Marr *et al.* 2003). Despite several reports of dead and dying introduced cockroaches in a hollow tree stump, and mortality of the introduced ant *Campanotus melichloros* in the immediate aftermath of the aerial operation, it is highly unlikely that the aerial operation caused broad and substantive non-target impacts on litter invertebrates. The sheer abundance of the foraging YCA in supercolonies meant that these ants largely monopolized the bait (Marr *et al.* 2003). Contractors engaged to assess impacts on canopy invertebrates, forest reptiles, and land birds found no evidence of non-target impacts (Stork *et al.* 2003).

The aerial baiting campaign was a success in terms of:

- Design and implementation of the island-wide survey as the basis for the aerial campaign.
- Approval under the EPBC Act (1999) of aerial baiting operations. All conditions specified in the approval were met.
- The recruitment of a reputable aviation company and skilled pilot/engineer to conduct aerial baiting.
- Proof of concept for the feasibility and efficacy of island-wide aerial baiting.
- Island-wide suppression of YCA supercolonies that met the pre-operational target of 99% knockdown.
- Assessment of non-target impacts.

11.3.3 Evaluation and lessons learned from the aerial campaign

At least 13 hard-won insights broadly relevant to rapid response to invasion by alien invertebrates emerge from the experience with YCA on Christmas Island. Except for the first three, these are in no particular order:

1) **The human dimension**. Successful responses to invasive species depend in large part on the passion and determination to succeed of the people involved. Top-down directives rarely engender zeal in scientists, managers, and field crews at the coalface of the invasion. We agree with Anderson (2005) that success is engendered from the bottom up. On Christmas Island, the coincidence of people with a 'love of place' sustained the effort, especially through moments of uncertainty and despair. Equally, the frailties of personality can be the Achilles heel of these efforts, and more than once threatened this programme. The consequences of these human failings assume greater proportion and immediacy in the crisis management of biological invasions.

2) **Solid science supported by good natural history**. Without a documented understanding of the impacts of the yellow crazy in rainforest on Christmas Island, it would have been nigh on impossible to convince management authorities and funding agencies of the seriousness of the problem. Embedding the scientific culture (e.g. design, analysis, reporting, synthesis, interpretation, and review) in invasive species management is essential and we believe it was crucial on Christmas Island. A sound understanding of the island's natural history was fundamental to crystallizing the implications of the invasion.

3) **Capacity for responsive funding**. Almost by definition, the crisis management of biological invasions is unpredictable and therefore not built into the operational budgets of management agencies. However, their capacity to respond to these demands in a timely fashion is crucial. The decision in February 2002 to implement the aerial campaign just 7 months later placed considerable financial demands on Environment Australia. In fact, one-third of the funding was sourced competitively through a Natural Heritage Trust

(NHT) grant, and it was pure coincidence that the timelines of this scheme suited the tight timelines of the aerial baiting operation.

4) **GPS and GIS technology.** Hand-held GPS units and GIS software (Christmas Island Geographic Information System) were both critical. Without GPS, sites could not have been located with accuracy and speed to conduct the IWS, to delimit supercolonies, and to transfer coordinates to the pilot for aerial operations. Without the CIGIS, data management, presentation, and interpretation would have been impossible.

5) **Public awareness**. Increased public awareness of impacts on the island through local, national, and international media helped focus the attention of the management authority on the emerging crisis. Furthermore, public awareness of the ongoing control effort increased both nationally and internationally the profile of the YCA in particular, and invasive ants and invasion on islands more generally. For example, the inclusion of the YCA in the list of 100 of the world's worst invaders (Lowe et al. 2000) was a direct result of its known and quantified impacts on Christmas Island.

6) **Clear demarcation of responsibility**. The aerial operational succeeded because there was a single authority responsible for management and funding. Invasion of the YCA into built environments, agricultural lands, and natural areas on mainland Australia has generated jurisdictional disputes on just who is the responsible agency for management (Commonwealth of Australia 2006). These disputes can cause significant delays in response and fuel a public perception of confusion and inaction.

7) **Bridging the science-management interface.** A cooperative programme of research and management of the YCA on Christmas Island through the Australian Research Council's Linkage programme brought scientific expertise to the operational programme. Essentially, university scientists were 'embedded' with natural resource managers to achieve the objectives of the programme. Once the magnitude of the challenge became apparent, traditional suspicions and demarcation of roles between scientist and manager blurred: scientists, who focused initially on impact analysis crossed over to support the planning and implementation of the operational programme, while managers rallied behind the scientific approach to achieve their goals. However, bridging the science–management interface can cause conflicts of interest within individuals. As scientists, we wanted to research and understand the nature of the YCA invasion and its cascading impacts. At the same time, we recognized the absolute need to destroy that which we wished to study.

8) **Competing resource demands**. Research and management of the YCA invasion was all consuming. This almost certainly had unintended consequences for other important management activities on the island. First, the YCA campaign diverted resources from other important programmes for invasive species (e.g. weeds, feral cats, and the introduced wolfsnake).

Second, other pressing issues of direct relevance to the management of the Christmas Island National Park (e.g. planning for a refugee detention centre, a satellite launching facility, and phosphate mining activities) all placed an extraordinary burden on managers in the lead up and conduct of the aerial operation.

9) **An independent steering committee.** A steering committee comprising scientists, managers, and policymakers guided, advised, reported, and evaluated the programme with independence from the management authority in a timely fashion, without any formal jurisdiction or direct funding. All were unpaid volunteers working primarily through frequent teleconferences across three time zones.

10) **Successful response sometimes requires a healthy dose of luck.** Serendipity played a significant role in the success of the programme. For example, the early detection of supercolonies was largely due to chance, while doing basic research on the last funded trip examining the role of the red land crab in island rainforest dynamics. This basic natural history and research, which began 15 years earlier, primed an appreciation for impacts following the removal of the native red land crab, a keystone species. We agree with Louis Pasteur: chance really does favour the prepared mind. Second, military-grade GPS became more widely available only in May 2000, increasing its precision fivefold and making it possible to navigate accurately beneath the rainforest canopy with hand-held units just in time for the first IWS. Previous attempts just a few months earlier had been a depressing and dismal failure. Third, an aviation company with the capacity to do job was available and a single pilot/engineer was willing to take on what was for them a relatively small job.

11) **Successful response requires quick thinking.** Things go wrong, and on Christmas Island, two incidents threatened to derail the aerial baiting operation almost before it began. First, the complexities of transferring coordinates from the CIGIS to the pilot's dual GPS system did not become apparent until after the aircraft was actually on the island. Even worse, all supercolonies had been mapped in the IWS on a metre (UTM) grid, but the pilot required coordinates in the form of degrees, minutes, seconds, and decimal seconds. Second, the bait arrived on the island too moist to flow through the hopper. In both situations, good old-fashioned nous was key to solving these problems and salvaging the project.

12) **Who you know is as important as what you know.** Networking with other groups proved pivotal. First, the notion of aerial baiting on a remote oceanic island stood on the shoulders of previous successes on New Zealand islands. Second, the fact that one steering committee member was involved directly in the complex effort to eradicate the red imported fire ant in Brisbane gave the control operation on Christmas Island access to their technology for aerial bait delivery and key contacts, including the aviation company.

13) **Isolation and tight timelines are not show-stoppers for invasive species management.** We think that the rapid conduct (Fig. 11.3) and success of the aerial operation (Fig. 11.5) dispels the notion that isolation stymies invasive species management (cf. Simberloff 2002). A can-do attitude, along with facsimiles, emails, and telephone conference calls were sufficient to overcome the tyranny of distance; restrictions imposed by the vagaries of shipping schedules were countered by good advanced planning.

11.4 Conclusions

Heuristic models that describe idealized responses to biological invasions are useful, but only to a point. These models (Fig. 11.1) describe adequately the sequence of events as they unfolded on Christmas Island, from initial detection and diagnosis of the invader, through to notification of stakeholders, interim management, more comprehensive assessment of threats, and to action and follow-up monitoring. The model even describes accurately that interim management responses and broader threat assessment occurred more or less simultaneously on Christmas Island. Perhaps it was inevitable that the Christmas Island response, and responses elsewhere (e.g. Anderson 2005; Coutts and Forrest 2007), even in the absence of specific or generic contingency plans, should all have followed this general pattern simply because it is logical to proceed in this way. What these models fail to do, however, is to educate practitioners—managers and scientists alike—of the cold, hard reality of confronting rapid, expansive, and high-impact invaders. We argue that models like these are only truly, and usefully, heuristic if they are accompanied by detailed case studies that flesh out the abstractions. 'Action' in our generic model (Fig. 11.1) belies the complexities of mobilizing a team and securing the funds to tackle the problem, overcoming myriad technical and organizational details, and even managing a large team whose very humanity makes the whole enterprise vulnerable.

We have identified ingredients to the success of the aerial baiting operation for YCA supercolonies on Christmas Island. Arguably, some of these, such as the drive and dedication of the people involved, good science, and the capacity for responsive funding should be seen as generic prerequisites. Others, such as sourcing the right aircraft and pilot, are obviously more idiosyncratic to this specific operation. Furthermore, we are quite willing to acknowledge lady luck in some aspects of the planning and implementation of the programme. While it would be unrealistic to expect that all emergency responses to invasive species will be successful, we should not fall to fatalism and the failure to act. Aim high (Simberloff 2002).

Events since the aerial campaign of 2002 offer further insights on invasive species management. First, the warm afterglow of success can be counterproductive because the very act of mitigating an urgent situation can foster a perception that the problem has been 'fixed'. This puts at risk the resources required for continued surveillance and maintenance management. By 2005, the YCA was resurgent in some areas treated during the aerial campaign, and new supercolonies had formed elsewhere (Fig. 11.3). Island managers have had to battle over many years to secure sufficient resources to sustain the gains of 2002, despite the fact that the ongoing nature of the problem, and the downturn in support, were both foreseeable (O'Dowd and Green 2002).

Second, the Christmas Island experience has had much broader, knock-on effects. It has directly influenced management of the YCA in the Northern Territory and Queensland in mainland Australia, and efforts to eradicate or control this ant on Pacific islands. Furthermore, this response on Christmas Island led to the successful listing of the YCA as a key threatening process under Australia's Environment Protection and Biodiversity Act (1999). Along with the listing of the red imported fire ant as a key threatening process, this led to a national threat abatement plan for invasive ants in Australia and its territories, which contains 72 actions, including regional cooperation to build capacity for prevention and rapid response to invasive ant species (Commonwealth of Australia 2006).

The aerial baiting campaign was but one battle in what will be a protracted war, but the prognosis for the suppression and containment of YCA supercolonies on Christmas Island is good. Environment Australia now has in place a 10-year plan (Fig. 11.3), with stable, ongoing funding for at least the first 4 years. Much of these resources will be devoted to maintenance management—biennial island-wide surveys, and ground based control—but funds have also been allocated for another aerial campaign, if needed, as well as the development of alternative baits, and research and development on novel approaches to control (O'Dowd and Green 2003).

If there is one overarching lesson from the aerial baiting campaign on Christmas Island and its aftermath, it is this: while funding and technology are fundamental in the battle against invasive species, it is passion, tenacity, and a steely will to succeed that will eventually tip the scales.

11.5 Acknowledgements

Natural resource and park managers at PANCI—D. Slip, S. Comport, M. Orchard, and M. Jeffery—provided crucial support to all aspects of this project. J. Barry (PAN) was instrumental in obtaining funding and in guiding the project through the EPBC referrals process. A. Andersen and the Crazy Ant Steering Committee provided strategic oversight and direction throughout the development and conduct of this programme. G. Foo, K. Retallick, and T. Zormann conducted both the IWS and initial ground baiting. A. bin Yon and E. bin Johari refuelled and reloaded the helicopter, and research students K. Abbott, N. Davis, and R. Marr helped provide the research underpinnings for the project. Chief pilot F. van Beek showed consummate skill and professionalism in the aerial campaign and J. McDermott committed his company to this effort. L. binte Bahari, J. bin Jantan, R. bin Sani, and A. bin Wahid provided field assistance. P. Morrow, J. Einam, and D. Meek facilitated production and delivery of the ant bait. K. Sandford-Readhead and the Brisbane City Council assisted in the conversion of GPS coordinates to compatible shapefiles. The Natural Heritage Trust and the Australian Research Council financially supported the control and research programmes, respectively. We thank the Director of National Parks, P. Cochrane, and former Environment ministers R. Hill and D. Kemp for their support of this programme.

12
Management of terrestrial vertebrate pests

John P. Parkes and Graham Nugent

12.1 Introduction

Successful management of terrestrial invasive vertebrates requires three elements:

1) Effective policies to ensure funds are allocated to the species and places of highest priority, and that the appropriate people fund the control—either those who cause the problem, if they can be identified, but more often those who benefit should pay. Appropriate long-term policies and funding mechanisms are particularly critical if sustained control is the management strategy being applied to the invasive species (Gibb and Williams 1994).
2) Effective instruments to deliver the control. These include both the availability of techniques and tools to manage the animals and the capacity of human resources to organize and do the work, especially where skills are required. There is a problem if we have a mouse plague but neither mouse traps nor someone who knows how to set them!
3) Knowledge of where and when to intervene with these tools to best effect (Choquenot and Parkes 2001). There is no point in having the perfect mouse trap but set it where there are no mice, or when mice are not a problem, or not setting enough when they are a problem!

These requirements are also governed by whether the strategic aim of management is to stop the pest arriving, to eradicate it, to stop it spreading, or to maintain some degree of control over it. In addition, under the sustained control strategy, they are governed by whether the impacts of the pest on the value to be protected is intermittent or chronic—and thus when and where to intervene (Parkes 1993).

Control tools have to meet these strategic needs, but must also do so in socially acceptable ways, without net adverse side effects, and at an affordable or justifiable cost. For example, a common weakness in pest control programmes is to design the programme entirely around what is possible with the tools available, rather than around the more important considerations of strategy (e.g. eradication versus sustained control), appropriate scales, and the relationships between the impacts of the pest at different densities and the rates of recovery or natural fluctuations of population densities.

174 | Invasive species management

However, in this chapter we focus on the tools and techniques available to exclude, remove, or kill terrestrial vertebrates, or prevent them from breeding, and explore how this toolkit fits the strategic needs of managers (from management at national borders, to management of established species by containment, eradication, or sustained control). We touch on obvious gaps in the toolkit, both at these strategic levels and for critical species, and describe briefly how they are being addressed around the world.

12.2 Tools to prevent new species arriving

Compared with invertebrates or plants, vertebrates have a well-known taxonomy and are usually large and obvious, making it more difficult for them to invade without being detected. Terrestrial vertebrates form a minor proportion of all invasive species, e.g. only eight of the 283 species that have become invasive in China (Xu et al. 2006b). However, their impacts, particularly to biodiversity values, are often disproportionately high, particularly on islands (Mulongoy et al. 2006).

Preventing invasion by vertebrates is easier than for invertebrates simply because they are larger and therefore less likely to be introduced accidentally, for example, 50% of new vertebrate occurrences in China were intentional compared with just 14% for invertebrates (Xu et al. 2006b). Therefore, the main tools to prevent invasion by new terrestrial vertebrates are the laws governing legal importation and border quarantine aimed at detecting illegal importation. Because it is seldom possible to prevent all incursions, laws also need to provide for some form of in-country surveillance and, more importantly, for a rapid response to remove accidental incursions or escapes before the species establishes a population (Chapters 1 and 2, this volume).

Managing the risk of unwanted incursion is easiest for island states (discussed in other chapters of this volume). Wildlife managers in most countries know what exotic terrestrial vertebrates are present as domestic, feral, or wild animals within their borders, although knowledge of which species are in the pet trade is often incomplete. They usually have some idea of which species are not present but may be particularly undesirable (e.g. Bomford 2003) and include them on a blacklist or impose conditions on how they may be held (for example, in zoos). They have access to information on the distributions and densities of many species elsewhere in the world and can make some judgement on whether they are likely to establish if they arrive or escape (Forsyth and Duncan 2001), and may have some knowledge of the likely pathway(s) by which the species might arrive. Thus the risks and pathways of arrival are probably easier to predict for vertebrates than for invertebrates or weeds, if only because there are fewer species.

The major risk of vertebrate invasion comes from a few invasive vertebrates that regularly find their way onto ships and aeroplanes, with the highest risk appearing to be from rodents (*Rattus* spp., *Mus musculus*), common and jungle mynas (*Acridotheres tristis* and *A. fuscus*), Asian house geckos (*Hemidactylus frenatus*), and brown tree snakes (*Bioga irregularis*) (Russell *et al.* 2008). The tools for dealing with this problem include: reducing the ability of the animals to embark at the source (e.g. for brown tree snakes in Guam; Engeman and Linnell 1998); control on the vessel (e.g. the 'de-rat' certification requirements under international and national maritime laws); and prompt reaction to eliminate any that do arrive (e.g. the prophylactic control against invasive rodents on high-risk islands using long-life bait stations; Morriss *et al.* 2008).

More broadly, however, there is a greater risk of new vertebrate species establishing in the wild as a result of the escape or release of exotic animals held legally or illegally in the pet trade. Managing these risks is not simple. Australia, for example, has catalogued all the exotic terrestrial species believed to be present (Vertebrate Pest Committee 2002), but the accuracy of the list is not known. A question is whether inaccurate or partial lists are better or worse than no lists at all. The former might give a false sense of security, while the latter might induce delays in responses. Other species imported or held illegally are unlikely to be reported. Despite these problems, such a catalogue allows some rational consideration of new species proposed for importation and management of any discovered to be present but not listed.

12.3 Tools to manage established wild populations

The same set of tools may be variously used to detect, contain, eradicate, or control established populations of vertebrates, although which tools or combination of tools and how they are applied are obviously strategy-specific.

12.3.1 Detection tools

With wild animals it is usually difficult to quickly detect initial incursions, or find the few survivors of an attempt at eradication, or to accurately delimit distributions. To be absolutely sure no animals are present we would have to search everywhere with an infallible detection device; this is seldom, if ever, possible. We therefore conduct sample surveys with devices that do not always detect animal presence, but when no animals are found it is unclear whether that is because no animals were present or because they were present but simply not detected. If, however, the probability of an animal being detected when it is actually present is known (e.g. Ball *et al.* 2005), we can apply the Bayesian methods originally developed for military search-and-destroy tactics or civilian search-and-rescue operations (Haley and Stone 1979). These estimate the probability and confidence that at least one individual remains in a search area despite the lack of detection, using the known detection probabilities of the search devices or systems (e.g. for the successful eradication of feral pigs on Santa Cruz Island in California; Ramsey *et al.* 2009, or the unsuccessful eradication of musk shrews on Ile aux Aigrettes; Solow *et al.* 2008). These methods also allow an explicit consideration of how much monitoring or control is required to achieve any desired level of confidence that animals are truly absent. Those levels are typically based either on the costs of being wrong (if these are unknown) or reflect some arbitrary level of managerial comfort (Ramsey *et al.* 2009). Managers have to compare the risk of wasting money on extra monitoring or control when in fact no pests remain, with the risk of falsely declaring the pest absent when in fact some do remain (e.g. Regan *et al.* 2006).

12.3.2 Exclusion

Pest populations can be prevented from spreading and their adverse impacts managed by establishing physical barriers, usually in the form of fences. Thousands of kilometres of dingo fence have been built in Australia (Allen and Sparkes 2001), and barriers are widely used in Africa to control large herbivores, especially elephants (Boone and Hobbs, 2004). Fences can also be used to help maintain control or locally eliminate populations within areas already occupied by the pest, by reducing the rate of reinvasion from unmanaged areas. This approach is particularly useful where the animals are managed for different purposes on adjacent lands (e.g. eradication of pigs from nature reserves on Hawaii that are adjacent to areas where they are managed for hunting; Katahira *et al.* 1993). Predator-proof fences capable of preventing animals as small as mice from regaining entry to areas from which they have been eradicated (McLennan 2006) have become something of a cornerstone to private conservation efforts in New Zealand. The principal limitations of fences are that they do not directly reduce the numbers of pests; they are expensive to build and maintain, and always eventually leak. The consequences of the latter in terms of the costs of detecting and dealing with breaches compared with unfenced sustained-control alternative strategies have yet to be revealed by events (Anon 2007).

12.3.3 Control tools

Tools to control vertebrate pests are older than written records, with snares and traps, slings and arrows, toxins, and biocontrol agents (the pet cat in Egyptian granaries) all used long before modern science became interested, and after best practice manuals made at least some, such as magic, redundant—well, not quite redundant but quaint and equally useless (e.g. homeopathic remedies; Eason and Hickling 1992). It has been said (reputedly by Ralph Waldo Emerson) that if you build a better mouse trap the world will beat a path to your door, and the development of new control tools remains a busy industry. The modern trends for developing control tools are generally to increase effectiveness, replace cruel methods with those that are less cruel, and to improve target specificity.

12.3.3.1 *Snares and traps*

The basic principles of snares and traps (for convenience we call them all traps) have not changed for millennia. Mascall (1590) described over 30 traps and one lure used to trap pests such as foxes in England in the 14th century, while Proulx (1999) categorized the plethora of traps used in North America in the 20th century— the basic difference being the later focus on humaneness (and hygiene), without ideally losing efficacy. In fact if you do take Emerson's advice and build a better mouse trap, someone will build an even better one, a process that has long been ongoing and is probably accelerating as technological possibilities expand. One of many such examples is a self-resetting electronic rat trap (US patent 5918409).

Traps, whether used for commercial fur harvesting or for pest control, and whether for holding or killing the animal, may have to meet international humaneness standards. Attempts to set these standards began in 1987 but, apart from standards for testing traps (e.g. International Organization for Standardization 1999), no consensus was reached. Two international standards may now be applied by countries for the use of traps: the above testing standards or the European Union Regulation No. 3254/91 of 1991, which essentially bans leghold traps from the Union and from countries wishing to export fur to the Union. Some countries have prohibited or are phasing out traps that do not meet national or international testing standards (Warburton *et al.* 2008). Modern food hygiene standards have also led to some innovative new traps. There is only one thing worse than a live mouse in the food factory and that is a dead one and the toxins used to kill it. European Union rules on these issues can close premises until the rodent and toxins are removed, so commercial pest control companies have developed systems that gas the rodent using CO_2 in an encased trap that sends off a radio signal to the pest control firm who come and collect the victim (Brigham 2006).

Trapping (both lethal and live-trapping) and its variants remains the main or only control tool for a surprisingly large number of vertebrate pests managed under sustained control strategies. Trapping seems to be the main control tool for many species with specialized diets (e.g. many predators and especially those that prefer live prey such as musk shrews, *Suncus murinus*); where non-target species cannot be easily avoided (e.g. for American raccoon, *Procyon lotor*, control in Hokkaido in the presence of the native raccoon dog, *Nyctereutes procyonoides albus*, Ikeda 2006); or where there are social constraints on lethal methods (e.g. mongoose, *Herpestes javanicus*, control on Amami Island, Okinawa, Japan (Ishii *et al.* 2006).

However, traps have some disadvantages. While they may be effective and efficient at small scales, they require someone to set and check them. They are expensive to apply at large scales or frequently, and individual animals can be, or become, trap-shy (e.g. Tuyttens *et al.* 1999). Thus, there are active research programmes in many countries to find alternative control methods for species currently managed by trapping. For example, in New Zealand, the stoat (*Mustela erminea*) is a major predator of native birds. It is mobile, generally at low densities, and has periodic (seasonal or every few years) impacts on its prey. Control of stoats still relies largely on trapping and is very expensive, even when applied for a short time when impacts are at their worst, so costs prohibit applications at the large scale needed to protect threatened bird populations (King and Murphy 2005). A major research effort into alternative controls (baits and toxins suitable for broad-scale distribution, and preliminary work on fertility control) was undertaken in New Zealand in the 1990s, as yet without a major breakthrough, but with incremental improvements to the older techniques (Murphy and Fechney 2003).

Live trapping is often inefficient compared with kill trapping or poisoning, but nonetheless has been the sole method used in several successful eradication attempts, usually because the lethal options threatened native non-target species. Coypu (*Myocaster coypus*) were removed from England by trapping (Gosling and Baker 1989). Mink (*Mustela vison*) have been removed from Hiiumaa Island in the Baltic by lethal trapping (Macdonald and Harrington 2003), and from the southern islands of the Hebrides by live-trapping (S. Roy, pers. comm.). A proposed attempt to eradicate North American beavers (*Castor canadensis*) from the seven million hectares they now occupy in Tierra del Fuego (where they were introduced in 1948; Anderson *et al.* 2006) would have to rely on trapping (and shooting) because these traditional tools are the only methods proven to effectively control and locally eliminate beavers in parts of their natural range. There is major uncertainty whether such simple tools will be as effective at huge scales, as they were developed for local extirpation (Parkes *et al.* 2008a).

However, most successful eradication attempts that have included traps have also used additional control methods, often because some animals cannot be caught in traps. For example, on Santa Cruz Island, California, 13% of 1421 feral pigs (*Sus scrofa*) were not caught in traps in a trial in a 2250-ha fenced area (Sterner and Barrett 1991), and only 16% of pigs were trapped in the final eradication over the whole island in 2006 (Parkes *et al.* 2008c). Where there are no technically feasible or socially acceptable alternatives to trapping, the resulting inability to catch any trap-shy animals seems to be a major constraint in eradicating some invasive species—musk shrews for the former reason (Seymour *et al.* 2005) and several pests in Japan for the latter reason (Ikeda 2006).

Research on lures has also advanced with modern chemistry, although the 'privie parts of a vixen mixed with galbanum' as recommended by Mascall (1590) still have their use in, for example, the use of the castor glands of beavers to attract them to traps. Successful control of insects using pheromones as lures has fewer counterparts in terrestrial vertebrate pest management, although it is an option for fish that rely on chemical cues to find spawning sites (e.g. lampreys in the Great Lakes; Li *et al.* 2003). Food is most commonly used to lure animals, but visual, auditory, and olfactory lures are also widely used.

12.3.3.2 Shooting

The slings and arrows of the past have evolved into firearms. Modern firearms and knowledge of ballistics allow several specialist applications to control pests, such as in urban areas where human safety when shooting unwanted animals is paramount, e.g. white-tailed deer (*Odocoileus virginianus*) in suburban North America (De Nicola *et al.* 1996). Other than the ongoing development of telescopic, low-light, and infrared sights, the main improvements in using firearms methods have been smarter applications using modern techniques such as radio-telemetry and GPS to track the hunters, dogs, helicopters, and target animals, or the use of Judas-animals for social species to enable hunters to locate con-specifics associating with the telemetered animals (e.g. Campbell *et al.* 2005).

One of the most spectacular advances was the use of helicopters to find animals and then to shoot them from the machine. This was first perfected as part of a commercial game harvesting industry for red deer (*Cervus elaphus*) and other ungulates in New Zealand in the 1970s (Nugent and Fraser 2005) that also acted as an effective control tool for these invasive species at least in non-forested areas (Parkes 2006a). Helicopter culling was the main method used in the eradication (success to be confirmed) of feral goats from Isabela and Santiago islands in the Galapagos (Lavoie *et al.* 2007) and of feral pigs from Santa Cruz Island in California. In the latter case, 77% of the 5036 pigs killed were removed by this method (Parkes *et al.* 2008b). Helicopter culling remains the main tool to sustain control on feral horses, donkeys, camels, pigs and goats across vast areas of Australia (Wilson *et al.* 1992).

As with trapping, shooting can be useful as a stand-alone tool for control but is seldom adequate for eradication because there are usually some areas of heavy cover or other refuges in which shooting is not effective or permitted. Nonetheless at least one current campaign relies entirely on shooting—an attempt to eradicate ruddy duck (*Oxyura jamaicensis*) from Great Britain (Genovesi 2005). Shooting on lakes has reduced the numbers but cannot yet kill 100% of the birds encountered on each shooting occasion, creating a risk that eradication will fail because the survivors become harder to kill and cause the funding agencies to lose heart, or the campaign falls into Zeno's paradox—the number removed each time gets smaller and smaller but can never reach zero. Eradication attempts can only succeed if the proportion removed annually exceeds the maximum possible rate of increase (Caughley 1977).

Hunting with dogs is another ancient human activity, and has long been used in pest control. Modern developments build on old team-hunting methods and dog training, and aim to ensure that no animals escape their first encounter with the hunters. This prevents the development of dog- and hunter-shy survivors, as occurred during the eradications of feral goats from Raoul Island. There team hunting was not used and the last few killed were aged females that had been present during most of the 20-year campaign (Parkes 1990b). The hunters involved with this campaign subsequently developed a 'wall of death' technique; on Santa Cruz Island, this involved a line of hunters and their dogs, in contact with and supported by a helicopter, working systematically across the landscape. To limit the number of pigs that escaped back through the hunting line the dogs were trained not to all simultaneously chase the same pig—only those that first encountered the pigs did so, except where pigs attempted to break back through the line. The helicopter was used to try to mop up any pigs that escaped. Up to 83% of pigs known to be present were killed using this method; the rest mostly by simultaneous helicopter hunting (Parkes *et al.* 2008c).

12.3.3.3 Poisoning

Natural toxins have long been used by humans to kill prey or emperors, along with use of natural antidotes, such as bezoar stones from goat stomachs, to avoid being poisoned. Modern trends in poison use are diverging. Some research aims to find a toxin, bait, and bait delivery system that can simultaneously control multiple pests at lowest possible cost, particularly for large-scale control and eradication campaigns. In New Zealand, for example, a broad-spectrum poison, 1080, is aerially applied over up to one million hectares annually, with introduced brushtail possums (*Trichosurus vulpecula*) the primary target (because they damage native ecosystems and carry bovine tuberculosis; Cowan 2005) (Fig. 12.1). Increasingly these operations are being designed to simultaneously target other sympatric pests such as ship rats (*Rattus rattus*), mice (*Mus musculus*) (Nugent *et al.* 2007) and stoats, some of which are killed by secondary poisoning (Murphy and Fechney 2003). Unanswered questions include:

1) Whether achieving high kills of the secondary pests (such as mice in the example above) is worthwhile where they have much faster breeding rates and smaller home ranges than the main target, and so require finer-scale, more frequent, and therefore more expensive control.
2) Whether it is better to find a universal bait or to mix different bait types (Morgan 1993). The main problem with broad-spectrum baits and toxins is the increased likelihood that some non-target species will also be killed. This can include native species, so managers need to ensure that kills of these are not so high as to outweigh the benefits of removing the pests (e.g. Powlesland *et al.* 1999). It can also include non-target exotic species such as domestic animals and deer (Nugent and Fraser 2005), which understandably provokes opposition from farmers and hunters, which may constrain where poisoning can be used. There is also often similar public opposition from those concerned about environmental contamination and direct threats to human health (Environmental Risk Management Authority 2007).

To avoid such constraint, other research is directed at finding species-specific baits or toxins—the Achilles heel approach (Marks 2001). At its simplest, all but the target pests can be physically excluded from toxic bait by using specially-designed bait stations. Likewise, the bait can be altered to make it less attractive to non-target species; 1080-cereal baits used to kill possums in New Zealand are often dyed green to reduce the likelihood of native birds feeding on them (e.g. Day *et al.* 2003), and increasingly may be coated with a repellent that deters deer but not possums, in order to reduce incidental by-kill of deer (Morriss *et al.* 2005).

The major alternative approach is the search for toxins that pose no incidental threat to humans or non-target animals. Even many commonly used acute or anticoagulant toxins have a wide range of toxicity across taxa (Eason and Wickstrom 2001), so the risks to both target and non-target animals can be manipulated to some extent by choice of toxin, careful selection of the dosages available in individual baits, and the number and distribution of baits available to each animal. Further, there are toxins that are effective against only one trophic level (e.g. para-aminopropiophenone for control of carnivores; Savarie *et al.* 1983), and even some that affect only one species or genus (e.g. toxins that target just *Rattus*; B. Hopkins,

Management of terrestrial vertebrate pests | 181

Fig. 12.1 Loading of a helicopter 'monsoon bucket' with baits for the control of brushtail possums in New Zealand. Large-scale aerial baiting was developed for possum control and has been applied worldwide to rodent eradications on islands. Photo: David Morgan.

pers. comm.). There are a number of biotechnological programmes aimed at designing and synthesizing completely new species-specific toxins and some of these have reached proof of concept stage but still require development of effective delivery systems.

12.3.3.4 Biocontrol

Attempts at biocontrol of terrestrial vertebrates has moved from introduction of predators, with, at best, limited benefits (Allen 1991), to, at worst, total disaster (e.g. the introduction of mustelids into New Zealand to control rabbits; King and Murphy 2005), to the introduction of new pathogens that are not already present as part of the natural disease organisms for any population of pests. There are a limited number of introduced pathogens that have imposed long-term control of vertebrates. Probably the most successful, in the sense of having long-term consequences on vertebrate assemblages and the ecosystem, has been the introduction of rinderpest into Africa in the 1880s (Sinclair 1995). This of course was not a deliberate attempt at biocontrol, but it did show that diseases could profoundly affect vertebrate abundance. The two main examples of the deliberate use of pathogens as biocontrols involve rabbit control, first with the importation of the New World virus *Myxoma*, a relatively benign infection in *Sylvilagus* spp., into Old World *Oryctolagus* populations in Europe and Australia (Fenner and Ratcliffe 1965), and second the importation into Australia and New Zealand of a new virus, rabbit haemorrhagic disease, which naturally evolved pathogenicity from an existing calicivirus found in rabbits (Cooke and Fenner 2002). Both diseases caused large initial reductions in rabbit abundance at most places, but the efficacy of both has waned, as the myxoma virus has evolved competing attenuated strains (Williams *et al.* 1995) and for unknown reasons for the calicivirus (Parkes *et al.* 2008b). Both of these viruses only lethally infect the target species, and to date few other candidate viruses with such specificity have been identified as potential candidates to control terrestrial vertebrates—the exception being another lagomorph calicivirus, European Brown Hare Syndrome virus (Frölich *et al.* 2003), which could be used (if required) to control hares (*Lepus europaeus*) in countries where they are an exotic problem. Other pathogens like rinderpest, anthrax, and canine distemper are not biocontrol candidates because the costs to desirable vertebrates would hugely outweigh any benefits.

12.3.3.5 Fertility control

Controlling the fertility of invasive species has been suggested as a more humane method of controlling their abundance (Fagerstone *et al.* 2002). However, most models show the benefits, in terms of population regulation, are less than when the same number of animals is killed, although a combination of lethal and fertility control is usually best (costs to achieve both being ignored; Cameron *et al.* 1999). A major restriction on the use of fertility control agents has been the lack of reliable one-shot oral delivery (i.e. in a bait) systems, requiring that animals be somehow injected or dosed individually, often more than once in their lifetime, which in turn usually requires their capture. This has proved possible for populations of feral horses in the USA (Turner *et al.* 1996), but is too impractical for most wild animals and often far more expensive than simply killing them.

To overcome these problems, a number of research programmes have aimed to develop genetically engineered, non-pathogenic agents that express proteins that act as immunological blocks to the target vertebrate's fertility (Tyndale-Biscoe 1994). Ideally these agents spread by themselves, but they can also be delivered alive or dead in baits. There are even efforts to combine vaccines that, for example, reduce possum fertility and protect them against tuberculosis infection (D. Collins, pers. comm.). Thus far, attempts to make this biotechnology work in the field against mice, rabbits, and foxes in Australia have failed (T. Peacock, pers. comm.). Mice failed because the engineered cytomegalovirus would not transmit in the field; rabbits failed because the proportion of females sterilized by the engineered benign *Myxoma* virus was too low to overcome compensatory fecundity in the non-sterile females (Twigg *et al.* 2000); and foxes failed because no candidate vector was found. However, research continues using both living vectors (engineered nematodes) and non-living, bait-delivered systems in New Zealand for use against possums (Cowan 2000), and there are models supporting development of a combined lethal and fertility control solution (e.g. Ramsey 2005).

12.4 Conclusions

Despite the large toolbox of methods for control of invasive alien terrestrial vertebrates, it is often inadequate when it comes to the needs peculiar to individual species, or for particular strategic needs. If we look across the 22 terrestrial vertebrates listed within the top 100 invasive species (Lowe *et al.* 2000) we can see that some have either no effective control tools or a limited number which restricts either the scale at which they can be managed or the feasibility of strategies such as eradication (Table 12.1). In addition, there are often non-technical societal constraints on strategies, especially eradication, e.g. introduced deer are widely seen as conservation pests in New Zealand but their eradication is opposed by hunters (Nugent and Fraser 1993).

Table 12.1 Terrestrial vertebrates listed among the top 100 invasive alien species and the availability of effective control tools to manage them (ranked, in our opinion, from absent = 0; to ineffective = +; to effective at small scales = ++; to highly effective at large scale = +++). Traps include all physical methods such as traps, snares, fences.

Species	Type of control tool			
	'Traps'	Shooting	Poison	Biocontrol
Bull frog (*Rana catesbeina*)	+	+	0	0
Cane toad (*Bufo marinus*)	+	+	0	**0**
Coqui frog (*Eleutherdactylus coqui*)	0	0	+	0
Common myna (*Acridotheres tristis*)	++	+	++	0
Red-vented bulbul (*Pycnonotus cafer*)	0	+	+	0
Starling (*Sternus vulgaris*)	+	+	++	0
Brown tree snake (*Boiga irregularis*)	++	0	++	0
Red-eared slider (*Trachemys scripta*)	++	0	0	0
Possum (*Trichosurus vulpecula*)	++	+	+++	**0**
Feral cat (*Felis catus*)	+	+	+++	0
Feral goat (*Capra hircus*)	++	+++	++	0

Table 12.1 (Con't.)

Species	Type of control tool			
	'Traps'	Shooting	Poison	Biocontrol
Grey squirrel (*Sciurus carolinensis*)	++	+	+	0
Macaque (*Macaca fascicularis*)	+	+	++	0
Mouse (*Mus musculus*)	+	0	++	0
Nutria (*Myocaster coypus*)	++	+	+	0
Feral pig (*Sus scrofa*)	++	+++	++	0
Rabbit (*Oryctolagus cuniculus*)	+	+	+++	+++
Red deer (*Cervus elaphus*)	+	+++	+	0
Red fox (*Vulpes vulpes*)	++	+	+++	0
Ship rat (*Rattus rattus*)	+	0	++	0
Mongoose (*Herpestes javanicus*)	++	+	+	0
Stoat (*Mustela erminea*)	++	0	++	0

As an example of the main problems, the common myna is a growing, but unresolved, pest in subtropical and tropical areas around the world (e.g. Freifeld 1999). The tools to control or eradicate them include live-traps, nest snares, shooting, and poisoned baits. However, all of these have drawbacks when dealing with a smart, social bird such as the myna. What is required is the development of a more effective primary control tool that achieves high percentage kills without teaching the birds avoidance behaviours (e.g. the use of delayed-action toxins such as DRC1339 rather than the acute chloral hydrate now commonly used), followed by a sequence of control tools (nest snares, traps, shooting) applied in order of least disturbance to kill survivors—at least when eradication is the aim (Parkes 2006b). New tactics and appropriate intervention strategies would lead to better policies and confidence that the problems can be managed by those responsible for dealing with mynas. In other words, we need a strategy to apply the three elements we raised in the introduction to this paper—funding policy, effective delivery instruments, and knowledge on where to apply them—not just a set of tools.

13
Management of invasive fish
Nicholas Ling

13.1 Introduction

Current understanding of the biology of fish invasions and the development of statistical tools for their prevention lags well behind comparable knowledge for other taxonomic groups such as birds and plants (Veltman *et al.* 1996; Goodwin *et al.* 1999). A recent analysis of fish introductions in Europe (Garcia-Berthou *et al.* 2005) warns that the probability of introduced fish becoming established far exceeds that proposed for other taxonomic groups, such as the 'tens' rule of Williamson and Fitter (1996b).

The introduction, establishment, and spread of invasive fish typically follows the same patterns as for other invasive organisms. Organisms must first be transported across a natural dispersal barrier—accidentally or deliberately—to an area outside their native range, and then released into a suitable habitat to allow establishment and spread. Successful introduction relies on surviving transport and subsequently establishing a viable self-sustaining population. Whether a non-indigenous fish species eventually becomes invasive depends on many factors, including the time since first release, interactions with existing indigenous and non-indigenous species, and perceived benefits of the species to humans. Depending on species, invasive fishes may damage indigenous biodiversity (Witte *et al.* 2000), alter food web structure (Simon and Townsend 2003), affect sports and commercial fisheries (Pycha 1980), degrade habitat quality through sediment resuspension and eutrophication (Zambrano *et al.* 2001), increase stream bank erosion (King 1995), and even cause physical injury to human water users (USGS 2004). However, unlike many organisms, fish are far less likely to be introduced accidentally. Because most fish species are reliant on continuous water immersion and relatively high water oxygen saturation, fish transportation can be difficult. Fish are very unlikely to be introduced as accidental stowaways in cargo, and the speed and reduced fouling of modern ships means fish are unlikely to be transported outside the hull, although they may be entrained within structurally complex hull structures, such as sea-chests, or slow-moving towed structures such as oil platforms (Foster and Willan 1979). Most fish introductions are deliberate and often based on beliefs of the economic, sporting, or aesthetic benefits they will afford. In some cases the perceived economic benefits may be real. Although the annual economic costs of exotic fishes in the USA has been estimated at US$5.4 billion, sports fishing contributes more than ten times that value to the US economy and a substantial proportion of that industry is based on game fishing for locally non-indigenous fishes. Exotic salmonids have established world-renowned game fisheries in New Zealand in the absence of suitable native game species, despite substantial ecological damage to indigenous fish and aquatic invertebrate communities (McDowall 2006). Cost:benefit tradeoffs are common for most globally important invasive fishes. Common carp is one of the most widespread freshwater species and is responsible for significant ecological and economic damage worldwide, yet it also contributed 2.8 million tonnes to global aquaculture production in 2000 (FAO 2003).

13.2 The role of humans

The intentional translocation of freshwater fish is not exclusively a modern phenomenon. Although most introductions worldwide have occurred since the mid-19th century, the introduction and redistribution of common carp in Europe dates as far back as the 12th or 13th centuries (Witkowski 1996). A large proportion of the freshwater fish fauna of the UK and Ireland is non-native and has been introduced from mainland Europe since the 15th century. The role of deliberate human transport in the introduction and spread of non-indigenous fishes is quite clear. In the USA, human population density is correlated with non-native fish diversity, although the relationship is complicated by the prolonged deliberate release of game fish species by state agencies into wilderness areas with comparatively low population density (McKinney 2001). The same is true in New Zealand, where the distribution of species such as perch, rudd, and tench, most commonly spread illegally by coarse fishing enthusiasts, is concentrated close to major population centres, whereas game fish salmonids are far more widely distributed due to historical legalized release by Acclimatization Societies and subsequent natural dispersal. Human access, as well as proximity to concentrations of human population, is important to the risk of initial release and also to re-invasion following the eradication of non-indigenous fishes. A UK study found that the probability of human introduction of fish to ponds was correlated with at least two of the following variables: distance to nearest road, nearest footpath, or nearest pond (*et al.* 2005a).

The introduction of game fish species for recreational purposes not only causes direct impacts on indigenous fauna by predation or competitive displacement, but may also encourage other destructive practices such as the transfer and spread of suitable bait fish to game fish waters. In New Zealand, the widespread transfer of salmonids to lentic water bodies was accompanied by the release of indigenous prey species, such as smelt and eleotrids, to waters where these species did not naturally occur. This has caused the demise or local extinction of genetically distinct populations of other indigenous species (McDowall 2006).

There is no universal definition of what defines an invasive species. While it is generally acknowledged that an alien species that establishes widespread feral populations may be considered invasive, not all such species appear to cause ecological damage or pose significant threats to indigenous biodiversity. For instance, although the goldfish (*Carassius auratus*) is one of the most widespread feral species worldwide, it is generally regarded as benign. By contrast, mosquitofish (*Gambusia* spp.) are typically regarded as destructive and nuisance species virtually everywhere they have been introduced, although sometimes the evidence of significant ecological damage is limited (Ling 2004).

13.3 Risk assessment

Much effort recently has focused on the development of risk assessment models for invasive fishes; the assessment protocol developed by Copp *et al.* (2005b) is an excellent example. Risk assessment procedures must operate at pre-border and post-border levels—firstly to establish the risk posed by fishes not already present in a country and secondly to address the risk posed by further transfer and release of already naturalized species. National laws restricting the importation of certain fishes are typically directed at the aquarium and ornamental fish trade. Given that the worldwide fauna stands at more than 20,000 species, such legislation usually lists excluded rather than allowed species, and such diversity means that quarantine inspectors may need expert knowledge to identify high-risk species in a multispecies importation. Taxonomic identification of some fishes may be difficult even for experts and taxonomic revisions may render older legislation quickly out of date (McDowall 2004). The situation is exacerbated if only common names are specified, because names like carp, cod, barb, and shark are applied to species across widely divergent taxonomic groups. Moreover, many aquarium species are imported as juveniles that may be difficult to identify and easily confused with allowed species. For instance, in New Zealand, the importation of goldfish is banned because juveniles are difficult to distinguish from the prohibited common carp.

Legislative control over the introduction, spread, and control of fish species (exotic or indigenous) is typically spread across numerous national or state agencies involved with biosecurity, aquaculture, sports fisheries, and conservation. The regulatory framework to control potentially invasive species is often diffuse and uncoordinated (Naylor *et al.* 2001). However, these authors hailed the New Zealand Hazardous Substances and New Organisms Act (1996) as a simple yet comprehensive measure to control introductions of new organisms whereby all exotic species are regarded as potentially invasive unless proven otherwise. Simberloff (2005) has further argued that the regulation of exotic species introduction and spread should be based on presumption of risk rather than assessment of risk because most current risk assessment procedures are 'narrowly focused, subjective, often arbitrary and unquantified, and subject to political interference'.

13.4 Economics of eradication and control

A cost:benefit analysis should be part of any consideration to undertake control. Where clearly defined economic costs accrue from persistence and spread of an invasive fish and the costs of eradication or control can be accurately defined, then the cost: benefit analysis is quite simplistic. However, where impacts do not have a direct monetary value, such as habitat modification or biodiversity loss, then the equation is considerably more difficult to compute. In an attempt to develop an objective assessment of the costs and benefits of invasive species control where economic costs may not be defined, Choquenot *et al.* (2004) argued for the application of a bioeconomic model that can be applied either to benefit maximization or cost minimization. Such bioeconomic models could be considered in the establishment of freshwater protected areas as proposed by Saunders *et al.* (2002). Freshwater protected areas offer considerable promise given that freshwater catchments may be protected from invasive species by natural biogeographic barriers.

13.5 Marine versus freshwater

By far the most significant adverse impacts have been caused by the introduction and spread of freshwater fishes rather than marine. Indeed, all of the eight fishes listed among the world's 100 most invasive organisms by the IUCN Invasive Species Specialist Group are primarily freshwater (Lowe *et al.* 2000), although both brown and rainbow trout are facultatively diadromous. Marine fishes therefore represent perhaps the least problematic of the non-indigenous marine species worldwide. Partly this reflects a lack of incentive for the introduction of marine species. Most countries have native marine fishes that afford suitable opportunities for sport or commercial harvest. Furthermore, the establishment of marine fishes is typically difficult given the number of individuals that would need to be released to establish a viable population. Anadromous salmonid species are one exception because juveniles can be raised in freshwater. Furthermore, many salmonids are not obligately anadromous and establish self-sustaining populations in landlocked waters. Despite this, efforts to establish self-sustaining anadromous salmonid populations in some countries, involving the release of millions of juveniles from hatcheries, has been time consuming and expensive.

13.6 Indigenous fish as invasive species

The redistribution of a country's indigenous fish fauna to areas outside their natural range can cause significant damage to native fauna and ecosystems. The North American freshwater fish fauna has undergone significant homogenization due to the introduction of a small number of cosmopolitan species for the enhancement of food or sport fisheries. Moreover, the primary cause of this homogenization is indigenous species introductions, rather than species extirpation. The most significant demonstration of this phenomenon is in those states where settlers regarded endemic local species as undesirable for food or sport (Rahel 2000). Similar faunal homogenization in the Iberian Peninsula has resulted from the spread of both exotic fish and translocated native species (Clavero and Garcia-Berthou 2006). Restricting the transfer and release of native fishes presents particular difficulties since it by-passes the rigorous biosecurity provisions at the international border. Many countries have legislation to prohibit the release of live aquatic life to waters where that species does not occur naturally, but law enforcement is difficult. Even if a species is already present, the introduction of individuals from other populations may significantly homogenize genetic diversity. Laws that prohibit interstate transport of fish and other aquatic life may help, however, river drainages are typically not confined within state, or even national boundaries, and natural spread throughout a river drainage can be expected. Significant natural barriers to fish passage assist in preventing natural spread and illegal releases. For instance, the importation to Tasmania from mainland Australia of any fish capable of living in Tasmanian waters is prohibited, as is any transfer of live aquatic life between the islands of New Zealand (with the exception of tropical or ornamental aquarium species). In many cases these laws were originally enacted to protect exotic sports fisheries from the spread of disease and inferior stocks although they now serve to protect natural biogeography and genetic diversity.

13.7 Routes of introduction and spread

13.7.1 Ballast water and vessel hull transport

Although most studies of ballast water have found few adult fish or larvae, ballast water is still a potentially important route of introduction for invasive fish. Ecologically and economically destructive invasions attributable to ballast water include the sea lamprey and round goby introduced to the Laurentian Great Lakes. Fish most likely to establish as a result of ballast water transport are small-bodied crevicolous species, especially gobies and blennies (Wonham *et al.* 2000). Such species are also most likely to be transported in hull structures such as sea-chests (Coutts and Dodgshun 2007) although the transport of their adhesive eggs, spawned in the port of origin, in sea-chests and on the outside of ships' hulls, is possibly a more likely mechanism than the transport of adult fish and is more likely to increase the number of propagules introduced, thereby enhancing the risk of establishment (Walsh *et al.* 2003). Post-larval fish may also be entrained with slow moving, structurally-complex towed structures such as oil platforms as long as those structures are not transported through waters hostile to their survival (Foster and Willan 1979).

13.7.2 Live fish importation and sale

The annual trade in ornamental and aquarium fish exceeds 1 billion individuals from around 4000 freshwater and 1400 marine species (Wittington and Chong

2007). The risks posed by the accidental or intentional release of non-native aquarium fish have been recognized in recent decades and many countries have enacted legislation to control the importation and trade in non-indigenous ornamental fishes. Examples are the Import of Live Fish Act 1980 and more recently The Prohibition of Keeping or Release of Live Fish (Specified Species) (Amendment) (England) Order, 2003, in the UK. Many countries ban the importation of species recognized as potentially invasive and the live release or keeping of high-risk or environmentally destructive species. Unfortunately, legislative controls are often ineffective at limiting the distribution and release of noxious species. Despite the listing of the alga *Caulerpa taxifolia* as an illegal noxious species by both Federal and California State authorities, it is still widely sold by aquarium stores in the state (Padilla and Williams 2004)

Markets selling live fish for human consumption, either imported or caught locally, also pose a risk for accidental or intentional release. While biosecurity and import legislation should effectively limit the risks posed by live fish importations, few regulatory restrictions may exist on the sale of live exotic fish that are already naturalized, unless those species are declared noxious, thereby banning their live possession or sale (Rixon *et al.* 2005). Weigle *et al.* (2005) assessed the risks of live release of imported marine and estuarine species associated with seven different importation and transfer industries: seafood companies; aquaculture operations; bait shops; aquarium shops; research and educational organizations; public aquariums; and coastal restoration projects. The risks for each activity contrasted strongly between the diversity of taxa imported (high for public aquaria and the aquarium/ornamental trade) and the number of individuals imported (many individuals from few taxa in the case of the seafood and live bait industries). The importation of imported live bait is clearly high risk given that the release of these organisms into the environment is virtually certain. Given the diversity of non-indigenous marine taxa typically stocked in public aquaria, this industry also poses significant risks of release since most large public aquaria use local coastal seawater supply. Unless effluent water is carefully treated or filtered, non-indigenous fish larvae or propagules of other organisms may be released. For instance, the aggressive invasion of the Mediterranean Sea by the alga *Caulerpa taxifolia* since the mid 1980s probably originated from the Aquarium of Monaco (Jousson *et al.* 1998).

The risks posed by the domestic ornamental and aquarium trade vary considerably based on the key taxa traded and the locality of sale. Given that the vast majority of aquarium fish species are of tropical origin, their survival, and certainly reproduction, in temperate latitudes is unlikely unless local peculiarities in water temperature allow. For example, three tropical aquarium species have been recorded as establishing breeding populations in New Zealand but are restricted to geothermally-heated locations in the central North Island (McDowall 1990). However, aquarium releases pose significantly greater threat in tropical and sub-tropical regions and have been identified as the prime culprit in the observed diversity of non-native marine species in south-eastern Florida. Sixteen non-indigenous marine species have been recorded, and their concentration close to human population centres implicates aquarium releases, although the typically low number of individuals recorded for each taxa means that physical removal of most species could be effectively undertaken given sufficient will (Semmens et al. 2004). This latter study highlights the importance of a rapid response to a potentially solvable problem. At least one tropical marine fish, the Indo-pacific lion fish, is now so firmly established in the south-eastern USA that eradication is unlikely (Whitfield et al. 2002). Quantitative models to estimate propagule pressure, based on uncertainty, number of fish traded, and fish buyer behaviour, that may serve as a useful method for aquarium trade risk assessment, have been developed by Gertzen et al. (2008).

13.7.3 Aquaculture for the aquarium trade

The majority of ornamental aquarium fish aquaculture is concentrated in tropical developing countries, particularly in South East Asia. Coincidently, equatorial developing nations also contribute more than 80% of world food aquaculture production. A recent evaluation of invasive fishes in ASEAN countries identified 79 introduced species although only 12 were considered to have become invasive, and only in some countries (Ponniah and Husin 2005). Information on species introduced for food aquaculture was much greater than that for ornamental species although knowledge of the extent of invasiveness and the biological impacts of alien species was typically poor.

At least 90% of freshwater ornamental aquarium fish traded worldwide are cultured rather than collected from the wild; however, only a small fraction of ornamental marine fish are currently farmed. Future expansion in the aquaculture of ornamental marine fish and invertebrates is likely given concerns about the overexploitation of coral reefs and the significant collateral mortality (up to 90%) caused by the capture of wild collected specimens by high-risk methods like cyanide fishing (Rubec et al. 2001). This industry poses a significant risk of release if aquaculture of such species occurs in areas where the organisms do not naturally occur (Tlusty 2002).

13.7.4 Aquaculture for food

Aquaculture is one of the fastest growing sectors of the global food economy. Throughout the period 1970–2000, growth in finfish aquaculture averaged greater than 10% per annum, with annual production reaching 14 million tonnes in 2000. In that year, aquaculture provided 73.7%, 65.3%, and 1.4% of global freshwater finfish, salmonid finfish, and marine finfish landings, respectively, exceeding US$28 billion (FAO Inland Water Resources and Aquaculture Service 2003). Aquaculture has become a leading vector for the introduction of invasive species worldwide and is also responsible for numerous other threats to ecosystem function and integrity. In a classic mismanaged example of aquaculture biological control, diploid black carp were permitted to be used for combating an outbreak of trematode infestation in channel catfish ponds in Mississippi in 1999. Previously only sterile triploid black carp had been approved for this purpose (Naylor *et al.* 2001). Inevitably, diploid black carp have been released and now appear to have established in the Mississippi and Atchafalaya Rivers (Jenkins and Thomas 2007). Escapes and unintentional releases of farmed sea-ranched salmon have been linked to negative ecological and genetic impacts on wild salmon stocks (Gross 1998). Up to 40% of wild salmon caught in the North Atlantic and 90% of salmon caught in the Baltic are of farmed origin (Hansen *et al.* 1997). The use of triploid farmed fish has been proposed as one way to minimize such effects (Cotter *et al* 2000). Salmon farming has also been implicated in the decline of wild salmon stocks by acting as sources for infestation by parasites. One single salmon farm increased infection pressure on wild migrating salmon by four orders of magnitude (Krkosek *et al.* 2005). Although salmon farmers routinely treat parasite infestations to reduce impacts on their stock, farms still provide significant and detrimental infection pressure on wild salmon, and the ecological effects of sea lice seem to far outweigh the ecological effects of pesticide treatments on farms (Woodward 2005).

13.8 Eradication and control

13.8.1 Early response

Immediate and early response to a suspected new introduction is critical to the success of any potential eradication. Suitable control tools need to be available, such as approvals for chemical eradication and private property access. An excellent example is the eradication of Australian marron crayfish (*Cherax tenuimanus*) and European gudgeon (*Gobio gobio*) from two locations in the Auckland region of New Zealand in 2005. The gudgeon were probably smuggled into the country by coarse fishing enthusiasts for use as live bait. A coordinated operation involving local and central government agencies resulted in rapid eradication of both species (MAF 2005) and no subsequent populations have been found to date. Early response to a possible introduction is clearly crucial in attempting to prevent further spread. One wonders what the outcome may have been had attempts been made to treat the initial observation of $1m^2$ of *Caulerpa taxifolia* off the Monaco Aquarium in 1984. Within 8 years the infestation was estimated at 3000 ha (Jousson *et al.* 1998).

13.8.2 Response tools

13.8.2.1 Preventing spread: physical barriers, electrical barriers, interstate/interisland biosecurity barriers

Freshwater fish generally disperse poorly. Little opportunity exists for fish to disperse from discrete water bodies such as isolated ponds and lakes, although eggs may be carried on aquatic vegetation transported by birds and vehicles. Adhesive eggs may also be spawned and transported on boat hulls. In contiguous watersheds, natural barriers like waterfalls, velocity barriers such as chutes and riffles, and of course the saline waters of estuaries and coasts may prevent spread.

Non-migratory marine species generally spread by entrainment of pelagic larval stages in coastal and oceanic currents, although some larvae have considerable ability to disperse by active swimming. Larvae of marine species may be transported in ballast water and adults in sea chests. One of the most common teleost families to spread in this manner is the gobies (Gobiidae). Migratory marine species either follow highly prescribed migration routes, such as diadromous eels (Anguillidae) and salmonids, or range widely over great oceanic distances (tuna, billfishes and pelagic sharks). In many cases it may be impossible to determine whether new records of marine fish are naturally dispersed vagrants or have been introduced by shipping or aquarium releases. Marine fish may also spread to new regions by the construction of canals. Since the construction of the Suez Canal in 1869, more than 60 Indo-Pacific marine fishes have invaded the Levantine Basin of the Eastern Mediterranean Sea and now dominate both the biomass and community structure of shallow sublittoral and littoral habitats (Goren and Galil 2005).

Constructed physical barriers such as dams, weirs, and culverts on streams and rivers are typically regarded as problems for fish passage and much research and expense has been invested in designing suitable structures to assist upstream fish passage, and in enacting legislation to prohibit the construction of impassable barriers (Roni *et al.* 2002). However, physical barriers may help to prevent the spread of nuisance fishes, while fish passes may assist such spread. Preventing the spread of invasive species in rivers where the upstream or downstream passage of migratory indigenous species is required therefore presents special problems. Some research has been undertaken to design structures that provide passage for native species but prohibit or trap invasive exotics (Stuart *et al.* 2006).

Recent technological applications in preventing the spread of invasive fish are electrical and bubble barriers, although these technologies are not new. Electrical barriers were extensively employed in the Great Lakes sea lamprey control programme in the 1950s (Smith and Tibbles 1980). An electrical barrier has recently been constructed on the Des Plaines River, Illinois, to prevent the spread of species such as round goby and ruffe from the Great lakes into the Mississippi drainage, and the spread of bighead, black, and silver carps in the opposite direction. Unfortunately the round goby had already invaded downstream of the barrier prior to its construction in 2002 (Corkum *et al.* 2004) and the ultimate success of this venture in preventing the spread of other invasive fish is uncertain because recent studies show that such barriers are only partially effective for some species (Dawson *et al.* 2006).

13.8.2.2 Chemical control

The chemical renovation of freshwaters is the most common and historically effective method in controlling or eradicating nuisance fish species. However, options for chemical eradication are limited to relatively small enclosed water bodies or small streams and rivers because of the quantity of chemical required and the potential for significant collateral ecosystem damage. Not only is cost an important consideration in treating a large water body but also the availability of sufficient quantities of the piscicide of choice. Annual worldwide supply of natural piscicides like rotenone is limited by its availability involving the harvest of source crops. The Strawberry Reservoir rotenone application (20.6 t) in Utah, USA, in 1990 was estimated to consume roughly one-third of worldwide rotenone production in that year and roughly four times the typical annual fisheries usage in the USA (McClay 2000). However, the recent decision by the European Union (EU) to ban rotenone use in plant protection products (EU 2008) is likely to increase its availability for alternate uses, especially if other countries follow the European decision in an attempt to reduce agricultural pesticide use and insecticide resistance. A variety of piscicides have been used to remove pest fish in order to enhance or recover the biodiversity values of a water body, to reduce or eliminate fish biomass in fish and non-fish aquaculture ponds, or to reduce pest fish impacts in larger freshwater systems.

Four chemicals (rotenone, antimycin, saponins, 3-trifluoromethyl-4-nitrophenol) have been extensively used for such purposes but at least 40 substances have been used on a more limited scale (Lennon *et al.* 1971) or investigated as potential piscicides (Clearwater *et al.* 2008). The key requirement for such chemicals is that they show strong selectivity towards fish but limited toxicity to other sectors of the biota in order to limit collateral ecosystem harm. In developed nations with robust legislation controlling the environmental use and application of toxic substances, pesticides typically must undergo a rigorous set of risk assessment procedures to qualify for registration, and then only for specified use or activity. In other circumstances, certain pure compounds or chemical mixtures may be authorized for piscicidal applications but may require specific, experimental, one-off, legislative approval by a local authorizing authority. This is usually described as 'off label' usage. National pesticide registration therefore sometimes alleviates the requirement for a protracted environmental impact assessment for each application. Formulations of fish control chemicals may vary depending on the target species or target environment (shallow versus deep lakes, static versus flowing water) and specific products or formulations are usually registered with national pesticide registration authorities. Rotenone is currently registered in several countries although allowed formulations may vary. For instance, powdered rotenone has recently been assessed by the Environmental Risk Management Authority and subsequently registered by the Agricultural and Veterinary Medicines Authority in New Zealand for piscicidal use but liquid formulations and rotenone carp baits have not. In the USA, antimycin, niclosamide, and 3-trifluoromethyl-4-nitrophenol (TFM) are also registered for use in controlling noxious fish. Numerous authors report problems with the dependability of supply of some fish control chemicals and the consistency of the products.

An important consideration in treating both static and flowing waters is to maximally dewater the system. This reduces the surface area for treatment and the volume of water to be treated, thus reducing the quantity of chemical required. It also dewaters structurally complex littoral habitats such as reed beds, root masses, undercut banks, and rocky shorelines that may be difficult to treat and which may provide refuge for fish in untreated pockets of water (Rayner and Creese 2006). Ground water recharge and subsurface springs also compromise eradication attempts. Generally, eradication is most successful in small, easily accessible, closed, shallow, lentic water bodies that are sparsely vegetated.

Chemical eradication of non-indigenous fish from flowing waters presents special difficulties. Dewatering the system by diverting stream flow or treating during low flow periods may create disconnected pools that need to be identified and treated separately. All other possible refugia such as minor tributaries must also be separately treated, and the reinvasion of fish from downstream must be prevented by constructing physical barriers or relying on natural barriers such as waterfalls. This approach has been successful in eradicating rainbow trout and brown trout in Australia to restore small streams for indigenous galaxiids. However, a high level of invertebrate mortality may result from rotenone treatment (Lintermans and Raadick 2003) and the ultimate success of these treatments may be compromised by subsequent illegal restocking of salmonids (Rayner and Creese 2006).

The two key variables affecting the response of fish to toxicants is concentration and contact time. Applications need to ensure complete mixing and adequate contact time which may be difficult to achieve in flowing waters. The toxicity of all fish control chemicals is markedly affected by environmental conditions such as water temperature, dissolved oxygen, hardness, pH, particulate organic matter, salinity, light exposure, substrate and plant adsorption. Sensitivity among fish species also varies greatly with salmonids typically more sensitive than coarse fish such as carps and catfish. Juvenile stages are usually more sensitive than adult, probably due to greater relative surface area for uptake and greater mass specific metabolic rate. However, fish eggs are usually less sensitive to the common piscicides than larvae, juveniles, and adults. For example, salmonid eggs are around 100 times less sensitive to rotenone than juveniles or adults so eradication efforts should be timed for post-hatch or pre-spawning (Ling 2003).

A significant disadvantage of chemical renovation of natural water bodies is the likely significant or total loss of non-target fish species. These then need to be reintroduced if a suitable source population is available and if sufficient numbers can be released to re-establish the species.

Rotenone

Rotenone, either in liquid or powdered formulations, has been used for many years primarily for the elimination of unwanted fish from enclosed static water bodies, but has also been used to poison flowing waters (streams and small rivers). The use of rotenone for fisheries management has been extensively reviewed in recent years (Finlayson *et al.* 2000; Ling 2003; Rayner and Creese 2006). The largest user of rotenone for fish control is the USA (roughly 5.5 t per annum) and usage is equally divided between powdered and liquid formulations (McClay 2002). Powders are now often favoured for the treatment of static waters because of their significantly lower cost, while liquid formulations are favoured for flowing water (McClay 2000). Some liquid rotenone formulations have generated public opposition due to the toxicity or smell of hydrocarbon components of the formulations. However, a recent European product (CFT legumine) is reportedly less noxious (CDFG 2007). Fish species sensitivity to rotenone varies considerably but fish are generally more sensitive than aquatic invertebrates, although significant invertebrate mortality or drift usually accompanies rotenone applications, given the relatively high concentration applied to ensure satisfactory mixing and rapid fish kills. Importantly, birds and mammals are relatively insensitive to rotenone and its use in the USA for nearly a century to eradicate nuisance fish from water bodies, including public water supply reservoirs, without significant ecological or public health concerns testifies to its relative safety in this regard (Ling 2003). A recent development is rotenone-impregnated baits for the control of grass carp and common carp (Fajt 1996). These baits have demonstrated some promise in fish control under experimental conditions but have sometimes been problematic in field applications causing significant collateral mortality of non-target species (Gehrke 2003).

Antimycin-A (Fintrol®)

The popularity of antimycin for fish control peaked in the 1970s, approaching that of rotenone, but it has since declined significantly due to cost and availability (Finlayson *et al.* 2002). US fisheries agencies used 94.7 t of rotenone in the 10-year period from 1988–1997, compared with only 50 kg of antimycin in the period 1991–2000. It is registered for use for fish control only in the USA and marketed as Fintrol®. It is highly toxic to scaled fish, less toxic to scaleless fish such as catfish, and relatively non-toxic to other aquatic life. The differential sensitivity of scaled and non-scaled fishes means that most of the antimycin use in the USA is by catfish farmers to selectively remove nuisance scaled fish species from their ponds (Finlayson *et al.* 2002). Antimycin baits have been experimentally tested on common carp and doses of around 1 mg/kg of fish are lethal but they have not been widely applied for fish control.

Natural saponins

Saponins are a diverse group of compounds mainly derived from plants. Two products—teaseed cake and mahua oil cake—have been widely used in Asian countries, primarily for the renovation of aquaculture ponds, either to remove fish prior to stocking or to eradicate finfish from shrimp ponds. Toxicity is much greater to finfish than to other aquatic organisms such as shrimps, and is enhanced at higher temperatures and salinities. Both teased cake and mahua oil cake are relatively impure horticultural by-products whose saponin content is variable. Pure saponin is toxic to fish at concentrations of around 0.5–1.0 mg/L but effective applications of the cake products are typically 25–100-fold greater. Saponins are not currently registered with any national pesticide regulatory authorities for invasive fish control.

TFM and niclosamide

TFM has been used in the USA since 1958 either singly or in combination with niclosamide (Bayluscide®) after 1963 to selectively poison sea lamprey ammocoete larvae in tributaries of the Great Lakes. TFM formulations include a liquid concentrate, which is applied by boat or backpack sprayers, and solid block, which is secured in a stream and allowed to dissolve slowly. TFM toxicity to lamprey larvae is significantly greater than to other fishes and aquatic life, however, unlike the other toxicants profiled here, TFM detoxifies only very slowly in the environment. Marking and Olsen (1975) found little or no loss of activity after 8 weeks. Niclosamide formulations include a wettable powder and granular products. Niclosamide toxicity varies with species. It is highly toxic to aquatic molluscs, more so than to most fishes, and is also used as a selective molluscicide. The sea lamprey control programme has been highly successful and by 1972, catches of spawning sea lamprey in Lake Superior had declined by 92% (Heinrich *et al.* 2003). Average annual TFM usage was around 34 t in the period 1995–1999 (McDonald and Kolar 2006).

13.8.2.3 Biocontrol measures

Most potential biocontrol measures are still considered experimental or hypothetical rather than effective and readily available control tools. Advancements in the biological control of aquatic species are considerably less developed and more problematic than counterparts in terrestrial biocontrol (Secord 2003).

Predatory fish

Although the introduction of a large predatory species may seem intuitively appealing for controlling nuisance fish this is unlikely to be effective given the relatively non-selective feeding of most fish. Introductions of large-bodied exotic piscivorous species have generally caused significant ecological problems (e.g. Nile perch in Lake Victoria), and many countries lack indigenous large-bodied piscivores. Natural predator–prey dynamics mean that a predatory species is unlikely to completely eliminate its prey thereby reducing the effectiveness of biocontrol. However, the introduction of the piscivorous European perch (*Perca fluviatilis*) to some New Zealand lakes has completely eradicated all other fish species, to the significant detriment of indigenous biodiversity.

Pheromones

Pheromone traps have proved to be highly effective in pest control and biosurveillance, particularly for insects, and this is a developing area of research for invasive fish control. Field trials of sea lamprey pheromone (Li *et al.* 2007) demonstrate the potential to use pheromones to enhance the capture of nuisance fish, and as a tool to detect pest-fish incursions, especially given the extreme sensitivity of some fish species to pheromonal compounds. Pheromones offer potential for sex-specific and species-specific removal, and for reproductive disruption (Sorensen and Stacey

2004). Application of pheromone technology is well advanced for sea lamprey but is also being investigated for round-goby (Murphy *et al.* 2001), goldfish, and common carp (Sisler 2005) although no teleost pheromone system has yet been fully elucidated and field tested (Sorensen and Hoye 2007).

Fish pathogens

The release of species-specific pathogens (parasites, bacteria, viruses) for controlling invasive fishes is currently only experimental. In Australia, limited success in controlling European perch has been achieved by the unintentional introduction of epizootic haematopoietic necrosis (Langdon and Humphrey 1987). Species-specific viral agents such as carp herpes virus (CyHV-3) offer some potential to significantly reduce common carp biomass (McColl *et al.* 2007) but this disease is not universally lethal, highly temperature dependent (Gilad *et al.* 2003), and the likely outcome is some degree of immunity and population recovery. Moreover the widespread importation or application of such a virus would encounter significant opposition in those countries where common carp are a legitimate and highly valued ornamental species. Indeed, most research on koi herpes virus worldwide is targeted at combating this disease in captive ornamental koi.

Habitat modification and restoration

Many invasive fishes typically thrive in degraded aquatic habitats and may actually cause habitat deterioration by encouraging eutrophication or increasing turbidity through sediment resuspension. Reversing the declining quality of aquatic habitats is unlikely to result in eradication of invasive species but may significantly reduce their impacts. For example, impacts of *Gambusia holbrooki* on a native Australian frog were found to be diminished in ponds that retained extensive riparian and littoral vegetation affording refuge for frogs and tadpoles, respectively (Hamer *et al.* 2002). Riparian revegetation increases stream shading, reduces sediment and nutrient loading, and may improve the habitat for native species over exotics. Harris (1997) argued that river restoration, including catchment management, flow allocation, pollution abatement, habitat reconstruction, and the restoration of connectivity, should form part of an integrated management strategy for common carp in Australia. Such a strategy has the potential to reduce carp impacts and restore indigenous fish populations but must be exercised with extreme caution and thorough species knowledge to ensure that any restoration efforts do not make matters worse rather than better.

Immunocontraceptive control and genetic modification

The induction of triploidy in fish using heat, cold, electrical or pressure shocks, or chemicals, has been commonly used to produce sterile, fast-growing fish for aquaculture (Tiwary *et al.* 2004) and to reduce the likelihood of breeding in fish released for biological control. However, the physiology of triploid fish is poorly understood (Maxime 2008) and some sterile triploids routinely reproduce by processes such as gynogenesis, whereby their eggs are activated by the sperm of other species. The gibel carp is a highly invasive freshwater species with gynogenetic triploid populations throughout Europe (Vetemaa *et al.* 2005). Although triploid species are believed to be able to reproduce only via parthenogenesis, gynogenesis, or hybridogenesis, a bisexually reproducing triploid amphibian has recently been discovered (Stöck *et al.* 2002). This highlights the need for extreme caution in the production and spread of triploid animals. Immunocontraceptive control and the genetic modification of fish to produce single-sex progeny (daughterless) have both been investigated in Australia for the control of common carp and other nuisance vertebrate species, but both are still in experimental planning or development (Thresher 2007). Immunocontraception relies on activating the fish's immune system to block fertilization. The required immunocontraceptive antigen could be delivered by a viral vector or by baits. Daughterless induction in fish relies on the heritable deactivation of the aromatase enzyme responsible for converting androgens into oestrogens, the result being that all offspring of mutants are sexually reproducing males capable of spreading the mutation throughout the population. Should the daughterless technology be successfully developed to provide a workable management option, considerable public resistance to the concept of the large-scale release of genetically modified organisms (GMOs) would need to be overcome and although these organisms are intragenic rather than transgenic (that is their own genetic material is modified without any addition of foreign DNA) they are likely to be subjected to the same stringent risk assessment required for the release of GMOs (Russell and Sparrow 2008). Moreover, genetic suppression of oestrogen production in females may potentially be reversed by the presence of environmental xeno-oestrogens (Jobling *et al.* 1998).

13.8.2.4 Physical removal

Complete eradication of fish from water bodies by physical removal is generally considered a hopeless task because the effective effort rises exponentially as the population is fished down. However, removing every last individual may not be necessary if one is able to hold the population at a sufficiently low density to hope for stochastic extinction to occur via the Allee effect (Courchamp *et al.* 1999c). Any control programme aimed at eradication also needs to consider the risk of illegal reintroduction which may render very expensive control measures ineffective. If a source population for reintroduction is locally available, or if the site has easy human access, then eradication efforts may ultimately be a waste of time, especially if the target species has some perceived recreational or other benefit. Physical removal often aims to reduce the ecological impact of invasive species rather than to achieve complete eradication but is probably the only effective measure in very large water bodies due to the impracticality or cost of chemical renovation.

One simple strategy to eradicate fish in small ponds and lakes is complete dewatering. The water body can be drained or pumped dry. Care must be taken to ensure that no wet refugia remain to harbour fish and the technique is best employed during times when eggs that might survive prolonged emersion in damp vegetation are not present. However, dewatering is of course entirely unselective, and restocking with desired species is necessary following a suitable period of conditioning to allow the recovery of invertebrate and plant communities. Partial dewatering is also effective to enhance the efficiency of physical removal by netting or other methods because capture efficiency increases as the stocking density is increased by dewatering.

A more drastic but equally simplistic measure is the use of explosives to eliminate nuisance fish in ponds and has been tried with limited success. Apart from the obvious risk of collateral habitat damage, the success of explosives depends on the ability to concentrate the explosive charge throughout the water body. This technique is more effective in small, shallow water bodies and relatively ineffective in deeper water.

More common methods for physical removal are passive capture techniques such as netting and angling, or active methods such as electrofishing (Fig. 13.2). Netting techniques and gear may be targeted to particular species and may sometimes be effective in removing nuisance fish without harming non-target species because of the selectivity of the techniques with respect to net type, mesh size, and the behaviour of target and non-target species. Some net types, and electrofishing, also offer the opportunity to release non-target species relatively unharmed whereas gill nets, trammel nets, and seines either cause significant damage or kill fish. Gill and trammel nets (Fig. 13.3) are also highly size selective (Hamley 1975), based on net mesh size, so that different meshes may be required to target different life stages of the species of concern. Trap nets and electrofishing are generally less size-selective depending on net design and electrofishing current parameters. Formicki *et al.* (2004) observed that the attachment of magnets to the entrance of fyke nets enhanced capture efficiency by 50% or more. It is well established that many fish are able to detect weak magnetic fields and this represents possibilities for controlling the movement of fish and improving the efficiency of fish removal efforts although this is completely untested.

Electrofishing is a very common method for surveying fish communities but can cause fish mortalities at all life stages. Although electrofishing equipment design may be altered to increase effectiveness against very small fishes (Copp 1989) no technique is likely to be completely effective in removing fish larvae and removal effort may be best directed at a time prior to spawning when larval density is low and young-of-the-year juveniles are large enough to be targeted. Dwyer *et al.*(1993) found that trout eggs were very sensitive to electroshocking at a critical early stage in their development. They caution that electrofishing over recently deposited redds can result in significant mortality. This therefore serves as a possible technique to destroy the redds of invasive salmonids and possibly other species also.

13.8.3 Case studies in the effectiveness of physical removal

The following examples show that complete removal or the minimization of ecosystem damage may be achieved by efforts at physical removal in very large water bodies.

Management of invasive fish | 201

Fig. 13.1 Koi carp now constitute approximately 70% of total fish biomass, at densities exceeding 2t/ha, in some waterways of the Waikato area in New Zealand. Koi are processed for fertilizer following a bow-hunting competition targeting these invasive fish. Photo: Brendan Hicks.

13.8.3.1 Nile perch in Lake Victoria

Nile perch were introduced illegally to Lake Victoria in East Africa, the world's second largest freshwater lake, sometime in the late 1950s and this was quickly followed by legalized releases throughout the 1960s (Pringle 2005). The species rapidly expanded its range throughout the lake and quickly came to dominate the fish biomass. By the late 1980s serious fears were held for the indigenous fish fauna comprising some 200 endemic haplochromine cichlid species (Ogutu-Ohwayo 1990). However, a highly valuable fishery for the species has subsequently developed with most of the harvest exported as quality fish fillets to Western Europe.

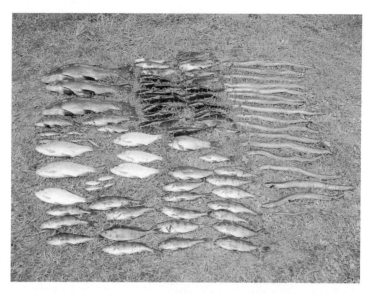

Fig. 13.2 Fish collected by boat electrofishing in a New Zealand urban lake (Lake Rotoroa, Hamilton). Eels (right) were the only indigenous species caught. Invasive species (clockwise from top left) are tench, catfish, goldfish, European perch, and rudd. Photo: Brendan Hicks.

Fig. 13.3 Juveniles of the invasive European perch collected by gill netting in an urban New Zealand lake (Karori Reservoir, Wellington). Four days of electrofishing and gill netting in this lake collected 4700 perch and no other fish species. Photo: Nicholas Ling.

Exports of fish fillet to the EU were valued at €210 million in 2005 and peaked at 56,000 t of fillets in 2004, although the fishery now seems to be declining (FAO 2006) and there are promising signs that many of the endemic cichlids that were once feared extinct are returning (Witte *et al.* 2000). The future of this fishery poses some significant challenges for the three countries that border Lake Victoria (Tanzania, Uganda, and Kenya) in order to promote both a sustainable commercial fishery and the conservation of biodiversity (Balirwa *et al.* 2003). The successful reduction in the biomass of Nile perch in Lake Victoria has been possible only because the fish has marketable value. However, many invasive fish species may not have commercial value and the costs of fishing down the population must be an ongoing expense on regulatory authorities. This cost must be weighed against the cost of non-intervention.

13.8.3.2 Common carp in Lakes Crescent and Sorrell, Tasmania

Common carp was discovered in the adjacent and interconnected Lakes Crescent and Sorrell in Tasmania in 1995. Previous discoveries of carp in Tasmania in the 1970s and 1980s had been eradicated using chemical control (rotenone) but poisoning was not pursued in these lakes due to their large size (combined area of 7615 ha) and the consequent cost of a rotenone operation (US$4.8 million in 1998), their value as sports fisheries, and the presence of an endemic galaxiid. An integrated control and eradication programme was quickly established based on physical removal by the Tasmanian Inland Fisheries Service. Objectives of the carp control programme were firstly to contain carp within the Lake Crescent/Sorrell system, followed by reduction and eventual eradication of carp. A barrier was constructed to prevent fish and larvae from invading downstream of the lakes, and physical removal has involved netting and electrofishing, particularly targeting spawning aggregations by following groups of radio-tagged male fish to locate spawning sites. Management of water levels in the lakes has been targeted at reducing opportunities for carp spawning (IFS 2004). Recently, exclusion of fish from suitable spawning habitat by the deployment of some 8 km of heavy-duty barrier netting has further substantially restricted opportunities for carp spawning in Lake Sorrell. This programme has been successful in substantially reducing the carp population in both lakes with eventual eradication likely. More than 10,000 carp have been removed since 1995. Current estimates of the populations of both lakes are around 100 individuals in Lake Sorrell and 10 in Lake Crescent.

13.9 Conclusions

The deliberate or accidental release of non-indigenous fishes, particularly freshwater species, has resulted in significant economic and ecological costs worldwide, yet our understanding of invasive potential and the development of effective response tools, especially for large water bodies, lags behind comparable knowledge for other organisms. However, significant advances have been made in recent years. The invasive fish risk identification and assessment protocol of Copp *et al.* (2005b), the quantitative invasion model of Gertzen *et al.* (2008), and the Carpsim software (Department of Primary Industries 2008) to simulate the expected outcomes of a range of control options are excellent examples of new tools applicable to the integrated pest management of fishes. The control of Nile perch impacts in one of the world's largest freshwater lakes and the potential for eradication of common carp from large lakes in Tasmania illustrate that control is feasible given the possibility to derive economic benefits from a non-indigenous fishery or the political and economic will to tackle the problem, respectively. The relatively poor dispersal abilities of freshwater fishes compared to other types of organisms should significantly assist in controlling the spread and enhance the potential for the eventual eradication of invasive fish, particularly if exciting new control options such as the genetic daughterless technology can eventually be applied. Unfortunately the increasing trade in live fish and expanding finfish aquaculture, both of freshwater and marine species, and for ornament and food, and the continued deliberate human-assisted dispersal of non-indigenous fishes for sport and commercial benefit will continue to homogenize finfish biodiversity, and greatly increase the ecological and economic damage attributable to this group of organisms. The global problem of invasive fishes will only get worse while many countries lack effective legislation to control the introduction and spread of non-indigenous species.

14
Marine biosecurity: management options and response tools

Richard F. Piola, Chris M. Denny, Barrie M. Forrest,

and Michael D. Taylor

14.1 Introduction

Marine alien invasive species (AIS) are now considered a major threat to the diversity and health of coastal regions worldwide (Carlton and Geller 1993; Vitousek et al. 1997; Cohen and Carlton 1998; Mack et al. 2000), overshadowing even the threat from the excess of other human activities whose impacts have traditionally received considerably greater attention (e.g. waste discharge, habitat reclamation). Human-mediated transport vectors such as shipping, aquaculture, and fishing act as a continuous source for inoculation of AIS into new regions, with the rate of species movements between different regions at unprecedented levels (Mack et al. 2000). Changing environmental conditions have also allowed for the successful invasion of new regions by species that had previously failed to establish (Dukes and Mooney 1999; Harris and Tyrrell 2001; Diederich et al. 2005; Grosholz 2005; Nehls et al. 2006). Although positive commercial and even ecological benefits of some AIS are recognized (e.g. Galil 2000; Sinner et al. 2000; Hayes and Sliwa 2003; Wonhom et al. 2005), the primary focus of scientists, regulatory agencies, and other stakeholders is on invasive species as a threat to ecological and socio-economic values (e.g. Hewitt et al. 2004). In the US and Canada alone, the projected economic impact from a few of the more notorious marine invasive species has been estimated to be in the order of approximately $US 2 billion per year (Pimentel et al. 2000; Colautti et al. 2006).

The biological invasion process can be broken down into a number of stages:

1) Entrainment of an organism by a human transport vector (e.g. maritime vessel) and the transport of the organism beyond its natural range (via vector movement).
2) The establishment of a viable population within a new location/region.
3) The spread of the organism away from its initial area of introduction (Fig. 14.1a).

The spread of the AIS and its interaction in the new environment may lead to adverse impacts on environmental, economic, social, and other values (Fig. 14.1a, Stage 4). Approaches for managing this invasion process fall into two broad categories: 'pre-border' management, which aims to *prevent* the arrival of new species during the transports stages (Stage 1); and 'post-border' management aimed at the *eradication and control* of the new species (Stages 2 and 3; Fig. 14.1b).

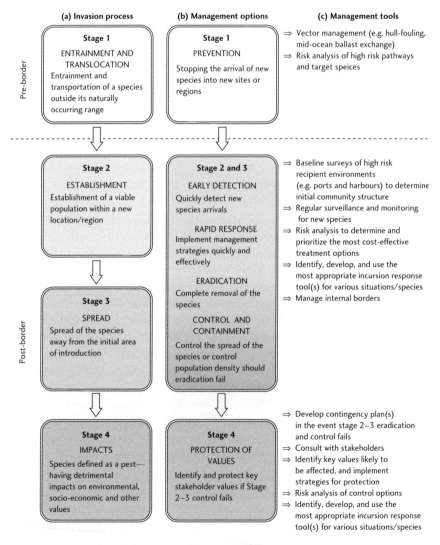

Fig. 14.1 Conceptual diagram outlining (a) the stages contributing to a successful biological invasion; (b) the pre- and post-border approaches for managing the invasion process; and (c) the management tools available.

As in terrestrial and freshwater systems, preventing the arrival of new species is likely to be the most successful and cost-effective management strategy for controlling marine AIS (Fig. 14.1b) (Simberloff 1996; Hewitt et al. 2004). This involves the effective management of common transport vectors, and the use of preventative risk analysis techniques to identify and target pathways and species that pose the greatest risk. Unfortunately, the highly connective nature of the marine environment, the ease of dispersal of reproductive propagules, and the presence of unmanaged (or difficult to manage) human-mediated pathways often makes prevention of AIS arrivals technically and financially prohibitive (Stoner 1992; Wittenberg and Cock 2001). As a result, practical and cost-effective 'post-border' incursion response procedures and tools are also vital for managing the establishment, spread, and impacts of marine pest species (Thresher and Kuris 2004; Wotton and Hewitt 2004).

Despite general acknowledgement of the threats posed to environmental, economic, social, and cultural values by marine pests, there has nonetheless been widespread uncertainty among scientists, government agencies, and marine user groups regarding how to deal with these threats. Without a structured approach to setting priorities, management efforts are largely ad hoc, and tend to lack focus and 'buy in' from affected stakeholders. This situation highlights the need for a comprehensive risk-based approach to identification and assessment of marine biosecurity risks, and establishment of management priorities that will enable limited budgets to be used most efficiently and effectively. An understanding of the feasibility, pros and cons of vector management, and treatment methods is an important component of this mix.

In this chapter we outline the common steps for managing marine species at all stages of the introduction process and detail the range of management options and tools available (e.g. vector management, risk assessment frameworks, surveillance and monitoring, control, and eradication tools). Particular emphasis is placed on the utility and development of specific response tools for use across a wide variety of commonly occurring marine habitats.

14.2 Pre-border management

14.2.1 Human-mediated invasion pathways

It has long been recognized that human activities in the marine environment have been a major pathway for the inadvertent spread of marine organisms well beyond their natural dispersal ranges (e.g. Chilton 1910; Elton 1958; Skerman 1960; Carlton 1985). There are only a few examples documenting natural movements of marine organisms across oceanic barriers, for example, those with long-lived planktonic larvae (Scheltema 1971) or rafting ability (Winston et al. 1996; Waters and Roy 2004). There are a large number of active pathways contributing to the global transport of marine AIS, including maritime shipping, aquaculture, fishing, the aquarium trade, and research and educational activities (Ruiz et al.

2000; Godwin 2003; Hewitt *et al.* 2004). The nature and magnitude of many of these pathways are primarily shaped by commerce and trade patterns across global, regional, and local scales. Depletion of resources, exploitation of new reserves, and advances in technology may all result in changes to patterns of trade. The establishment of new source regions of goods and commerce has the potential to greatly increase and change invasion regimes within recipient locations (Taylor *et al.* 1999; Meyerson and Mooney 2007). Since the majority of world trade relies on ship transport (Minchin and Gollasch 2003), the primary pathways for marine AIS introduction are those associated with international and domestic shipping, via ballast water discharge (Carlton and Geller 1993; Ruiz *et al.* 2000), and fouling on vessel hulls (Gollasch 2002; Godwin 2003) and other submerged areas such as sea-chests (Coutts and Dodgshun 2007).

Ballast water is used in commercial vessels to stabilize the vessel en route, and is uploaded or discharged depending on the amount of cargo onboard (Fig. 14.2). Ballast water can contain an assortment of organisms at various stages of development, including plankton, crustaceans, fish, larvae, eggs, or cysts (Williams *et al.* 1988; Carlton and Geller 1993). The uptake and release of ballast water (and its associated biota) has the potential to transport and introduce entire assemblages of marine organisms from one bioregion to another (Carlton and Geller 1993). Ballast water may also include sediments that accumulate in the bottom of the tanks, providing a suitable habitat for the survival of species with resistant life stages or resting cysts (e.g. dinoflagellates) as well as adult stages of benthic organisms. Transfer via ballast water has been implicated in the introduction and

Fig. 14.2 Ballast water overflowing from inspection ports on a tanker vessel undertaking mid-ocean ballast water exchange. Photo: T. Dodgshun, Cawthron Institute.

further spread of many invasive marine species including the northern Pacific seastar *Asterias amurensis* and the European green crab *Carcinus maenas* (Cohen *et al.* 1995; Taylor *et al.* 1999; Ross *et al.* 2003).

Fouling on maritime vessels has long facilitated dispersal of marine biota (Fig. 14.3). Over two thousand years ago, fouling on ocean-going Polynesian vessels radiating throughout the Pacific were probably responsible for transferring many species beyond their native ranges (Carlton and Hodder 1995). From the 1300s, European vessels began to transport hull-fouling and boring organisms from the north-eastern Atlantic throughout the world (see Carlton and Hodder 1995 and references therein). In fact, modern observational and experimental studies examining the dispersal of fouling taxa on a replica 16th century wooden sailing vessel, concluded that such vessels may have significantly altered the distribution of marine and estuarine organisms globally (Carlton and Hodder 1995). The development and widespread use of tributyltin (TBT) antifouling paints in the 1960–1970s resulted in a dramatic decrease in fouling on treated vessels (Evans *et al.* 2000; Lewis 2001), but the total prohibition of TBT in 2003 (IMO 2001), and its replacement with less effective biocides such as copper (Floerl *et al.* 2004; Piola and Johnston 2006a,b; Dafforn *et al.* 2008) may result in high numbers of fouling invaders (Minchin and Gollasch 2003).

Aquaculture is another important vector for the transport of alien marine species, both historically and today (Minchin 2007). Since the 1800s some organisms, such as oysters, have been transported between distant bioregions for commercial

Fig. 14.3 Scientists inspect the hull of a heavily fouled yacht washed ashore in the Port of Nelson, New Zealand. Hull-fouling assemblages present included several known invasive pest species. Photo: *The Nelson Mail*.

purposes (Carlton and Mann 1996; Minchin 2007). Short distance translocations are also commonplace. For example, industries such as mollusc aquaculture collect spat in different areas to where adults are ultimately grown and harvested (Minchin 2007). With the continued development of faster and more reliable transport, the distribution of cultivated species is becoming more globally widespread (Minchin 2007). Non-target species associated with cultured organisms may also be unintentionally introduced via aquaculture, e.g. epibiota on mollusc shells (Critchley and Dijkema 1984) or fouling taxa attached to gear and equipment (Forrest and Blakemore 2006). Additionally, a multitude of transport modes (e.g. air, water, road, rail) with ever decreasing journey times allows the rapid deliberate and unintentional dissemination of live aquatic products (and their associated pest parasites and diseases). Hence pinpointing transfer pathways can be extremely difficult owing to the complex array of transport networks (Minchin 2007).

The aquarium and ornamental marine species trade is emerging as an important source for AIS introductions globally (Padilla and Williams 2004). Thousands of species and millions of individuals of aquarium fish alone are transported worldwide annually (McDowall 2004). The popularity of e-commerce and Internet-based trading of aquarium species (e.g. the invasive marine alga *Caulerpa taxifolia*, see Walters *et al.* 2006) will undoubtedly see this trend continuing. Intentional and inadvertent releases of aquarium species have long been recognized as a leading source of AIS in freshwater systems (Courtenay and Robins 1973; Courtenay and Stauffer 1990) and, more recently, as a vector for the introduction of marine algae and fish species into natural systems (Whitfield *et al.* 2002; Semmens *et al.* 2004; Walters *et al.* 2006). The introduction of invasive species via the aquarium trade is difficult to manage, however, as the aquarium industry remains largely unregulated, and has traditionally received little attention from ecologists, conservationists, and policy makers (Padilla and Williams 2004).

14.2.2 Management of human-mediated pathways

Given the technical and financial constraints associated with controlling marine pests after they have established in a new location, it is clearly preferable to prevent the initial introduction as a first line of defence (e.g. Bax *et al.* 2001; Meyerson and Reaser 2002; Simberloff 2003b; Branch and Steffani 2004; Hewitt *et al.* 2004). The key method for the prevention of marine AIS introductions involves the analysis, identification, and management of known transport vectors (Fig. 14.1c). There has been considerable effort globally to identify risks associated with international vessel traffic (Carlton 1985; Coutts *et al.* 2003; Coutts and Taylor 2004; Verling *et al.* 2005), and to develop target lists of high-risk species (or 'next pests'), their likely distribution ranges, and potential high risk entry locations. These approaches may all be used as a means of prioritizing and focusing pre-border management efforts (Gollasch and Leppäkoski 1999; Hayes and Sliwa 2003).

Currently, pre-border management of maritime pathways mainly focuses on exotic species transported via ships' ballast water. Qualitative and quantitative risk assessment techniques have been developed to analyse and manage the various elements of the ballast water cycle in order to aid pre-border interception of AIS through the identification of high/low risk vessels (Hayes and Hewitt 1998, 2000) and high/low risk transport routes (e.g. GloBallast risk assessments, see Clarke et al. 2004). Such techniques can be highly flexible, operating at several levels of complexity, depending upon the availability of data (Hayes and Hewitt 1998, 2000).

At a more applied level, International Maritime Organization (IMO) guidelines currently require vessels to conduct mid-ocean ballast water exchange (BWE), and to not discharge any unexchanged water unless exempted on the grounds of safety (IMO 1997). The premise of this approach is that BWE will substantially reduce the risk of new introductions by displacing pest species during exchanges, or uploading oceanic species that are unlikely to survive in the recipient coastal zone. In general, however, the effectiveness of BWE has been shown to be highly variable and organisms from the port of origin invariably remain in the ballast tanks (Taylor et al. 2007). Adopted by the IMO in 2004, the International Convention for the Control and Management of Ships' Ballast Water and Sediments, introduces new standards for the management of ships' ballast water, and progress is now being made on the development of various ballast water treatment technologies (Herwig et al. 2006; Tang et al. 2006; Gregg and Hallegraeff 2007).

Although pre-border prevention of species transfer via hull fouling is technically feasible, widespread implementation and enforcement of viable management tools (e.g. hull cleaning) is largely impractical. As a result, hull fouling management usually focuses on the use of risk analysis for identifying specific high-risk vectors or routes (e.g. specific countries or regions) and target pest species. Such analyses involve developing a target list of potentially invasive species, based upon pre-defined selection criteria (e.g. Hewitt and Hayes 2002; Hayes and Sliwa 2003). Such criteria may be as simple as selecting species based upon their invasiveness or impacts elsewhere, but may also involve examining species' attributes in an attempt to characterize their invasibility and potential distribution ranges (Forrest et al. 2006). It is also useful to characterize the attributes of receiving environments to determine habitats at greatest risk from invasion, which can help determine values at risk, and hence priorities for management.

14.3 Post-border management

Given the complexity of marine ecosystems, the lack of effective or affordable prevention tools, and the large number of managed and unmanaged vectors able to facilitate species transport, it is unrealistic to expect a 100% effective pre-border management strategy (Wotton and Hewitt 2004). This raises the question as to whether post-border management, which has a track record of various successes in freshwater and terrestrial systems (e.g. Genovesi 2005; Allen and Lee 2006), might also be feasible in the marine environment.

A recent synthesis of biological invasions in New Zealand by Allen and Lee (2006) provides a number of examples where the efficacy of post-border management of pests in terrestrial and freshwater systems has been demonstrated. These include: successes in the restoration and recovery of native vegetation through control of introduced herbivores such as rabbits, goats, and deer (Coomes *et al.* 2006); control programmes for introduced predators (e.g. stoats, rats) of native birds or their eggs (McLennan 2006); spraying programmes for invasive aquatic and terrestrial weeds (e.g. Swales *et al.* 2005); the release of biological control agents (e.g. for insects) and commercial harvest for introduced mammals (e.g. goats, deer, and pigs, Parkes 2006a). These examples and many others highlight a wide range of control strategies. In contrast to freshwater and terrestrial systems, the marine environment is highly inter-connected and expansive, relatively inaccessible, and is often hostile to work in. Intuitively, it is apparent that many of the methods developed for freshwater and terrestrial systems are unlikely to be directly transferable to the marine environment. And in fact, there is a longstanding view that management of marine pest incursions post-border will generally be futile (e.g. Sanderson 1990; Brown and Lamare 1994; Thresher and Kuris 2004). There is, however, emerging evidence that management may be feasible under certain circumstances.

14.3.1 Early detection and rapid response

Surveillance and monitoring programmes for the early detection of new incursions form the first stage of an effective post-border management strategy. Once a threat has been detected, the next stage involves determining the most effective response actions, which include complete eradication, containment measures to slow or prevent the spread of the organism, or population control to reduce pest densities to levels that minimize adverse effects (Fig. 14.1). The level of response will depend on a combination of factors, including:

- Potential impacts of the organism on the environment, economy, and people.
- The feasibility and risks of response options.
- The ability to target the invasive species.
- Public concern or interest (Wotton and Hewitt 2004).

Baseline biological surveillance and monitoring is crucial for effective rapid response to new incursion events in the marine environment. Surveillance programmes aim to collect baseline information on the identity and numbers of species within areas deemed to be of high risk of incursion (e.g. ports and marinas) in order to identify new arrivals (Hewitt and Martin 2001). Large-scale baseline monitoring programmes of this nature, such as the multinational GloBallast programme, are being increasingly promoted in a range of countries such as South Africa, India, Brazil, China, Ukraine, and Iran (Global Ballast Water Management Programme 2004). In Australia, baseline biological surveys have been completed

Marine biosecurity: management options and response tools | 213

for over half the nation's ports (Convention on Biological Diversity 2005). The increased effort placed on pest surveillance and delimitation surveys has resulted in the development of some sophisticated approaches (Hayes *et al.* 2005; Gust and Inglis 2006), but these are still based on sampling and detection at defined levels of confidence and can by no means guarantee finding all individuals. The fact is that many first incursions are found simply by accident or enquiry, rather than by active surveillance (e.g. Hay and Luckens 1987; Coutts and Forrest 2007).

Most reports of successful eradications of marine invaders acknowledge that a major factor in their success is due to the populations in question being only recently established or spatially restricted (Culver and Kuris 2000; Bax *et al.* 2002; Wotton *et al.* 2004; Anderson 2005). Once an introduced marine organism becomes geographically dispersed, management options become increasingly limited, and invariably require expensive, long-term commitment (Sinner *et al.* 2000). For this reason, surveillance efforts must be coupled with incursion response systems that can be rapidly deployed upon first detection of new species (Wotton and Hewitt 2004). Swift response to the early detection of the black mussel *Mytilopsis adamsi* in Darwin, Australia resulted in the total eradication of this species in the region (Bax *et al.* 2002). Similarly, when the invasive alga *C. taxifolia* was first observed in California in 2000, a rapid management response lead to its complete eradication within 2 years (Anderson 2005). If newly established populations are allowed time to reproduce the chances of eradication are greatly reduced. Continued inaction following the initial introduction of *C. taxifolia* into the Mediterranean Sea in 1984 resulted in it becoming a major environmental and economic problem in the region (Meinesz 1999).

Clarity regarding the overall desired management outcomes is critical prior to the implementation of any post-border management, since this will influence the scope, time-frame, and cost of the programme. For example, if the desired outcome is eradication, then effective ongoing pest surveillance and vector management are critical to success, while intensive management activities may only be a short-term requirement (Coutts and Forrest 2007). Alternatively, a programme aiming to control pest populations (e.g. to manage densities to a level that avoids adverse effects) is likely to require a long-term commitment including ongoing funding, with issues such as pest detection and management of re-invasion less of a priority (Forrest 2007). In some circumstances, immediate containment or other interim management actions may be necessary prior to the determination of final outcomes (Wotton and Hewitt 2004).

Following the identification of a new incursion, it is crucial that a contingency plan(s) be implemented in order to determine the most appropriate action(s) for a rapid response. Ideally, such contingency planning would occur prior to an invasion taking place (at least in the case of high-risk species). Such plans should define particulars such as which stakeholders require notification, funding, and appropriate management options. In order for such management plans to be effective, however, it is vital that decision makers be aware of all the tools available at their disposal. In Australia and New Zealand for example, management options for the control of undesirable marine organisms have been well described (McEnnulty *et al.* 2001; Stuart 2002), and the National Introduced Marine Pest Information System (NIMPIS) database provides a rapid response toolbox detailing control and eradication attempts for selected marine species (NIMPIS 2002a). The range of tools available for marine systems are described in section 14.3.2.

14.3.2 Response tools

Nearshore marine environments at risk from AIS encompass a diverse range of habitats and ecosystems, and therefore require a varied range of response tools. Given that the majority of vectors for AIS transport are anthropogenic in nature (shipping, aquaculture, fishing, etc.), it is not surprising that most introduced species establish in environments subject to high levels of human development and disturbance (e.g. ports and harbours) rather than in less impacted areas such as open coast (Wasson *et al.* 2001). Within these developed environments AIS are typically more prevalent on artificial surfaces and structures rather than natural substrata (Glasby *et al.* 2007). It is not surprising, therefore, that most existing incursion response tools focus on the control of marine AIS within modified habitats, and in particular on artificial surfaces and structures. Despite this, relatively pristine habitats are also susceptible to invasion (Wyatt *et al.* 2005), which stresses the need for incursion response tools for the control of AIS in such areas. In the following sections we discuss the diversity of tools available for managing new species incursions within artificial and natural habitats, including discussion of management tools for human vectors (e.g. recreational vessels) that have a significant role in the spread of alien species post-border. In many cases, a number of different response tools are employed simultaneously in an attempt to maximize the chance of success in eradicating or controlling AIS. Table 14.1 provides a summary of presently available treatment methods, detailing their appropriate application, stakeholder and community acceptability, chance of success, legal considerations, benefits, and limitations.

14.3.2.1 Physical removal

The most effective method of treating vessels infected with introduced fouling species is to remove them from the water (e.g. dry-docking or slipping) and to scrape the hull clean of all fouling biota. Regular application of antifouling paints can then be used to minimize the recurrence of fouling assemblages (Floerl *et al.* 2005a). Removing a vessel from the water can however be time consuming and expensive, often leading to delays in the treatment of infected vessels. One common alternative to land-based treatment is in-water defouling, where divers remove fouling biota *in situ,* using mechanical brush systems or scrapers. Unfortunately, this technique can actually enhance the recruitment of some taxa onto the recently cleaned surface if all traces of existing biota are not removed (Floerl *et al.* 2005b). Additionally, viable organisms and/or fragments of defouled material released into the surrounding environment may survive and establish, increasing the risk of an introduction occurring (Hopkins and Forrest 2008). As a result, in-water cleaning has been restricted or banned in some countries. An improved mechanical brush system that simultaneously removes and collects biofouling from vessel hulls is currently in development, with trials indicating its effectiveness at removing and collecting up to 90% of biofouling from treated vessels (Hopkins and Forrest 2008). A prototype underwater vacuum device and cutting system was also developed and trialled for the removal of the invasive ascidian *Didemnum vexillum* from vessel hulls, however this proved ineffective except as a means of biomass reduction (Coutts 2002).

High pressure (>2000psi) spraying is another available technique for the physical removal of unwanted fouling species. This was found to be effective at dislodging microscopic gametophytes of the Asian kelp *Undaria pinnatifida* from marine farming equipment and associated mussel shells (Forrest and Blakemore 2006). In contrast, Canadian aquaculture farmers had less success in trialling water blasting to remove the invasive sea squirts *Ciona intestinalis* and *S. clava* from mussel lines, with damage incurred by both the mussels stocks and farming equipment (Heasman pers. comm.). Although the use of high-pressure water jets is generally considered an acceptable (environmentally friendly) and successful management option under the right circumstances, it can be expensive and time consuming to implement, and appropriate procedures are needed to prevent the re-release of AIS back into the marine environment.

Physical removal may be a cost-effective eradication tool within natural habitats, particularly in small discrete areas where a species distribution is limited. Over larger areas however, these methods become expensive and time-consuming, and may need to be repeated to ensure complete removal. Consideration must also be given to any potential effects that mechanical and physical removal methods may have on the habitats in question and associated flora and fauna. Manual and mechanical removal of numerous algal pest species has been attempted with varying degrees of success. For example, small outbreaks of *C. taxifolia* (up to 200m^2) have been eradicated by divers manually removing the plants (Cottalorda *et al.* 1996; Meinesz 1999; Meinesz *et al.* 2001; Creese *et al.* 2004). Diver-operated suction devices have also been trialled for the control *C. taxifolia* in Australia (Creese *et al.* 2004), Croatia (Zuljevic and Antolic 1999a,b) and the Spanish Mediterranean (Meinesz *et al.* 2001). Removal by hand was successful in eradicating the seaweed *Ascophyllum nodosum* from San Francisco Bay, primarily due to its early detection and the relatively small area of infected shoreline (Miller *et al.* 2004). In contrast, monthly removal of *U. pinnatifida* by divers across an 800m^2 area in Tasmania (Australia) was ultimately unsuccessful in eradicating the alga due to the persistence of 'hot spots' of growth (Hewitt *et al.* 2005). Mechanical harvesting proved a viable control measure for the alga *Sargassum muticum* in England, however this method was costly, time consuming, labour intensive, and caused considerable physical and ecological damage to the shoreline (Critchley *et al.* 1986). Containment and disposal of collected materials also proved problematic. The early detection of

Table 14.1 Summary of treatment methods for the control and eradication of marine pest species on artificial and natural substrates and habitats, detailing their acceptability, chance of success, legal considerations, application issues, benefits, and limitations.

Type	Treatment	Artificial habitats	Natural habitats	Structures and/or habitats	Acceptability and chance of success
PHYSICAL REMOVAL	Physical removal (i.e. hand picking; scraping)	Yes	Yes	Wharf piles Vessels Seaweed beds Seabed	*ACCEPTABILITY*: high *SUCCESS*: low
	In-water hull cleaning (regular scraping and brushing)	Yes	N/A	Vessels	*ACCEPTABILITY*: low *SUCCESS*: moderate
	In-water hull cleaning (rotating brushes that collect fouling)	Yes	N/A	Vessels	*ACCEPTABILITY*: moderate *SUCCESS*: moderate
	High pressure spraying	Yes	Yes	Buoys Vessels Intertidal areas	*ACCEPTABILITY*: high *SUCCESS*: high
	Suction devices	Yes	Yes	Vessels Seabed Seaweed beds	*ACCEPTABILITY*: moderate *SUCCESS*: low

Table 14.1 (Con't.)

Application issues	Benefits	Limitations
Requires good underwater visibility May require repeated treatments Limited to a small area Labour intensive	Selective (low collateral impact) Does not require complex equipment	Not all targeted species may be collected Diver safety issues All target individuals must be collected Unsuited to cryptic species
Difficult to treat niche areas	Quick (i.e. 30m vessel in 4 hours) Can be done *in situ*	Discharge of all fouling material directly into the environment Removes adults and/or stimulates the release of propagules or fragments into the water column Diver safety issues
Specialized brushes, pumps, and collection bags required Not all fouling may be removed/collected Brushes may not be able to access 'nook and crannies' on a hull	Quick (i.e. 30m vessel in 4 hours) Can be done *in situ* Fouling material collected on the surface (90%)	Discharge of fine particulate to the environment May remove adults or stimulate the release of propagules or fragments into the water column Diver safety issues
Simple to apply Applied above-water Effectiveness depends on water pressure	Can be applied quickly and cheaply (i.e. standard off-the-shelf technology)	May fragment/redistribute species May be expensive Time consuming
Specialized equipment required Labour intensive Only practical over small areas Not all fouling may be removed/collected	Can target specific areas	May fragment/redistribute species Can miss species Diver safety issues

Table 14.1 (Con't.)

Type	Treatment	Artificial habitats	Natural habitats	Structures and/or habitats	Acceptability and chance of success
WRAPPING and SMOTHERING	Wrapping and Encapsulation	Yes	Yes	Wharf piles Jetties Pontoons Vessels Buoys Seabed (smothering)	*ACCEPTABILITY*: high *SUCCESS*: high
	Wrapping and Encapsulation with chemicals (e.g. acetic acid/chlorine)	Yes	Yes	Jetties Pontoons Vessels Wharf piles Seabed (smothering)	*ACCEPTABILITY*: moderate *SUCCESS*: high
	Smothering with plastic and/or geotextile sheeting	Yes	Yes	Seaweed beds Rip-rap Seabed	*ACCEPTABILITY*: high *SUCCESS*: low
	Smothering with dredge spoil	No	Yes	Seaweed beds Rip-rap Seabed	*ACCEPTABILITY*: moderate *SUCCESS*: moderate
CHEMICAL	Chemical treatments	Yes	Yes	Wharf piles Jetties Pontoons Vessels Buoys Seabed	*ACCEPTABILITY*: moderate–low *SUCCESS*: moderate

Application issues	Benefits	Limitations
Plastic dispenser required Relatively simple to deploy and can be left in place Slow acting (i.e. days/weeks) May inconvenience port operations	100% effective if applied correctly Cost-effective Structures /habitats can be treated *in situ* Can remain on for long periods and may act as a secondary treatment	Unselective May emit offensive odours Disposal issues (plastic and collected biota) Diver safety issues
Fast acting Safety gear required Requires attention to ensure effective concentration	Can quickly treat structures that are in heavy use (i.e. vessels, pontoons) Minimize any larval release	Potentially hazardous Collateral damage Can be expensive Disposal issues Potentially corrosive
Labour intensive Specialized equipment may be required Only practical over small areas Requires good underwater visibility	Environmentally friendly	Collateral damage Diver safety issues
Specialized equipment required Only practical over small areas	Cost-effective	Unselective May result in secondary impacts if dredge spoil is contaminated (e.g. heavy metals) May provide suitable substrate for further invasions
Fast acting Safety gear required Requires attention to ensure effective concentration	Can quickly treat structures that are in heavy use (i.e. vessels, pontoons) Minimize any larval release	Potentially hazardous Unselective Can be expensive Disposal issues Potentially corrosive

Table 14.1 (Con't.)

Type	Treatment	Artificial habitats	Natural habitats	Structures and/ or habitats	Acceptability and chance of success
PHYSICAL TREATMENT	Heat treatment	Yes	Unlikely	Wharf piles Seafloor Vessels	ACCEPTABILITY: high SUCCESS: moderate
	Desiccation	Yes	N/A	Vessels Aquaculture equipment Scientific equipment Buoys, ropes, chains, tyres	ACCEPTABILITY: high SUCCESS: high
	Freshwater	Yes	N/A	Vessels Aquaculture equipment Scientific equipment Buoys, ropes, chains, tyres Aquaculture seed-stock	ACCEPTABILITY: high SUCCESS: moderate

populations and removal of individuals before they reach reproductive maturity is crucial to the success of algal removal by physical means. Failure to achieve this was one of the major factors resulting in the failure to control *U. pinnatifida* in southern New Zealand (B. Forrest, pers. comm.). Further, given the ability of many algal species to regenerate from small fragments, physical removal may actually enhance a populations spread (Critchley *et al.* 1986; Glasby *et al.* 2005).

Trawling and dredging techniques, which involve pulling large equipment behind a vessel to collect pest organisms in and on the surface of the sediments, have been trialled as a means of controlling benthic pest species but are generally of limited success (see McEnnulty *et al.* 2001 and references therein). Trawling has been suggested as a possible control method for the Asian date mussel *Musculista senhousia,* which forms large colonies in intertidal mud flats in estuaries and sheltered bays (McEnnulty *et al.* 2001). However, since *M. senhousia* is also found as a fouling organism on pylons and other artificial structures (Willan 1987), dredging

Application issues	Benefits	Limitations
Specialized equipment required Not effective on non-uniform surfaces Labour intensive Only practical over small areas	Environmentally friendly	Unselective Difficulties in achieving sufficiently high water temperatures Only suitable for early life stages Diver safety issues
Applied above-water Specialized gear may be required to remove vessels (i.e. dry dock) May inconvenience port operations	Cost-effective Environmentally friendly	Removal of some structures (i.e. vessels) may be expensive Some organisms can survive for extended periods out of water
Logistical issues involved if large amounts of freshwater are required	Cost-effective Environmentally friendly	Some species (i.e. mussels) can survive for extended periods in freshwater

can comprise only one component of a successful eradication effort. In Japan, scallop dredges are used to periodically remove *A. amurensis* from areas of the sea floor in scallop harvesting plots (Ito 1991). While seastars reinvade the cleared areas, a significant number of scallops can be harvested before reinvasion becomes a problem (McLoughlin and Bax 1993). Aside from the obvious physical damage caused by dredging, environmental impacts of this type of control would be high in areas where resuspended sediments are highly polluted (McEnnulty *et al.* 2001). Dredging and trawling have also been demonstrated to change the characteristics of some soft sediment habitats in a way that inhibits the further settlement and attachment of many non-target sessile invertebrates (Stead 1971a,b).

14.3.2.2 Wrapping and smothering

A variety of materials including plastic sheeting, rubber, jute matting, and dredge spoil have been used on a range of artificial and natural surfaces to control a range of benthic pest species such as algae, ascidians, and seastars. When applied correctly, such approaches prevent light availability for photosynthesis (in the case of plants) and impede water flow, resulting in anoxic conditions and the eventual mortality of encapsulated biota (McEnnulty *et al*. 2001; Coutts and Forrest 2007).

To date, one of the most successful and cost-effective methods available to treat artificial structures *in situ* is to encapsulate them with a physical barrier such as impermeable plastic (polyethylene). Wrapping wharf piles in impermeable plastic has been widely and successfully implemented during eradication attempts of *D. vexillum* (Pannell and Coutts 2007) (Fig. 14.4) and trialled for the snowflake coral *Carijoa riisei* (Montgomery 2007). In both cases, this technique successfully eliminated all encapsulated biota, except in a few instances where the wrapping became damaged or failed to completely prevent water exchange (such as on complex wharf structures). This technique has also proven successful for clearing *S. clava* and *D. vexillum* from floating pontoon structures (Coutts and Forrest 2005, 2007). In areas where wharves and pontoons are in high demand and require rapid treatment, chemicals such as acetic acid and bleach (chlorine) can be added to the encapsulated water within the wrapping to accelerate mortality. For example, the addition of 4% acetic acid within wrapped pontoons resulted in 100% mortality to the ascidian *S. clava* in 10 minutes (Coutts and Forrest 2005). Similarly, wharf piles infected with the Asian kelp *U. pinnatifida* have been successfully sterilized using bromine compounds applied inside PVC sleeves (Stuart 2002).

Fig. 14.4 A marina pontoon wrapped in geotextile fabric during efforts to control an infestation of the pest ascidian *Didemnum vexillum* in Tarakohe Harbour, New Zealand. Wrapping restricts water exchange to fouling communities growing on the pontoon, resulting in the development of anoxic conditions and eventual mortality. Photo: A. Coutts, Cawthron Institute.

Wrapping has also been used as a method to treat vessel hulls infected with fouling pest species. Vessels infected with *D. vexillum* were successfully treated *in situ* using a plastic encapsulation technique and the addition of chemicals (Coutts and Forrest 2007). Similarly, the black-striped mussel *Mytilopsis sallei* and Asian green mussel *Perna viridis* detected on the hulls of fishing vessels in Darwin Harbour were successfully eradicated by wrapping the vessel hulls in PVC sheaths and adding chlorine. At the time of writing, the largest reported vessel to be treated using the wrapping techniques was a 113m frigate in New Zealand (Denny 2007), although this attempt was unsuccessful due to strong currents in the area and difficulties in maintaining the integrity of the plastic wrap around certain vessel structures (e.g. propeller blades).

On natural substrates, plastic sheeting has again been used to smother the alga *C. taxifolia* (Zuljevic and Antolic 1999b; Meinesz *et al.* 2001; Creese *et al.* 2004) and *D. vexillum* (Coutts and Forrest 2007; Pannell and Coutts 2007). Jute matting, which is cheaper and more environmentally friendly than plastic, has also successfully been used for smothering *C. taxifolia* over small areas of seabed (Glasby *et al.* 2005); however, deployment proved more difficult over larger areas due to its positive buoyancy and the amount of weight required to anchor it in place. In the French Mediterranean, mats soaked in copper sulphate have been placed over beds of *C. taxifolia,* with the chemicals leaching from the mats resulting in increased mortality (Uchimura *et al.* 2000).

Dredge spoil and sediment have also been used in attempts to smother benthic pest species. Studies in the USA found that covering the starfish *Asterias forbesi* with a layer of mud or sand resulted in death as the individuals were unable to escape (Loosanoff 1961). In contrast however, the alga *Sargassum muticum* was found to be far more resistant to burial and decayed more slowly than similar macroalgal species, suggestive of the fact that burial initiated a self-protective response from the plant (Morrell and Farnham 1982). It must be noted that control programmes involving physical burial of invasive taxa with sediments must be carefully designed because of the potential to cause significant environmental damage and alter the habitat in a way that facilitates other introduced species (McEnnulty *et al.* 2001).

Crucial to the success of encapsulation and smothering techniques is the repeated monitoring of affected structures and the complete treatment and removal of every individual. Other factors important to success include: the size and topographic complexity of the infected area; the hydrodynamics of the location; and maintaining the smothering for a prolonged period (McEnnulty *et al.* 2001; Creese *et al.* 2004; Coutts and Forrest 2007). For algal and/or colonial species, it is also important to limit the amount of fragments generated during deployment as these may settle and establish nearby (Creese *et al.* 2004). Wrappings on artificial structures such as wharf piles and pontoons are typically able to be left in place for extended periods of time (>12 months), providing further protection from re-infection (i.e. during reproductive periods). Additionally, should the outside of wrappings become re-infected, their removal provide a secondary treatment option. There are several environmental and public safety issues that must be considered during the use of plastic encapsulation and smothering methods, although these are not insurmountable (Table 14.1).

14.3.2.3 Physical treatment

A variety of heat-based methods have been developed for the treatment of natural habitats and artificial structures. Eradication of the sabellid polychaete *Terebrasabella heterouncinata* at a Californian aquaculture facility involved immersing abalone shells in warm seawater to kill the polychaetes (Culver and Kuris 2000). Heated water was successfully used to eradicate gametophytes of the Asian kelp *U. pinnatifida* fouling the hull of a sunken vessel in the Chatham Islands, New Zealand (Wotton *et al.* 2004). This work comprised sterilizing sections of the vessel hull by attaching a plywood box lined with industrial electrical elements to the side of the hull and heating the encapsulated water to 70°C. In addition, a flame torch was used for inaccessible areas (e.g. near the seafloor) and for areas with heavy fouling. A similar *in situ* heat treatment method based on surface generated steam supply has been applied to marina pontoons and natural rocky reef habitats, albeit with limited success (Blakemore and Forrest 2007). Laboratory-based studies have also suggested heat treatment as a feasibility method for disinfecting ballast water (Mountfort *et al.* 1999).

Heat treatment, freshwater baths, and air-drying have all demonstrated potential for managing the transfer of marine pest species via human-mediated pathways such as aquaculture. Exposure of mussel seed-stock to hot water at 55°C for approximately 5 seconds was effective in achieving complete mortality of *U. pinnatifida* gametophytes, whilst having little effect on mussel survival (Forrest and Blakemore 2006). Similarly, fresh water immersion resulted in 100% mortality of *U. pinnatifida* on infected seed mussel ropes, without affecting mussel health (Forrest and Blakemore 2006). As mussels and oysters can survive for extended periods out of water, desiccation has been found to be a cost effective and environmentally friendly method to control fouling species. Mussel infrastructure (i.e. moorings, warps, floats, and backbones) can be removed from the water, desiccated, and later returned to the same location (Forrest and Blakemore 2006).

Heat treatment is generally difficult to implement in open marine conditions, unless fouled habitats can be isolated and have uniform surfaces (e.g. vessel hulls, wharf pylons). It is also unlikely to be effective in controlling organisms with thick coverings or shells (e.g. oysters; Nel *et al.* 1996). Heat treatment has the added disadvantages of being impractical over large areas and damaging to non-targeted species, although adverse impacts on the natural environment are likely to be short-term (see Table 14.1). Logistic constraints in the procurement and use of large volumes of freshwater in some marine environments (e.g. aquaculture farms, isolated locations) may also limit its efficacy as a pest control solution.

14.3.2.4 Chemicals

Chemicals have been trialled with varying success for the control and eradication of a range of AIS. In aquaculture, solutions of saturated salts (brine) and hydrated lime have been successfully used on mussel (Fig. 14.5) and oyster farms to prevent translocation of pest species such as the algae *Codium fragile* ssp. *tomentosoides*, *Sargassum muticum* and *Cladophora* spp., and ascidians *Molgula* spp., *S. clava*, and *C. intestinalis* (Shearer and MacKenzie 1961; Minchin 1996; MacNair and Smith 1998; Carver *et al.* 2003; Atlantic Canada Aquaculture Industry Research and Development Network 2006; Mineur *et al.* 2007). Similarly, a combination of chlorine baths and sun drying has been effectively used to remove the alga *U. pinnatifida* from infected mooring ropes and chains (Stuart and Chadderton 1997).

During the successful eradication of the black mussel *Mytilopsis adamsi* in Darwin (Australia), marinas within which the mussels were found were sealed off from surrounding waters and dosed with ~190t of liquid sodium hypochlorite and 7.5t of copper sulphate, killing the mussels in <18 days (Ferguson 2000; Bax *et al.* 2002). Infected vessels also had their internal water plumbing treated by adding copper sulphate solution and detergent to pipes with standing water. Acetic acid has also been proposed as a method to manage biofouling pests associated with shellfish aquaculture seed-stock (Forrest *et al.* 2007).

Fig. 14.5 A 500 kg bag of seed mussels being lowered into a chemical bath containing 0.5% sodium hypochlorite (bleach) solution during industry-scale trials evaluating methods to reduce the spread of pest species via aquaculture transfers. Photo: C. Denny, Cawthron Institute.

A range of chemicals and toxicants have also been trialled for the management of pest species within natural habitats. The successful eradication of *C. taxifolia* in Southern California lagoons was achieved by covering colonies of the alga (ranging in size from 1–500 m^2) with PVC tarpaulins and applying bleach (in liquid and solid tablet form; Anderson 2005). Applying copper ions directly to the thalli of *C. taxifolia* via *in situ* electrolysis has also proven effective in killing the alga (Gavach *et al.* 1996). Herbicides have been employed in the control of estuarine emergent plant species such as the cryptogenic reed *Phragmites australis* in Chesapeake Bay, USA (Ruiz *et al.* 1999) and the introduced rice grass *Spartina anglica* in Australia and New Zealand (Kriwoken and Hedge 2000). Herbicides have proven less effective in the control of algal species, failing to prevent the spread of the introduced alga *Sargassum muticum* in southern England (Critchley *et al.* 1986), and proving ineffective and labour-intensive to administer during *in situ* trials for the control of *U. pinnatifida* (McEnnulty *et al.* 2001). The broad spectrum insecticide carbaryl has been used (or considered) for the control of various crustacean pest species, such as the European green crab *Carcinus maenas* (Carr and Dumbauld 1999) and a thalassinid burrowing shrimp species in intertidal oyster beds in Washington State, USA (McEnnulty *et al.* 2001). Advantages of carbaryl are its tendency to be short-lived in the environment with no bio-accumulation (Dumbauld *et al.* 1997) and short-term effects on non-target populations (Brooks 1993). The alkaline properties of lime, and its ability to corrode calcium carbonate, have led to its use as a control agent of seastars, including *A. amurensis* (McEnnulty *et al.* 2001). Following the broadcast application of lime, seastars are exposed to the corrosive particles as they settle or crawl over it, dying within 2 weeks. The spray application of lime solution has also proven effective in controlling pest tunicate species such as *Styela clava* and *Ciona intestinalis* on mussel aquaculture crops in Prince Edward Island, Canada.

Using chemicals to modify the characteristics of a habitat has also been used for AIS control. Salt (NaCl) has been used successfully for the control of *C. taxifolia* at a variety of spatial scales. A 4ha area of *C. taxifolia* in a coastal lagoon in Australia was controlled by raising the salinity of the entire lagoon through the addition of 1000t of salt (Hilliard 1999). Similarly, salt dispensed from a barge was effective in treating *C. taxifolia* in shallow water soft sediment habitats in Australia (Glasby *et al.* 2005). This method is less successful in deeper waters (>6m depth) however since the salt disperses before reaching the substratum. Small-scale infestations of *C. taxifolia* (4 m^2) in Sydney Harbour, Australia, were eliminated by scuba divers spreading salt by hand (4 cm thick, ~50 kg/m^2) (Creese *et al.* 2004). This type of treatment is usually only applicable to small, relatively enclosed bodies of water such as lagoons/lakes and again is non-selective, often resulting in the death of non-target species.

In general, while chemicals have been used with some success in aquatic environments their utility is compromised by a range of factors. These include:

- A lack of selectivity resulting in mortality of non-target species.
- The large doses required for effective control.

- The often extended time periods required for the chemical to be in contact with the organism.
- Problems associated with the effective application (e.g. suitable delivery systems, dilution, containment within areas of interest).

Additionally, the use of toxic chemicals in aquatic environments is often cost-prohibitive, since they often require repeat treatments to be effective (McEnnulty et al. 2001). Since most chemicals used for controlling AIS are general biocides, non-target biota in treatment areas are also killed, though communities generally return to former levels within less than 1 year (Bax et al. 2002). Some toxicants, such as copper, may continue to persist in the environment long after the target invader has been eliminated (Gavach et al. 1999), affecting non-target species by direct toxicity or through bio-accumulation.

14.4 Discussion

Despite a number of attempts worldwide, very few marine pest species have been successfully eradicated (but see Culver and Kuris 2000; Bax et al. 2002; Kuris 2003; Miller et al. 2004; Wotton et al. 2004; Anderson 2005). However, it is possible in many cases to effectively manage new species introductions despite the unique and difficult challenges associated with marine pest incursions. Significantly, each attempt to eradicate or control marine pest incursions builds on our knowledge of marine biosecurity and adds to the development of management options and treatment tools. These include the identification of critical success factors that dictate the overall outcomes of such programmes, primarily:

- Early identification and detection of the invader.
- Expert knowledge about the biology and ecology of the invader.
- Sufficient resources to fund a programme to its conclusion.
- The existence (or ready development) of effective control procedures for the target pest organism.
- Monitoring and review during and after the incursion response.
- Implementation of protocols to prevent reinvasion (Myers et al. 2000; Bax et al. 2001; Wotton and Hewitt 2004; Wotton et al. 2004; Anderson 2005; Coutts and Forrest 2007).

With regards to eradication, it is critical to be able to detect and remove all target organisms, or at least reduce pest densities to levels that cannot sustain a viable population. Failure to achieve the latter has proven to be the major stumbling block for many attempted eradications in the marine environment (e.g. Coutts and Forrest 2007).

Effective management strategies for preventing new incursions must begin with pre-border strategies to assess the risks posed by different species, the likely vectors for their arrival, and attributes of 'at risk' recipient environments. However, the lack of completely effective pre-border management tools means that the continued incursion, establishment, and spread of marine invasive species is inevitable (Wotton and Hewitt 2004). Ongoing surveillance and monitoring is therefore vital to detect new species arrivals, and to allow the management response to commence immediately. If the opportunity to rapidly respond is missed and the invasive species becomes widely distributed, it will be difficult, if not in most cases impossible, to eradicate. In the few documented examples of successful marine pest eradications, the target pest was always detected at an early stage and its distribution was restricted to a localized area or habitat (Culver and Kuris 2000; Bax *et al.* 2002; Kuris 2003; Miller *et al.* 2004; Wotton *et al.* 2004; Anderson 2005).

In instances where exclusion and eradication of new invaders fails, and containment at the point of incursion is no longer an option, prompt and clear decisions must be made on the best management end point for a given situation. A key consideration is whether or not spread can be prevented and, if not, whether incursion response is desirable (Forrest *et al.* 2006). For example, in the case of aquaculture it may be more desirable and cost-effective to simply manage pest densities to a level that avoids adverse affects to stock and equipment, even when repeated incursions are inevitable and eradication is unfeasible. Additionally, it may be worthwhile characterizing the 'manageability potential' of the target pest to estimate the degree of management success that can be expected. For example, introduced organisms that are large and conspicuous, and have highly specific habitat and environmental requirements and short dispersal ranges, are inherently more manageable than small or cryptic organisms that are habitat generalists with long planktonic life-stages (Forrest *et al.* 2006). Using a similar approach, key attributes of the receiving environment can be assessed to determine the feasibility of surveillance and/or incursion response. For example, it is much easier to manage a sheltered, accessible environment with clear water and a relatively two-dimensional bathymetry, compared to an exposed remote environment with turbid water and a complex heterogeneous bathymetry.

Once a marine pest has become established in a new location, one important approach to managing spread is the identification of 'internal borders' (Forrest and Gardner, in review). Internal borders define post-border management intervention points around which relatively localized management approaches may be feasible, and involve applying a similar set of criteria and tools used for preventing and managing pest incursions at a national scale to protect key values at smaller geographic scales (e.g. vector management, pest surveillance, incursion response, containment, etc.). Critical to this is an understanding of the natural habitat barriers or broad-scale oceanographic features that may prevent or restrict the natural dispersal of pest organisms (Forrest *et al.* 2009). Knowledge of the natural dispersal potential of a pest organism may be used to identify instances where even the most robust management of human transport pathways and activities would be futile in preventing its spread.

For some marine pest species, current management strategies appear ineffective and ongoing introductions are almost inevitable. In order to manage new incursions of the more intractable pest species the development of novel solutions will be needed (see Box 14.1). For example, the seastar *A. amurensis* is a notorious predatory pest. Life history characteristics that make it a successful invader—asexual and sexual reproduction, high fecundity, extended planktonic larval stage, wide environmental tolerances, and its ability exploit a wide range of prey types and habitats (NIMPIS 2002b; Ross *et al.* 2003)—reflect the inadequacy of current management tools. Given the very high densities of *A. amurensis* larvae observed in some port environments (among the highest ever reported for seastar larvae in the port of Hobart, Australia; Bruce *et al.* 1995) and the association of this species with shipping vectors such as ballast water (Ross *et al.* 2003), the pest is highly likely to spread to new regions. Effective post-border management tools for these types of species do not exist, and solutions will need to be found in the development of new methods such as molecular probes for detection of propagules (e.g. Deagle *et al.* 2003), or the development of semiochemical and other technologies for pest attraction (e.g. Ingvarsdóttir *et al.* 2002). For any new management methods, however, there is clearly a need to balance management efficacy against the risk of collateral impacts on the wider environment including social factors. In many instances, this may mean that promising but potentially high risk solutions will be publicly and politically unacceptable (Thresher and Kuris 2004).

Box 14.1 Novel solutions for early detection of marine invasive species, and rapid response

---

Ballast water
Ballast water is a major mechanism for the transfer of AIS from source regions to new locations worldwide. Planktonic stages of many marine organisms are entrained in ships' ballast water, transported to new bioregions, and expelled when ballast is discharged. In order to effectively manage the invasion risk posed by ballast water, we need to know which unwanted species are being transported, by which ships, and at what densities. Species-specific identification of planktonic organisms in ballast water can be problematic, however, particularly for larval and juveniles stages.

Recent advances in genetic and molecular methodologies may be the answer to effective ballast water management. A polymerase chain reaction (PCR)-based test for detecting DNA of the northern Pacific starfish *Asterias amurensis* has successfully been developed (Deagle *et al.* 2003). This powerful technique overcomes many of the limitations of ballast water sampling, by successfully detecting single larvae in large amounts of mixed plankton. It is also very species-specific—able to distinguish larvae of *A. amurensis* from other *Asterias* species. Along similar lines, a fluorescent *in situ* hybridization assay has also been developed for the detection of *A. amurensis* in ballast water (Mountfort *et al.* 2007). This fluorescein-labelled species-specific probe targets *A. amurensis* larvae, allowing for their easy detection in ballast water samples without the necessity of expensive equipment. With further research and development, these techniques will provide rapid and cost-effective tools for a suite of marine pest species in ballast water and environmental samples.

Hull fouling management

Vessel hull fouling ranks alongside ballast water discharge as a primary pathway for the spread of AIS worldwide; however, practical management solutions to address this problem remain elusive. Antifouling coatings are effective at preventing growth on vessels hulls, but efficacy is limited on vessels that are frequently idle or subject to poor maintenance regimes (Floerl *et al.* 2005a). When fouling occurs, vessels are often removed to land for cleaning (dry-docking) or, perhaps more commonly, have their hulls cleaned in-water (in the case of small craft or large vessels outside their dry-docking schedule). Many concerns exist regarding conventional in-water cleaning (e.g. mechanical removal using brushes and scrapers). Entire organisms and/or viable fragments (e.g. colonial organisms or algae) may survive and establish, or the physical disturbance associated with removal may trigger the release of viable gametes and propagules (ANZECC 1996). Several methods are currently under development to reduce biosecurity risks posed by in-water hull cleaning. Trials have begun on diver-operated rotating brush systems, which incorporate suction and collection capabilities, that are able to remove and reclaim ~90% of fouling material from vessels hulls *in situ* (Hopkins and Forrest 2008). Encapsulation techniques are also being developed, whereby vessel hulls are wrapped in plastic *in situ* in order to eliminate fouling organisms through the creation of anoxic conditions (Coutts and Forrest 2005; Denny 2007). Mortality can be further accelerated through the addition of chemical agents to the encapsulated water (Coutts and Forrest 2005).

Pest detection and response

A relatively new strategy for the control of invasive species in aquatic environments involves the use of semiochemicals. Semiochemical is a generic term for chemical substance that carries a message (e.g. pheromones, allomones, kairomones). Semiochemicals such as pheromones are responsible for eliciting strong behavioural responses (e.g. settlement, gamete formation and reproduction) across a range of marine organisms including polychaetes, decapods, and echinoderms (Bartels-Hardege *et al.* 1996; Hamel and Mercier 1996; Ingvarsdóttir *et al.* 2002; Watson *et al.* 2003). The role of semiochemicals in host location by salmonid parasitic sea lice *Lepeoptheirus salmonis* were investigated as a control tool for infestations in aquaculture stocks (Ingvarsdóttir *et al.* 2002). Isolated chemicals from salmon-conditioned water was shown to be significantly attractive to sea lice in a slow-release system, and it is hoped these may form the basis for *in situ* lures for the control of sea lice in the field. Similar techniques, in combination with other control measures such as trapping, may prove useful for the control of other pest species such as crabs and sea stars (Sutton and Hewitt 2004). Further, compounds such as sex pheromones have also been suggested as manipulating chemicals for controlling the life history events of marine pest species such as *A. amurensis* (McEnnulty *et al.* 2001).

People involved in the management and control of pest species within the marine environment face a unique range of challenges and problems, many of which are not prevalent in freshwater and terrestrial pest management (e.g. the expansive inter-connectivity nature of the marine environment, accessibility issues, etc.). For this reason, it is often thought that post-border management of marine pest species is largely futile (e.g. Sanderson 1990; Brown and Lamare 1994; Thresher and Kuris 2004). Nevertheless while the successful management of any marine pest incursion must begin with effective pre-border strategies, there is also a suite of knowledge and tools available for effective post-border management of marine pest species, at least under certain circumstances. An effective marine biosecurity system will conceivably consist of vector management, surveillance, incursion response, and control measures that target particular pests or suites of functionally similar species (e.g., biofouling organisms), coupled with generic approaches (e.g. vector management) that aims to reduce human-mediated transport of all pest organisms (Forrest *et al.* 2009). The reality of marine post-border management is that there will be some successes and many failures, but this is not to say that we should focus all of our attention on pre-border management. There is a fundamental role for science in refining the knowledge and tools on which post-border management priorities and decisions are based. In particular, major gains will be made in the development of novel response tools that are publicly acceptable, cost-effective, and can be targeted towards specific pests, or groups of pest organisms, across a range of spatial scales.

14.5 Acknowledgments

Our sincere thanks to Grant Hopkins for his valuable comments on a draft manuscript of this chapter. Funding for this work was provided by the New Zealand Foundation for Research Science and Technology NIWA/Cawthron Outcome Based Investment programme, Effective Management of Marine Biodiversity and Biosecurity (EMMBB).

15

Management of interacting invasives: ecosystem approaches

Leigh S. Bull and Franck Courchamp

> It is usually easy enough to shoot the goats, wild cattle or sheep from small islands, but unfortunately this very often creates only fresh conservation problems
>
> R.H. Taylor, 1968

15.1 Introduction

Too often the success of an invasive species management programme is measured solely by the decrease or eradication of that species. However, this way of thinking distracts from the ultimate goal of these programmes, which is not just the removal of the alien species but rather the restoration of the ecosystem's biodiversity. Several decades ago, Taylor (1968) alluded to the fact that the former does not necessary lead to the latter; the incorrect management of an invaded ecosystem can in fact result in potential problems following the removal of a species. Ecosystems, be they invaded or pristine, consist of a community of organisms and their physical environment that interact as an ecological unit (Lincoln *et al.* 1998). Because of these interactions, any alteration to the species composition can have flow-on effects throughout the ecosystem (Chapin *et al.* 2000). It is this potential for flow-on effects, as Taylor (1968) suggested, that researchers, managers, and conservationists must consider before attempting any control or eradication programme. While a number of invasive species eradications have had the desired positive effects on native biodiversity, there are instances in which such actions have had either no effect, an unexpected, or even opposite impact on an ecosystem (Mack and Lonsdale 2002; Zavaleta 2002; Courchamp and Caut 2005). Such outcomes have largely been as a result of not considering the importance of the interactions between species (both native and introduced) within the ecosystem.

Unfortunately, multiply-invaded ecosystems are now the rule rather than the exception. These ecosystems are generally more difficult to manage than those that have been invaded by a single species, because as the numbers of interacting invaders increase in an ecosystem, and as aliens in late stages of invasion eliminate native species, they are more likely to replace the functional roles of the native species (Zavaleta *et al.* 2001). In the cases of these multiply-invaded ecosystems, the majority of past management actions for alien species has been the formation of separate control or eradication programmes, and most often for the most visibly destructive species (Courchamp and Caut 2005). While such single-species eradications may be successful in terms of their removal of the target species, this in itself may lead to unexpected and detrimental impacts on the ecosystem or species which the original intent was to conserve. The most common secondary outcome is the ecological release of a second (plant or prey) alien species which was previously controlled by the removed species (herbivore or predator) (Zavaleta 2002). Such outcomes can be anticipated, or ideally avoided, by first obtaining knowledge about species interactions occurring within the ecosystem and the general ecological rules that they follow (Courchamp *et al.* 1999b; Zavaleta 2002; Courchamp *et al.* 2003a; Courchamp and Caut 2005).

Species abundance and composition within an ecosystem (either natural or modified) exist largely due to the interactions between species that regulate these factors. In any ecosystem, populations of producers, consumers, and predators are in part controlled by one another through food web and other biotic interactions, including competition and provision of habitat (Hairston *et al.* 1969; Fretwell 1987; Polis and Strong 1996). Such complex interactions necessitate a deeper understanding of the system in order to predict properly the result of management actions such as the removal of one species from the ecosystem (Courchamp *et al.* 2003a). Despite every invaded ecosystem being in some way unique, they all follow, at least qualitatively, the same set of basic ecosystem rules (Zavaleta 2002). With these basic ecological rules in mind, managers and eradication experts can make great gains towards anticipating, planning for, preventing, and mitigating the unexpected (Zavaleta 2002).

Any ecological release of a species from some pressure, such as competition or predation, brought about by the removal of a species from an ecosystem, has the potential to change subsequent species interactions and species abundance. Therefore, before implementing an eradication or control programme, it is important to consider how food-web interactions (both vertical and horizontal) may be limiting populations of producers, consumers, or predators within the ecosystem (Zavaleta *et al.* 2001). Two ways in which food-web interactions may be working is through top-down regulation by higher-level consumers or predators, and by bottom-up regulation of populations by food availability or resource limitation (Zavaleta *et al.* 2001).

Bottom-up regulation of predators by prey implies that removing an alien prey should reduce both alien and native predators (Polis 1999; Zavaleta 2002). In comparison, the removal of an exotic predator from a single-invaded ecosystem, can release native prey from strong *top-down* regulation, thereby potentially increasing prey abundance. Similarly, removal of alien herbivores (in the absence of predators) exerting top-down pressure on native plants can lead to rapid recovery of native plant populations (Zavaleta *et al.* 2001; Zavaleta 2002). However, in an ecosystem where different trophic levels have been invaded, the scenario becomes more complicated, as will be shown later in the chapter.

The aim of this chapter is to illustrate the importance of considering ecological interactions between species (including those between invasive species) when planning a sequence of management actions in either natural or modified ecosystems. Different types of species interactions are discussed in the following sections, particularly with respect to the community and ecological dynamics that are behind these interactions and how this knowledge can be used by managers and researchers to reduce the likelihood of unexpected or unwanted outcomes in the management of invasive species. Most studies investigating the impacts of species interactions have looked at invasive mammalian species on island ecosystems, so the case studies presented reflect this bias. The case studies are used to demonstrate key principles regarding invasive species interactions and the ways in which they can be managed successfully. The main tools and techniques that can aid in the successful management of an invaded ecosystem are also discussed.

15.2 Cases when removal of alien species does not lead to ecosystem recovery

15.2.1 When the alien species has an important functional role

Instances exist whereby an alien species has been present in an ecosystem for sufficient time that it dominates or has replaced native species and habitats. In some cases, this can lead to positive association between a native and alien species, thus further complicating the management of invaded ecosystems. In such instances, consideration must be given to how the removal of such an invader could in fact remove an ecosystem function necessary to the survival of other (perhaps threatened) biota (Zavaleta *et al.* 2001). For example, Carter and Bright (2002) describe how the exotic but non-invasive Japanese red cedar (*Cryptomeria japonica*) plantations on Mauritius provide refuges for native birds against predation by introduced macaques (*Macaca fascicularis*). Given that nest predation by macaques is significantly lower in cedar than in native forest, the removal of the Japanese red cedar would indirectly increase the impacts of another alien species (macaques) on endemics with high conservation value (Carter and Bright 2002). Another example concerns alien pollinators or seed dispersers that have become the most important source of pollination or dispersal following the loss of native ones. In instances where an alien species is non-invasive, its removal may not be of significant benefit to the ecosystem.

15.2.2 When the alien species has a long lasting effect

There are instances whereby alien plants have indirect negative effects on a native species even after the removal of the exotic ones (Zavaleta 2002). For example, the invasive species from the genus *Tamarix*, and the iceplant (*Mesembryanthemum crystallinum*) have been shown at some sites to salinize soil to such an extent that native organisms are not able to recolonize after their removal (Vivrette and Muller 1977; El-Ghareeb 1991; Bush and Smith 1995; Shafroth *et al.* 1995).

Management of interacting invasives | 235

15.2.3 When the alien species interacts with other aliens

15.2.3.1 Interactions resulting from conspicuous aliens

Hyperpredation

When a decline in a native prey species is observed in an ecosystem also containing an alien predator and prey, generally the initial response is to attempt to remove the most visibly devastating species—the predator. However, such actions have the potential to lead to a further decline in the native prey species; the availability of abundant exotic prey can inflate alien predator populations, which then increase the predators' consumption of native species, subsequently driving the indigenous prey to very low numbers and potentially to extinction (Zavaleta et al. 2001; Zavaleta 2002; Courchamp and Caut 2005). This process, termed **hyperpredation,** illustrates how introduced prey can have an important indirect effect in such ecosystems. A prey species introduced into an environment in which a predator has also been introduced is likely to allow a high enough increase of this predator that native prey, less adapted (in terms of behaviour and life history traits) to high levels of predation, could suffer a population decline (Courchamp et al. 1999b, 2000, 2003a; Courchamp and Caut 2005).

The interaction between introduced cats (*Felis catus*), rabbits (*Oryctolagus cuniculus*), and native birds through the hyperpredation process on Macquarie Island (a Tasmanian State Reserve) resulted in the decline of burrow-nesting petrels, as well as the extinction of an endemic parakeet (*Cyanoramphus novaezelandiae erythrotis*) and a banded rail (*Rallus philippensis*) (Taylor 1979; Brothers 1984). Despite cats being introduced to the island 60 years before rabbits, the dramatic impact of cat predation on the bird populations dated back only 10 years after the introduction of rabbits (Taylor 1979). The presence of the rabbit population not only maintained, but significantly increased, the cat population during winter (when seabirds are absent from the island), therefore resulting in increased predation pressure on the land bird species. Rabbits are more adapted to cat predation and were thus able to support the increase. However such an increase was fatal to several native bird populations (unadapted to mammalian predation pressures) which were extirpated by an over-sized cat population that no longer depended on the presence of birds to survive (Courchamp et al. 1999b, 2000).

Diet studies of the predator in question should be conducted in order to assess not only the importance of the impact on the local population, but also potential hyperpredation processes (Courchamp and Caut 2005). Not surprisingly, the presence of hyperpredation has consequences on the management actions required: should a control programme be aimed at the predator only or at the introduced prey and predator simultaneously? Removing only an introduced predator population without controlling the introduced prey is not recommended for several reasons. First, eradicating the predator may be difficult to achieve since the introduced prey constitute a constant source of food (Fig. 15.1a). Second, removing the predation pressure would increase the difficulties of later coping with the introduced prey, which are often characterized by high reproductive rates. On the other hand,

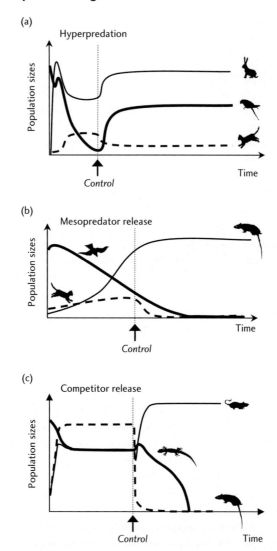

Fig. 15.1 Three examples of how the overlooked presence of other invasive species can hinder the success of a control programme aimed at protecting a local species. (a) The hyperpredation process hinders the control of the predator (here a cat, to use an example from the main text), as alien alternative prey (rabbits) are abundant which boosts the predator's population growth rate, thereby precluding full protection of the local prey (parakeet). (b) A mesopredator release may occur if the superpredator (cat) is removed, with the result that the shared prey (seabird) that was to be protected can be lost by a demographic explosion of the mesopredator (rats). (c) A competitor release is here equivalent to mesopredator release as the shared prey (lizard) may also be eliminated by a poorly-designed control protocol resulting in an explosion of a lower competitor population (mouse) as the higher competitor (rat) is eliminated.

Fig. 15.2 Hyperpredation relationships on Santa Cruz Island, involving a colonizing predator (bald eagle), an introduced prey (pig), and two native prey (insular fox and striped skunk). Photo: Gary Roemer.

controlling only the introduced prey is unsatisfactory in the long term because predators could place high predation pressure on the indigenous prey. Through the use of models, Courchamp et al. (2003b) determined that a regime of combined control of both introduced species (predator and prey) is the best restoration strategy in such cases. They noted that if the control of introduced prey is not sufficient, the indigenous prey will be destroyed, even if the predator population is being controlled, while too high a prey control would cause the predator to shift to local prey. Courchamp and Caut (2005) advocated the simultaneous commencement of both control programmes; beside being the most ecologically viable option, costs may be reduced (if transportation, or hunting and trapping can be shared) and efficiency increased (e.g. through the additive effects of primary and secondary poisoning of predators (Robertson et al. 1994; Torr 2002).

The interaction between introduced pigs (*Sus scrofa*) and native golden eagle (*Aquila chyrsaetos*) populations on Santa Cruz Island, California, provides an example of bottom-up regulation as well as the presence of hyperpredation (Fig. 15.2) (Roemer et al. 2002; Courchamp and Caut 2005). The consequences of these processes on endangered island grey fox (*Urocyon littoralis*), as well as the management implications for native predators, are discussed next.

The endemic island grey fox occurs only on the Channel Islands, and has evolved over the past 20,000 years to such an extent that the largest six islands have their own endemic subspecies (Roemer et al. 2001). Severe declines had been noted in the fox populations, and were initially attributed to predation by golden eagles visiting the islands. While it was originally thought that golden eagles did not stay on the islands long enough to constitute a threat to the local prey, the discovery of a nest on Santa Cruz Island revealed this was not the case: the presence of fox remains in the nest confirmed eagle predation. Despite these findings, the question remained: how could the eagle threaten fox survival through predation, if there was not enough local prey on the island to allow the continuous presence of the eagles? The remains of piglets also found in the nest provided the answer. Feral pigs (introduced in the 1850s) were present on Santa Cruz Island, and by producing piglets all year round, they provided visiting eagles with enough resources for them to colonize the island. Even the irregular and low predation rate of eagles on a species such as the fox that is ill-adapted to avian predation (both in terms of behaviour and life-history strategies), was sufficient to drive the fox populations towards extinction (Roemer *et al.* 2002). Roemer *et al.* (2002) were able to show that in the absence of pigs, any introduction of eagles, however large, will eventually lead to colonization failure and fox population persistence. However, when pigs are present, a single eagle pair will be able to colonize the island and build a population that is so large that foxes will go extinct, while pigs will remain at moderate densities. With the use of models to mimic different control strategies (control of pigs only, of eagles only, or of both species, with different strength) and comparison of their relative efficiency, Courchamp *et al.* (2003b) revealed that the eradication of pigs (in absence of eagle control), the intended course of action on Santa Cruz, would lead to the extinction of the fox. In theory, the most efficient solution would be to remove both eagles and pigs. Each removal project was faced with its own difficulties; the removal of the large pig population was logistically difficult, whereas the protected status of the golden eagle meant that it had to be removed from the island via live trapping methods (Courchamp *et al.* 2003b). This scenario illustrates the many challenges, not always apparently obvious, that conservation workers face in the attempt to conserve threatened species.

15.2.3.2. Interactions resulting from inconspicuous aliens

Release from introduced herbivores

Introduced herbivores are sometimes removed from an ecosystem where they cause damage that may threaten local flora and fauna. However, such management actions have shown mixed results with regards to the restoration of the native vegetation (Coblentz 1978, 1997; Van Vuren and Coblentz 1987). Surveys following the eradication of rabbits and goats (*Capra hircus*) from Round Island, Mauritius, revealed that in the short term the general predicted effects of eradication were upheld: increases in plant biomass and tree recruitment (Bullock *et al.* 2002). However, unpredicted effects (such as differential population responses of reptiles and increasing rates of establishment and influence of non-native plants) also occurred, leading Bullock *et al.* (2002) to predict that new ecological communities (not necessarily dominated by local plants) are likely to develop on Round Island as a consequence.

Zavaleta *et al.* (2001) predict that the greatest potential for negative impacts on native vegetation exists when herbivore eradication removes the disturbance that is necessary to suppress establishment of late successional (tree or shrub) aliens. For example, on San Cristobal Island, Galapagos, the removal of feral cattle from degraded grasslands containing suppressed populations of exotic guava (*Psidium guajava*) led to the rapid growth of this plant into dense and extensive thickets (Eckhardt 1972).

The removal of introduced pigs and goats on Sarigan Island provides another example of the devastating effects that the removal of herbivores can have on the flora. Due to logistical difficulties, the management programme for the island included only a minimal pre-eradication study. While the programme was successful in removing the introduced ungulates, it failed in its ability to detect the presence of the introduced vine *Operculina ventricosa*, which appeared to be a preferential food item for the goats. The release from grazing pressure enabled the introduced plants to fully express their competitive superiority over the native plants, resulting in their rapid invasion of the community as quickly as 2 years after the removal of the alien grazers (Fig. 15.3) (Kessler 2002). Future monitoring will be required to determine what effect the vine will have on the regeneration and expansion of the native forest and its fauna. A more thorough pre-eradication study incorporating simple fenced exclosure plots would have helped managers to

Fig. 15.3 Control of herbivores without taking into account introduced plants may lead to undesired chain reactions, as occured on Sarigan Island, with the invasion of *Operculina ventricosa* following goat removal. Photo: Curt Kessler.

detect any potential unwanted results (i.e. the release of the vine from top-down regulation) arising from the eradication of the herbivores.

The effects of alien herbivore removal on native vegetation, under certain circumstances, might also have indirect negative effects, because of the presence of other alien animals (Zavaleta et al. 2001). Rabbit removal on Macquarie Island led to major increases in cover of the native tussock grass *Poa foliosa* (Copson and Whinam 1998). The expansion of this tussock species, the preferred habitat of the introduced ship rat (*Rattus rattus*), could expand the range of rats on the island and consequently bring them into contact with burrow-nesting bird colonies on the island, which have escaped rat predation so far (Copson and Whinam 1998).

Donlan et al. (2002) used both large- and small-scale experimental manipulations to investigate the impact and recovery of an island plant community following the removal of exotic herbivores from the San Benito Islands, Mexico. The hypotheses tested were:

- With herbivore removal, plant community structure changes due to the release of top-down regulation.
- The response by the plant community is predictable from the hierarchy of herbivore preference.

Removal of European rabbits, donkeys, and goats began on San Benito West in early 1998, while removal of rabbits on San Benito East was postponed until late 1999 to facilitate the comparison between the two islands. During the course of the San Benito West eradication programme, the food preferences of rabbits and exclosure plot studies were conducted on San Benito East. Results from the food-preference trials accurately predicted changes in the perennial plant community: the changes in relative abundance of the plant species were positively correlated with the preference hierarchy on San Benito West (herbivores removed), and negatively correlated on San Benito East (herbivores present). Despite the relationship between herbivore food preference and changes in plant cover providing strong evidence of a top-down effect, recovery of the ecosystem was shown to depend on the bottom-up effects of resources such as water availability (Donlan et al. 2002).

In an attempt to preserve and restore an area of tropical dry forest on Hawaii, an area of 2.3 ha (Kaupulehu preserve) was fenced in 1956 to exclude cattle and feral goats. Cabin et al. (2000) examined the effects of this long-term ungulate exclusion from the Kaupulehu preserve and the recent control of rodents (*Rattus rattus, R. exulans,* and *Mus musculus*) by comparing the flora present to that of an adjacent area subjected to continuous grazing since the fence was constructed. Compared to the adjacent area, the preserve had a relatively diverse flora with substantially greater coverage of native overstorey and understorey species. However, Cabin et al. (2000) noted that the dominant herbaceous cover of alien fountain grass (*Pennisetum setaceum*) and predation by rodents had thwarted the regeneration of the native canopy trees within the preserve. These results once again indicate that in restoration programmes, the removal of an alien herbivorous species should be viewed as the first critical step in the recovery of the ecosystem.

The mesopredator release effect

When an alien predator and an alien prey species co-occur, removal of the predator can lead to the release of the prey from top-down regulation (Zavaleta *et al.* 2001). This process of the rapid expansion of a prey population once top-down control by a predator has disappeared is termed **mesopredator release** (Fig. 15.1b), and can lead to negative effects if the increased alien prey population competes with or consumes native biota (Zavaleta 2002).

On Amsterdam Island, the attempted reduction of the cat population was abandoned as it was alleged to have caused a compensating rise in the number of rats and mice (*Mus musculus*), just as is predicted by the mesopredator release theory (Holdgate and Wace 1961). However, through the collection of long-term data (1972–2007) on the productivity of the threatened Cook's petrel (*Pterodroma cookii*) breeding on Little Barrier Island, New Zealand, and the sequential removal of cats (1980) and Pacific rats (2004) from the island, Rayner *et al.* (2007) tested the predictions of the mesopredator release hypothesis. This study did in fact find that the removal of cats resulted in an increase in the predatory impacts of Pacific rats, and more importantly, a decline in the fecundity of the Cook's petrel (Rayner *et al.* 2007). Furthermore, the removal of both cats and Pacific rats resulted in an increase of Cook's petrel breeding success to a level above that recorded when both introduced predators were present on the island. A further finding of conservation importance was the altitudinal variation in the impact of rats on the Cook's petrel productivity. The observed spatial variation in the mesopredator release was attributed to the interactions between environmental gradients, resource availability, and the nutritional requirements of Pacific rats. As noted by Rayner *et al.* (2007), local variation in the outcomes of mesopredator release has significant implications for island restoration, and provides further support regarding the importance of ecosystem level understanding to predict the potential impacts of introduced species management on oceanic islands.

The presence of a third predator in the prey–mesopredator–superpredator system complicates matters further. The managers of Bird Island, Seychelles, conservation programme were aware of the potential dangers of removing a superpredator from an ecosystem also containing a mesopredator, and as such took the cautious approach of simultaneously removing both introduced cats and rats in order to protect the local bird colonies. Unfortunately, the presence of the introduced crazy ant (*Anoplolepis longipes*), in very low numbers on the island, was overlooked (Feare 1999). It appears that the ant larvae could be an important prey item of the introduced rodents, and the rat eradication led to a demographic explosion of the ants (Feare 1999). This resulted in the ants extending their range over the island and impacting heavily on land crab and bird colonies. This example once again highlights the importance of pre-eradication, particularly diet, studies in order to obtain a thorough picture of the trophic web interactions occurring in an ecosystem.

When alien predators and prey co-occur, eradication of only the alien prey can cause the predator to switch to native prey. In New Zealand, because rats are a major component of the stoat (*Mustela erminea*) diet, efforts to reduce both species was attempted through the control of the rats only (Murphy and Bradfield 1992; Murphy *et al.* 1998b). These actions not only failed to eliminate the stoat populations, but the reduced availability of alien prey resulted in a diet switch by the stoat to incorporate more native birds and eggs.

The competitor release effect

Having earlier covered the potential affects of control attempts on non-target species through trophic interactions, this section looks at the consequences relating to competitive interactions. The control of an invader has the potential to release any species interacting with that invader from its pressure, be it predation or competition (exploitation or interference) (Courchamp and Caut 2005). Control of a superior competitor may lift the pressure of competition from an inferior competitor, subsequently leading to an increase in its population; such a process is termed the **competitor release effect** (see Fig. 15.1c).

While not always tested, there are numerous instances in which the removal of a target alien species has coincidentally facilitated the emergence in the community of another long-suppressed non-indigenous species (Mack and Lonsdale 2002). For example, following the biological control of the weed St. John's wort (*Hypericum perforatum*) at several sites using chrysolina leaf beetles (*Chrysolina quadrigemina*), other non-indigenous invading species became more abundant (Huffaker and Kennett 1959; Tisdale 1976). Similar results have been observed with respect to invasive aquatic macrophytes. For example in southeast Florida, the use of herbicides and grass carp to control the widespread invader *Hydrilla verticillata* has coincided with an increase in the equally-unwanted invader *Hygrophila polysperma* and its replacement of hydrilla as the number one non-native aquatic weed in some southeast Florida canals (Duke *et al.* 2000).

There are a number of ecosystems to which both rats and mice have been introduced. In such instances, rats generally dominate and are generally viewed as strong competitors of mice (Ruscoe 2001). Techniques to control rodents (trapping and poisoning) often lack specificity regarding rodents; therefore their use in ecosystems containing multiple introduced rodent species are often viewed as beneficial due to non-target rodent species mortalities. However, in a number of ecosystems containing both rats and mice, the successful eradication of rats has corresponded with a dramatic increase in mouse numbers, often to levels exceeding those prior to the eradication programme (Brown *et al.* 1996; Witmer *et al.* 1998). For example, rats and rabbits were successfully eradicated from Saint Paul Island, Indian Ocean, but the control programme did not focus on the small mouse population that was known to occur there (Micol and Jouventin 2002). The complete removal of the rat population released the mice from their competitors, causing such a demographic explosion that mouse numbers far exceeded the habitat carrying capacity. While it may be argued that mice are less harmful than rats, mouse outbreaks can be very problematic; mice have been shown to be active predators of invertebrates, reptiles, and even the chicks of large birds such as the albatross (Newman 1994; Smith *et al.* 2002; Wanless *et al.* 2007).

Caut *et al.* (2007) were able to demonstrate how some control strategies that overlook the competitor release effect may fail to restore the ecosystem through an unexpected increase of the inferior competitor, even if that species is being controlled too. Using mathematical models to mimic the effects of controlling introduced species in the presence of their competitors, Caut *et al.* (2007) found that it was possible for a competitive release effect to occur even when both introduced competitors were being controlled simultaneously (as is the case in most rat eradication programmes). The competitive release of the inferior competitor (mouse) is due to the indirect positive effect of control (the removal of their competitor, the rat) exceeding its direct effect (their own removal). Furthermore, both control levels and target specificity were found to have a direct influence on the extent of the competitor release process: the stronger and more specific the control, the greater the effect (Caut *et al.* 2007).

Because the intensity of the competitor release is directly proportional to the control effort, indiscriminate intensification of the control will exacerbate this process, with high potential impact on native prey species (Caut *et al.* 2007). Furthermore, while most control programmes aim for high target specificity (Simberloff and Stiling 1996b; Murphy *et al.* 1998a), this recent study highlights the role of control specificity in terms of the likelihood of a competitor release effect. While conservation managers appear to be faced with a dilemma regarding control intensity and specificity, Caut *et al.* (2007) suggest the following:

- Obtain an understanding of the invaded ecosystem as a whole in order to assess potential processes (including competitor release) that may occur during or following control.
- Use as many specific methods as there are species to be controlled. If resources are limited, as is the case in most instances, controlling the inferior competitor should be the first priority so that the combination of control and competition (or predation) eliminates it; after which the remaining resources can then be used to target the superior competitor without the danger of releasing the inferior competitor (Caut *et al.* 2007).

It is important to note that these should only be taken as guidelines, and that assessment of the data gathered during pre-eradication studies should provide a much better basis on which to construct an optimal strategy.

15.3 Mitigating actions

Several tools have already been mentioned in this chapter, which will reduce the likelihood of unexpected outcomes in eradication and control programmes.

15.3.1 Pre-eradication studies

Pre-eradication studies provide useful insights into potential ecosystem responses to invasive species removal, thus helping managers to reliably avert or plan for the undesired side effects of eradication (Blossey 1999; Zavaleta 2002). Such studies should at the very least incorporate both surveys of the species present in the ecosystem, and diet studies in order to obtain an understanding of the interactions and trophic web links that exist between the species present and an estimation of possible outcomes of the eradication of the target species (Fig. 15.4). For instance in the case of a possible herbivore removal programme, food-preference trials can be used to accurately predicted changes in the plant community (Donlan *et al.* 2002),

All such studies should remain simple, standardized, and easily replicable so that they can be repeated during and after the programme, on other islands, or by new researchers/staff in the future. Methods should also be properly recorded in order to help both replication and future analyses.

15.3.2 Exclosure experiments

Erecting fenced exclosure plots prior to the removal of an herbivore from an ecosystem will provide an indication of how the vegetation may respond post-eradication (Cabin *et al.* 2000; Donlan *et al.* 2002; Zavaleta 2002).

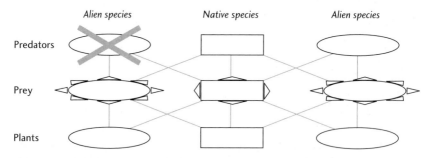

Fig. 15.4 In a simple, three-level, nine-species ecosystem, the presence of more than one alien species can mean alien removal will trigger other alien species releases. Here, species are represented by a square if they are native and by ellipse if they are alien. The grey lines are some of the energy flows among them. The cross represents an eradication. Species which undergo a sudden population increase (resp. decrease) are surrounded by arrows going outwards (resp., inward). Real ecosystems will contain far more species, with less straightforward outcomes. Depending on which species is removed, different outcomes are possible, but for many configurations most outcomes are globally negative, with an increase of one or several alien species. The configuration displayed here illustrates how the eradication of an alien species to protect a local (prey) species can in fact have the opposite effect.

15.3.3. Control strategies

15.3.3.1 Hyperpredation

Control options for cases where hyperpredation is likely include:

- No control: the alien predator population will remain large mainly due to the presence of alien prey, and the native prey population will go extinct with time due to cat predation.
- Control of alien predators only: the presence of alien prey precludes the eradication of predators, allowing only a partial recovery of the native prey population.
- Simultaneous control of both aliens: for the same control effort as above, eradication of alien prey helps achieving the predator extinction, allowing total recovery of the native prey population.

It is important to note that if the control of the introduced prey is not sufficient, the native prey may be destroyed, even if the predator population is being controlled. Thus, the control effort is a critical factor contributing to the success of a programme: control effort must be above the intrinsic growth rates of both introduced species (Courchamp *et al.* 1999b). These parameters should be obtained during pre-eradication studies so that the required timing and control efforts can be established.

15.3.3.2 Mesopredator release

The hyperpredation process has already highlighted the importance of controlling both introduced predator and prey species; however, the level of control is crucial to avoid mesopredator release in systems containing native prey (e.g. bird, lizard), introduced mesopredator (e.g. rats), and introduced superpredator (e.g. cat):

- High superpredator and low mesopredator control effort: when the superpredator disappears the native prey may go to extinction through mesopredator release.
- Low superpredator and high mesopredator control effort: this combination of control will enable the survival of the prey population, although it should be noted that sometimes a very low number of superpredators can lead a prey population to extinction (Vazquez-Dominguez *et al.* 2004).
- High control effort for both predators must be regarded as a potentially good option, provided the timing and forces are set according to the specificities of the ecosystem of concern.

The fate of the native prey species will depend on the superpredator control level; the prey will disappear if this control level is too high (Courchamp *et al.* 1999a).

15.3.3.3 Competitor release

Assuming both competitors are controlled by a common, non-specific method, the extent of the competitor release process is directly influenced by both the levels of control and the target: the stronger and more specific the control, the greater the effect (Caut et al. 2007).

- Removal of superior competitor: when the inferior competitor is not controlled at the same time, its population reaches carrying capacity.
- Superior competitor is controlled, but not eradicated: the inferior competitor population here again increases its population through a competitor release process.
- Simultaneous control of both competitors: the inferior competitor increases when the indirect positive effect (removal of the superior competitor) exceeds the direct negative effect (its own removal). This occurs when the control effort is high and either the control specificity is high or the intensity of competition is high (Caut et al. 2007).

15.3.3.4 Post-eradication monitoring

Post-eradication monitoring will provide information on the eradication success, both in terms of complete elimination of the target species and of the recovery of the invaded ecosystem (Blossey 1999). A good example of the utility of such monitoring is given by a recent study where it allowed the monitoring of the recovery of a native species (Pascal et al. 2005). Post-eradication monitoring will also enable managers to catch unanticipated side effects, or know whether and when to implement contingency plans for dealing with undesired outcomes (Mack and Lonsdale 2002; Zavaleta 2002).

The information collected during pre- and post- monitoring allows us to:

- Contribute information to a knowledge base on effects of alien removal in an ecosystem context that better prepares us for future control and eradication programmes.
- Learn about the impacts of alien species on both alien and native components of the biota—information that is crucial to the determination of whether ecosystem restoration is likely to be achieved.
- Provide vital impact evidence that will help to support future eradication campaigns. As rightly claimed by Towns and colleagues (2006), 'With well-conceived projects that include rigorous measurement of the responses to rat eradications, it may be possible to generate models of rat effects for a range of species and locations. Without such models, the motives and justification for rat eradications will continue to be questioned'.

Also, it is important—when logistically possible—to couple pre- and post monitoring with a comparison to another island that is free of the alien species: the before-after control-impact (BACI) protocol. This protocol is used to assess the impact of an event on variables that measure the state of an ecosystem. The design involves repeated measures over time, made at one or more control sites and one or more impacted sites, both before and after the time of the event that may cause an impact (Manly 2000).

Fig. 15.5 Rats are not only a major invasive alien species, they are also one that interacts substantially with other introduced species; as predators, competitors, or prey. As they are often the target of management programmes, it is essential to consider their trophic relationships with other species, in order to avoid potential chain reactions following their removal. Photo: Jean-Louis Chapuis.

15.4 Conclusions

The removal of a single alien species from an ecosystem does not necessarily lead to biodiversity restoration. In some cases it even has the potential to lead to opposite effects, through the release of other alien species, for example. Such processes can have a dramatic impact on an ecosystem and its native biota. In these cases, the eradication of alien species does not lead to biodiversity restoration. Pre-eradication studies are essential for establishing the likelihood of such incidences.

Understanding what species are present in an ecosystem, which species they interact with, and where they are located in the trophic web will place conservation managers in a good position for making decisions regarding the potential impacts of control efforts on that ecosystem (Fig. 15.5). Furthermore, this will enable not only success in removing unwanted species, but also the increased likelihood of achieving the primary goal of most programmes—biodiversity restoration.

References

Abbott K. (2005). Supercolonies of the invasive yellow crazy ant, *Anoplolepis gracilipes*, on an oceanic island: forager activity patterns, density and biomass. *Insectes Sociaux* **52**:266–273.

Abbott K. (2006). Spatial dynamics of supercolonies of the invasive yellow crazy ant, *Anoplolepis gracilipes*, on Christmas Island, Indian Ocean. *Diversity and Distributions* **12**:101–110.

Abbott K. and Green P.T. (2007). Collapse of an ant-scale mutualism in a rainforest on Christmas Island. *Oikos* **116**:1238–1246.

Abdelkrim J., Pascal M., and Samadi S. (2007). Establishing causes of eradication failure based on genetics: case study of ship rat eradication in Ste. Anne Archipelago. *Conservation Biology* **21**:719–730.

Agami M. and Waisel Y. (1985). Inter-relationships between *Najas marina* L. and three other species of aquatic macrophytes. *Hydrobiologia* **126**:169–173.

Allen L.R. (1991). The eradication of feral goats from an island national park. *Australian Vertebrate Pest Control Conference* **9**:22–25.

Allen L.R. and Sparkes E.C. (2001). The effect of dingo control on sheep and beef cattle in Queensland. *Journal of Applied Ecology* **38**:76–87.

Allen R.B. and Lee W.G. (2006). *Biological Invasions in New Zealand*. Ecological Studies **186**, Springer-Verlag, Berlin Heidelberg.

Allessio Leck M., Parker V.T., and Simpson R.L. (eds.) (1989). *Ecology of Soil Seed Banks*, pp. 3–87. Academic Press, San Diego, CA.

Alliance for the Chesapeake Bay (2003). *Citizen's guide to the control of invasive plants in wetland and riparian areas*. http://www.alliancechesbay.org/pubs/projects/deliverables-251-1-2005.pdf.

Altieri M.A. and Liebman M. (eds.) (1988). *Weed Management in Agroecosystems: Ecological Approaches*, pp. 25–39, 216–218. CRC Press, Boca Raton, FL.

Anderson C.B., Griffith C.R., Rosemond A.D., Rozzi R., and Dollenz O. (2006). The effects of invasive North American beavers on riparian plant communities in Cape Horn, Chile. Do exotic beavers engineer differently in sub-Antarctic ecosystems? *Biological Conservation* **128**:467–474.

Anderson L.W.J. (2005). California's reaction to *Caulerpa taxifolia*: a model for invasive species rapid response. *Biological Invasions* **7**:1003–1016.

Anon (2004). Comb jelly Jonahs. *Biocontrol News and Information*, **25**:29N–33N.

Anon (2007). Understanding reinvasion risk! Xclude it! *Xcluder Pest Proof Fencing Company Newsletter*, 2, 3. Cambridge, New Zealand.

ANZECC. (1996). *Working together to reduce impacts from shipping operations: Code of practice for antifouling and in-water hull cleaning and maintenance*, pp. 10. Australia and New Zealand Environment and Conservation Council, Canberra.

APHIS web facts. *Excluding foreign pests and diseases*. http://permanent access.gpo.gov/lps3025/exclude.html (accessed March 2008).

Ashley M., Storrs M., and Brown M. (2002). Caring for country: community-based management of *Mimosa pigra* on Aboriginal lands in the Northern Territory, Australia. In Julien M., Flanagan G., Heard T., Hennecke B., Paynter Q., and Wilson C. (eds.), *Research and Management of* Mimosa pigra: *papers presented at the 3rd Internationals Symposium on the Management of* Mimosa pigra, pp. 106–109. CSIRO, Darwin, Australia.

References

Ashton P.J., Scott W.E., Steÿn D.J., and Wells R.J. (1979). The chemical control programme against the water hyacinth *Eichhornia crassipes* (Mart.) Solms on Hartebeespoort Dam: Historical and Practical Aspects. *South African Journal of Science,* **75**:303–306.

Atkinson I.A.E. (1985). The spread of commensal species of *Rattus* to oceanic islands and their effect on island avifaunas. In P. J. Moors (ed.) *Conservation of island birds*, pp. 35–81. International Council for Bird Preservation, Cambridge, UK.

Atlantic Canada Aquaculture Industry Research and Development Network. (2006). *Green Algae Project Objectives*, pp. 6–7. Provincial Research and Development Updates, Prince Edward Island, Canada.

Bainbridge D.A. (1990). Soil solarization for restorationists. *Restoration and Management Notes* **8**:96–98.

Bainbridge D.A. (2007). *A Guide for Desert and Dryland Restoration: New Hope for Arid Lands*. Island Press, Washington, DC.

Baker H.G. (1965). Characteristics and modes of origin of weeds. In Baker, H.G. and G.L. Stebbins (eds). *The genetics of colonizing species*, pp. 147–72. Academic Press, New York, NY.

Baker S. (2006). The eradication of coypus (*Myocaster coypus*) from Britain: the elements required for a successful campaign. In Koike F., Clout M.N., Kawamichi M., De Poorter M., and Iwatsuki K. (eds.) *Assessment and control of biological invasion risks*, pp. 142–147. Shoukadoh Book Sellers, Kyoto and IUCN, Switzerland.

Balciunas J.K., Grodowitz M.J., Cofrancesco A.F., and Shearer J.F. (2002). *Hydrilla*. In van Driesche R.G., Lyon S., Blossey B., Hoddle M.S., and Reardon R. (eds.) *Biological control of invasive plants in the eastern United States*, pp. 91–114. USDA Forest Service, Morgantown, WV.

Balirwa J.S., Chapman C.A., Chapman L.J., Cowx I.G., Geheb K., Kaufman L., *et al.* (2003). Biodiversity and fishery sustainability in the Lake Victoria Basin: an unexpected marriage. *Bioscience* **53**:703–715.

Ball S.J., Ramsey D., Nugent G., Warburton B., and Efford M. (2005). A method for estimating wildlife detection probabilities in relation to home-range use: insights from a field study on the common brushtail possum (*Trichosurus vulpecula*). *Wildlife Research* **32**:217–227.

Barko J. and Smart R.M. (1981). Sediment-based nutrition of submersed macrophytes. *Aquatic Botany* **10**:339–352.

Barko J., Smart R.M., McFarland D.G., and Chen R.L. (1988). Interrelationships between the growth of *Hydrilla verticillata* (L.F.) Royle and sediment nutrient availability. *Aquatic Botany* **32**:205–216.

Barnett D.T., Stohlgren T.J., Jarnevich C.S., Chong G.W., Ericson J.A., Davern T.R. and Simonson S.E. (2007). The art and science of weed mapping. *Environmental Monitoring and Assessment* **132**: 235–252.

Bartels-Hardege H.D., Hardege J.D., Zeeck E., Muller C., Wu B.L., and Zhu M.Y. (1996). Sex pheromones in marine polychaetes V: a biologically active volatile compound from the coelomic fluid of female *Nereis (Neanthes) japonica* (Annelida Polychaeta). *Journal of Experimental Marine Biology and Ecology* **201**:275–284.

Barton J. (2004). How good are we at predicting the field host-range of fungal pathogens used for classical biological control of weeds? *Biological control* **31**:99–122.

Baskin C.C. and Baskin J.M. (1998). *Seeds: Ecology, Biogeography and Evolution of Dormancy and Germination*. Academic Press, New York, NY.

Bates J.D., Svejcar T., Miller R.F., and Angell R.A. (2006). The effects of precipitation timing on sagebrush steppe vegetation. *Journal of Arid Environments* **64**:670–697.

Bax N., Carlton J.T., Mathews-Amos A., Haedrich R.L., Howarth F.G., Purcell J.E., *et al.* (2001). The control of biological invasions in the world's oceans. *Conservation Biology* **15**:1234–1246.

Bax N., Hayes K., Marshall A., Parry D., and Thresher R. (2002). Man-made marinas as sheltered islands for alien marine organisms: establishment and eradication of an alien invasive marine species. In Veitch C.R. and Clout M.N. (eds.) *Turning the Tide: The Eradication of Invasive Species*, pp. 26–39. IUCN SSC Invasive Species Specialist Group. IUCN, Gland, Switzerland and Cambridge, UK.

Bazzaz, F.A. (1986). Life history of colonizing plants: some demographic, genetic, and physiological features. In Mooney H.A. and Drake J.A. (eds.) *Ecology of biological invasions of North America and Hawaii*, pp. 96–110. Springer-Verlag, New York.

Beggs J. (2001). The ecological consequences of social wasps (*Vespula* spp.) invading an ecosystem that has an abundant carbohydrate resource. *Biological Conservation* **99**:17–28.

Benech Arnold R.L., Sánchez R.A., Forcella F., Kruk B., and Ghersa C.M. (2000). Environmental control of dormancy in weed seed banks in soil. *Field Crops Research* **67**:105–122.

Benedictow O.J. (2004). *The Black Death 1346–1353: The complete history*. Boydell Press, Suffolk.

Bertolino S. and Genovesi P. (2003). Spread and attempted eradication of the grey squirrel (*Sciurus carolinensis*) in Italy, and consequences for the red squirrel (*Sciurus vulgaris*) in Eurasia. *Biological Conservation* **109**:351–358.

Bigler F., Babendreier D. and Kuhlman U. (eds.) (2006). Environmental Impact of Arthropod Biological Control: Methods and Risk Assessment. CABI Publishing, Wallingford, UK.

Blakemore K.A. and Forrest B.M. (2007). *Heat treatment of marine fouling organisms*. Report No. 1300, Cawthron Institute, Nelson, New Zealand.

Blossey B. (1999). Before, during and after: the need for long-term monitoring in invasive plant species management. *Biological Invasions* **1**:301–311.

Blossey B. and Nötzold R. (1995). Evolution of increased competitive ability in invasive non indigenous plants: a hypothesis. *Journal of Ecology* **83**:87–889.

Bomford M. (2003). *Risk assessment for the import and keeping of exotic vertebrates in Australia*. Bureau of Rural Sciences, Canberra, Australia.

Bomford M. and O'Brien P. (1995). Eradication or control for vertebrate pests? *Wildlife Society Bulletin* **23**:249–255.

Boone R.B. and Hobbs N.T. (2004). Lines around fragments: effects of fencing on large herbivores. *African Journal of Range & Forage Science* **21**:147–158.

Bos M.M., Tylianakis J.M., Steffan-Dewenter I., and Tscharntke T. (2008). The invasive yellow crazy ant and the decline of forest ant diversity in Indonesian cacao agroforests. *Biological Invasions* **10**:1399–1409.

Bossenbroek J.M., Kraft C.E., and Nekola J.C. (2001). Prediction of long-distance dispersal using gravity models: zebra mussel invasion of inland lakes. *Ecological Applications* **1**: 1778–1788.

Boylen C.W., Eichler L.W., and Sutherland J.W. (1996). Physical control of Eurasian water milfoil in an oligotrophic lake. *Hydrobiologia* **340**:213–218.

Branch G.M. and Steffani C.N. (2004). Can we predict the effects of alien species? A case-history of the invasion of South Africa by *Mytilus galloprovincialis* (Lamarck). *Journal of Experimental Marine Biology and Ecology* **300**:189–215.

Briese D.T. (1997). Biological control of St. John's wort: past, present and future. *Plant Protection Quarterly* **12**:73–80.

Brigham A.J. (2006). HACCP and effective pest control in 'high risk' areas. In Feare C.J. and Cowan D.P. (eds.). *Advances in vertebrate pest management IV*, pp. 255–266. Filander Verlag, Fürth, Germany.

Brooks K.M. (1993). Impacts on benthic invertebrate communities caused by aerial application of carbaryl to control burrowing shrimp in Willapa Bay, Washington. In Abstracts of technical papers presented at the 85th annual meeting of the National Shellfisheries Association, Portland, Oregon, May 31–June 3 1993. *Journal of Shellfish Research* **12**:146.

Brothers N.P. (1984). Breeding, distribution and status of burrow-nesting petrels at Macquarie Island. *Australian Wildlife Research* **11**:113–131.

Brown J.K. and Kapler-Smith J. (eds.) (2000). *Wildland fire in ecosystems: Effect of fire on flora*. General Technical Report RMRST-GTR-42-Vol.2. U.S.D.A. Forest Service, Rocky Mountain Research Station, Ogden, UT.

Brown K.P., Moller H., Innes J., and Alterio N. (1996). Calibration of tunnel tracking rates to estimate relative abundance of ship rats (*Rattus rattus*) and mice (*Mus musculus*) in a New Zealand forest. *New Zealand Journal of Ecology* **20**:271–275.

Brown M.T. and Lamare M.D. (1994). The distribution of *Undaria pinnatifida* (Harvey) Suringar within Timaru Harbour, New Zealand. *Japanese Journal of Phycology* **42**:63–70.

Bruce B.D., Sutton C.A., and Lyne V. (1995). Laboratory and field studies of the larval distribution and duration of the introduced seastar *Asterias amurensis* with updated and improved prediction of the species spread based on a larval dispersal model. Final report to the Fisheries Research and Development Corporation. CSIRO Division of Marine Research, Hobart, Australia.

Buchan L.A.J. and Padilla D.K. (1999). Estimating the probability of long-distance overland dispersal of invading aquatic species. *Ecological Applications* **9**:254–265.

Buckingham G.R. (1996). Biological control of alligator weed, *Alternanthera philoxeroides*, the world's first aquatic weed success story. *Castanea* **61**:232–243.

Buckingham G.R. (2002). Alligator weed. In van Driesche, R.G., Lyon, S., Blossey, B., Hoddle, M.S., and Reardon, R. (eds.) *Biological control of invasive plants in the eastern United States*, pp. 5–15. USDA Forest Service, Morgantown, WV.

Buddenhagen C.E. (2006). The successful eradication of two blackberry species *Rubus megalococcus* and *R. adenotrichos* (Rosaceae) from Santa Cruz Island, Galapagos, Ecuador. *Pacific Conservation Science* **12**:272–278.

Buhler D.D., Liebman M., and Obrycki J.J. (2000). Theoretical and practical challenges to an IPM approach to weed management. *Weed Science* **48**:274–280.

Bullock D.J., North S.G., Dulloo M.E., and Thorsen M. (2002). The impact of rabbit and goat eradication on the ecology of Round Island, Mauritius. In Veitch C.R. and Clout M.N., (eds.) *Turning the tide: the eradication of invasive species*, pp. 53–63. IUCN SSC Invasive Species Specialist Group, IUCN, Gland, Switzerland and Cambridge, UK.

Burbidge A.A. and Morris K.D. (2002). Introduced mammal eradications for nature conservation on Western Australian islands: a review. In Veitch C.R. and Clout M.N. (eds.) *Turning the tide: the eradication of invasive species*, pp. 64–70. IUCN SSC Invasive Species Specialist Group, IUCN, Gland, Switzerland and Cambridge, UK.

Burke M.J., and Grime J.P. (1996). An experimental study of plant community invasibility. *Ecology* **77**: 776–790.

Burrows N.D., Algar D., Robinson A.D., Singara J., Ward B., and Liddelow G. (2003). Controlling introduced predators in the Gibson Desert of Western Australia. *Journal of Arid Environments* **55**:691–713.

Bush D.E. and Smith S.D. (1995). Mechanisms associated with decline of woody species in riparian ecosystems of southwestern US. *Ecological Monographs* **65**:347–370.

Byers J.E., Reichard S., Randall J.M., Parker I.M., Smith C.S., Lonsdale W.M., et al. (2002). Directing research to reduce the impacts of nonindigenous species. *Conservation Biology* **16**:630–640.

Cabin R.J., Weller S.G., Lorence D.H., Flynn T.W., Sakai A.K., Sandquist D., et al. (2000). Effects of long-term ungulate exclusion and recent alien species control on the preservation and restoration of a Hawaiian tropical dry forest. *Conservation Biology* **14**: 439–453.

Cacho O. (2004). When is it optimal to eradicate a weed invasion? In Sindel B.M. and Johnson S.B. (eds.) *Weed Management: Balancing People, Planet and Profit*. Proceedings of the 14th Australian Weeds Conference, pp. 49–54. Weed Society of New South Wales, Sydney.

Cacho O.J., Spring D., Pheloung P., and Hester S. (2006). Evaluating the feasibility of eradicating an invasion. *Biological Invasions* **8**:903–917.

Cameron E.Z., Linklater W.L., Stafford K.J., and Veltman C.J. (1999). Birth sex ratios relate to mare condition at conception in Kaimanawa horses. *Behavioural Ecology* **10**:472–475.

Campbell K. and Donlan C.J. (2005). Feral goat eradications on islands. *Conservation Biology* **19**:1362–1374.

Campbell K.J., Baxter G.S., Murray P.J., Coblentz B.E., and Donlan C.J. (2005). Development of a prolonged estrus effect for use in Judas goats. *Applied Animal Behaviour Science* **32**: 737–743.

Capperino M.E. and Schneider E.L. (1985). Floral biology of *Nymphaea mexicana* Zucc. (Nymphaeaceae). *Aquatic Botany* **23**:83–93.

Caring for Country Unit website http://www.nlc.org.au/html/care_menu.html

Carlton J.T. (1985). Transoceanic and interoceanic dispersal of coastal marine organisms: the biology of ballast water. *Oceanography and Marine Biology Annual Review* **23**:313–371.

Carlton J.T. (1999). Invasions in the sea: six centuries of reorganising the earth's marine life. In Sandlund O.T., Schei P.J., and Viken A. (eds.) *Invasive Species and Biodiversity Management*, pp. 195–212. Kluwer Academic Publishers, The Netherlands.

Carlton J.T. and Geller J.B. (1993). Ecological roulette: The global transport of nonindigenous marine organisms. *Science* **261**:78–82.

Carlton J.T. and Hodder J. (1995). Biogeography and dispersal of coastal marine organisms: experimental studies on a replica of a 16th-century sailing vessel. *Marine Biology* **121**: 721–730.

Carlton J.T. and Mann R. (1996). Transfers and world-wide introductions. In Kennedy V. S., Newell R. I. E., and Eble A. F. (eds.) The eastern oyster *Crassostrea virginica*, pp. 691–705. Sea Grant College, College Park, MD.

Carr L. and Dumbauld B. (1999). *1999 monitoring and control plans for* Carcinus maenas *in Willapa Bay and Grays Harbour, Washington*. Washington Department of Fish and Wildlife, Washington, DC.

Carrion V., Donlan C.J., Campbell K., Lavoie C., and Cruz F. (2007). Feral donkey (*Equus asinus*) eradications in the Galapagos. *Biodiversity and Conservation* **16**: 437–445.

Carter S.P. and Bright P.W. (2002). Habitat refuges as alternatives to predator control for the conservation of endangered Mauritian birds. In Veitch C.R. and Clout M.N. (eds.) *Turning the tide: eradication of invasive species*, pp. 71–78. IUCN SSC Invasive Species Specialist Group, IUCN, Gland, Switzerland and Cambridge, UK.

Carver C.E., Chisholm A., and Mallet A.L. (2003). Strategies to mitigate the impact of *Ciona intestinalis* (L.) biofouling on shellfish production. *Journal of Shellfish Research* **22**:621–631.

Caughley G. (1977). *Analysis of vertebrate populations*. Wiley, London.

Caut S., Casanovas J.G., Virgos E., Lozano J., Witmer G.W., and Courchamp F. (2007). Rats dying for mice: modelling the competitor release effect. Austral Ecology. **32**: 858–868.

CBD (1992). *Convention on Biological Diversity*. http://www.cbd.int/convention/convention.shtml

CBD (2002). Report of the Sixth Meeting of the Conference of the Parties to the Convention on Biological Diversity, UNEP/CBD/COP/6/20, 27 May 2002.

CDFG (2007). *Lake Davis pike eradication project: Final environmental impact report/environmental impact statement (EIR/EIS)*. Resources Agency California Department of Fish and Game. http://www.dfg.ca.gov/lakedavis/EIR%2DEIS/ (accessed August 2008).

References

Center T.D., Frank J.H., and Dray F.A. (1997). Biological control. In Simberloff D, Schmitz D.C., and Brown T.C. (eds.) *Strangers in Paradise. Impact and Management of Nonindigenous species in Florida,* pp. 245–266. Island Press, Washington DC.

Chong G.W., Reich R.M., Kalkhan M.A., and Stohlgren T.J. (2001). New approaches for sampling and modeling native and exotic plant species richness. *Western North American Naturalist* **61**: 328–335.

Champion P.D. and Clayton J.S. (2000). Border control for potential aquatic weeds. Stage 1, Weed risk model. *Science for Conservation* **141**, Wellington Dept. of Conservation, New Zealand.

Champion P.D. and Clayton J.S. (2001). Border control for potential aquatic weeds. Stage 2. Weed risk assessment. *Science for Conservation* **185**, Wellington Dept. of Conservation, New Zealand.

Chapin F.S., Zavaleta E.S., Eviner V.T., Naylor R.L., Vitousek P.M., Reynolds H.L., *et al.* (2000). Consequences of changing biodiversity. *Nature* **405**:234–242.

Chicoine T.K., Fay P.K., and Nielsen G.A. (1985). Predicting weed migration from soil and climate maps. *Weed Science* **34**:57–61.

Chikwenhere G.P. and Phiri G. (1999). History of water hyacinth and its control efforts on Lake Chivero in Zimbabwe. In Hill, M.P., Julien, M.H., and Center, T.D. (eds.) *Proceedings of the First IOBC Global Working Group Meeting for the Biological and Integrated Control of Water Hyacinth,* pp. 91–97. Zimbabwe, Plant Protection Research Institute, Pretoria, South Africa, 16–19 November 1998.

Chilton C. (1910). Note on the dispersal of marine crustacea by means of ships. *Transactions NZ Institute* **42**:131–133.

Chong G.W., Reich R.M., Kalkhan M.A., and Stohlgren T.J. (2001). New approaches for sampling and modeling native and exotic plant species richness. *Western North American Naturalist* **61**: 328–335.

Chong G.W., Otsuki Y., Stohlgren T.J., Guenther D., Villa C., and Waters M.A. (2006). Evaluating plant invasions from both habitat and species perspectives. *Western North American Naturalist* **66**:92–105.

Choquenot D. and Parkes J. (2001). Setting thresholds for pest control: how does pest density affect resource viability? *Biological Conservation* **99**:29–46.

Choquenot D., Nicol S.J., and Koehn J.D. (2004). Bioeconomic modelling in the development of invasive fish policy. *New Zealand Journal of Marine and Freshwater Research* **38**:419–428.

Cilliers C.J. (1999). Biological control of parrot's feather, *Myriophyllum aquaticum* (Vell.) Verdc. (Haloragaceae), in South Africa. *African Entomology Memoir* **1**:113–118.

CITES Convention on International Trade in Endangered Species of wild fauna and flora. http://www.cites.org/eng/disc/what.shtml

Civeyrel L. and Simberloff D. (1996). A tale of two snails: is the cure worse than the disease? *Biodiversity and Conservation* **5**:1231–1252.

Clarke C., Hilliard R., Junqueira, A.de O.R., Neto A.de C.L., Polglaze, J., and Raaymakers S. (2004). Ballast Water Risk Assessment, Port of Sepetiba, Federal Republic of Brazil, December 2003: Final Report. *GloBallast Monograph Series* No. 14.

Clarke P.J., Latz P.K., and Albrecht D.E. (2005). Long-term changes in semi-arid vegetation: invasion of an exotic perennial grass has larger effects than rainfall variability. *Journal of Vegetation Science* **16**:237–248.

Clavero M. and Garcia-Berthou E. (2006). Homogenization dynamics and introduction routes of invasive freshwater fish in the Iberian Peninsula. *Ecological Applications* **16**:2313–2324.

Clearwater S.J., Hickey C.W., and Martin M.L. (2008). Overview of potential piscicides and molluscicides for controlling aquatic pest species in New Zealand. *Science for Conservation* **283**. Department of Conservation, Wellington.

Clout M.N. and De Poorter M. (2005). International initiatives against invasive alien species. *Weed Technology* **19**:523–527.

Clout M.N., Denyer K., James R.E., and McFadden I.G. (1995). Breeding success of New Zealand pigeons (*Hemiphaga novaeseelandiae*) in relation to control of introduced mammals. *New Zealand Journal of Ecology*, **19**: 209–212.

Clout M.N. and Russell J.C. (2006). The eradication of mammals from New Zealand islands. In Koike F., Clout M.N., Kawamichi M., De Poorter M., and Iwatsuki K. (eds.) Assessment and control of biological invasion risks, pp. 127–141. Shoukadoh Book Sellers, Kyoto and IUCN, Switzerland.

Coates P. (2006). *Strangers on the land: American perceptions of immigrant and invasive species.* University of California Press, Berkeley, CA.

Coblentz B.E. (1978). The effects of feral goats (*Capra hircus*) on island ecosystems. *Biological Conservation* **13**:279–286.

Coblentz B.E. (1997). Goat eradication on Isabella Island. *Conservation Biology* **11**:1047–1048.

Cock M.J.W. (2002). Risks of non-target impact versus stakeholder benefits in classical biological control of arthropods: selected case studies from developing countries. In *International Symposium on Biological Control of Arthropods*, pp. 25–33, January 14–18, 2002, Honolulu, Hawaii, USA, USDA Forest Service Publication FHTET-03-05, Morgantown, WV.

Cock M.J.W. and Seier M.K. (2007). The scope for biological control of giant hogweed, *Heracleum mantegazzianum*. In Pyšek P., Cock M.J.W., Nentwig W., Ravn H.P. and Wade W. (eds.) *Ecology and Management of Giant Hogweed* pp. 255–271. CABI Publishing, Wallingford, UK.

Cock M.J.W., Ellison C.A., Evans H.C., and Ooi P.A. (2000). Can failure be turned into success for biological control of mile-a-minute weed (*Mikania micrantha*)? In Spencer N.R. (ed.) *Proceedings of the X International Symposium on Biological control of Weeds*, 4–14 July 1999, Montana State University, Montana, USA, Montana State University, Bozeman, Montana, USA. 155–167.

Codex FAO/WHO world food standards Codex Alimentarius. http://www.codexalimentarius.net/web/index_en.jsp

Cohen A.N. and Carlton J.T. (1998). Accelerating invasion rate in a highly invaded estuary. *Science* **279**:555–557.

Cohen A.N., Carlton J.T., and Fountain M.C. (1995). Introduction, dispersal and potential impacts of the green crab Carcinus maenas in San Francisco Bay, California. *Marine Biology* **122**:225–237.

Colautti R.I. and MacIsaac H.J. (2004). A neutral terminology to define 'invasive' species. *Diversity and Distributions,* **10**:135–141.

Colautti R.I., Bailey S.A., van Overdijk C.D.A., Amundsen K., and MacIsaac H.J. (2006). Characterised and projected costs of non-indigenous species in Canada. *Biological Invasions* **8**:45–59.

Commonwealth of Australia (2006). *Background document for the threat abatement plan to reduce the impacts of tramp ants on biodiversity in Australia and its territories.* Department of the Environment and Heritage, Canberra, ACT Australia.

Connell, J.H. (1983). On the prevalence and relation importance of interspecific competition: Evidence from field experiments. *American Naturalist* **122**:661–696.

Convention on Access to Information, Public Participation in Decision Making and access to Justice in Environmental Matters United Nations ECE, 1998. Aarhus. http://www.unece.org/env/pp/documents/cep43e.pdf (accessed September 2008).

Convention on Biological Diversity. (2005). *Towards the Development of a Joint Work Plan for the Management of Marine Invasive Alien Species*, pp. 37. Workshop of the Joint Work Programme on Marine and Coastal Invasive Alien Species. Annex II: Draft Joint Work Plan for the Management of Marine Invasive Alien Species, Montreal.

Cook G. D., Setterfield S.A., and Maddison J. P. (1996). Shrub invasion of a tropical wetland: Implications for weed management. *Ecological Applications* **6**:531–537.

Cooke B.D. and Fenner F. (2002). Rabbit haemorrhagic disease and the biological control of wild rabbits, *Oryctolagus cuniculus*, in Australia and New Zealand. *Wildlife Research* **29**: 689–706.

Coombs E.M., Clark J.K., Piper G.L., and Cofrancesco, Jr., A.F. (2004). *Biological Control of Invasive Plants in the United States*. Oregon State University Press, Corvallis, OR.

Coomes D.A., Mark A.F., and Bee J. (2006). Animal control and ecosystem recovery. In Allen R.B. and Lee W.G. (eds.) *Biological Invasions in New Zealand*. Ecological Studies **186**: 379–353, Springer-Verlag, Berlin Heidelberg.

Cooney R. (2004). *The precautionary principle in biodiversity conservation and natural resource Management: an issues paper for policy-makers, researchers and practitioners*. IUCN, Gland, Switzerland and Cambridge, UK.

Cooney R. and Dickson B. (eds.) (2005). *Biodiversity and the Precautionary Principle: Risk and Uncertainty in Conservation and Sustainable Use*. Earthscan, London, UK.

Copp G. (1989). Electrofishing for fish larvae and 0+ juveniles: equipment modifications for increased efficiency with short fishes. *Aquaculture Research* **20**:453–462.

Copp G.H., Wesley K.J., and Villizi L. (2005a). Pathways of ornamental and aquarium fish introductions into urban ponds of Epping Forest (London, England): the human vector. *Journal of Applied Ichthyology* **21**:263–274.

Copp G.H., Garthwaite R., and Gozlan R.E. (2005b). *Risk identification and assessment of non-native freshwater fishes: concepts and perspectives on protocols for the UK*. Science Series Technical Report 129. CEFAS, Lowesoft, UK.

Copson G. and Whinam J. (1998). Response of vegetation on subantarctic Macquarie Island to reduced rabbit grazing. *Australian Journal of Botany* **46**:15–24.

Corkum L., Sapota M.R., and Skora K.E. (2004). The round goby, *Neogobius melanostomus*, a fish invader on both sides of the Atlantic Ocean. *Biological Invasions* **6**:173–181.

Correll R. and Marvanek S. (2006). Sampling for detection of branched broomrape. In Preston C., Watts J.H., and Crossman N.D. (eds.) *Managing Weeds in a Changing Climate*. Proceedings of the 15th Australian Weeds Conference. Weed Management Society of South Australia, Adelaide. 618–621.

Cottalorda J.M., Robert P., Charbonnel E., Dimeet J., Menager V., Tillman M., *et al.* (1996). Eradication de la colonie de *Caulerpa taxifolia* de´- couverte en 1994 dans les eaux du Parc National de Port-Cros (Var, France),. In Ribera M. A., Ballesteros E. Boudouresque C. F., Gomez A., and Gravez V. (eds.) *Second International Workshop on Caulerpa taxifolia*, pp.149–156. Univ. Barcelona Publ, Spain.

Cotter D., O'Donovan V., O'Maoiléidigh N., Rogan G., Roche N., and Wilkins N.P. (2000). An evaluation of the use of triploid Atlantic salmon (*Salmo salar* L.) in minimising the impact of escaped farmed salmon on wild populations. *Aquaculture* **186**:61–75.

Courchamp F. and Caut S. (2005). Use of biological invasions and their control to study the dynamics of interacting populations. In Cadotte M.W., McMahon S.M., and Fukami T. (eds.) *Conceptual Ecology and Invasions Biology*, pp. 253–279. Springer Verlag, Berlin.

Courchamp F., Langlais M., and Sugihara G. (1999a). Cats protecting birds: modelling the mesopredator release effect. *Journal of Animal Ecology* **68**:282–292.

Courchamp F., Langlais M., and Sugihara G. (1999b). Control of rabbits to protect island birds from cat predation. *Biological Conservation* **89**:219–225.

Courchamp F., Clutton-Brock T., and Grenfell B. (1999c). Inverse density dependence and the Allee effect. *Trends in Ecology and Evolution* **14**:405–410.

Courchamp F., Langlais M., and Sugihara G. (2000). Rabbits killing birds: modelling the hyperpredation process. *Journal of Animal Ecology* **69**:154–164.

Courchamp F., Chapui J.-L., and Pascal M. (2003a). Mammal invaders on islands: impact, control and control impact. *Biological Reviews* **78**:347–383.

Courchamp F., Woodroffe R., and Roemer G.W. (2003b). Removing protected populations to save endangered species. *Science* **302**:1532.

Courtenay W.R. and Robins C.R. (1973). Exotic aquatic organisms in Florida with emphasis on fishes: a review and recommendations. *Transactions American Fisheries Societ,* **102**: 1–12.

Courtenay W.R. and Stauffer J.R. (1990). The introduced fish problem and the aquarium fish industry. *Journal of the World Aquaculture Society* **21**:145–159.

Cousens R. (1985). A simple model relating yield loss to weed density. *Annals of Applied Biology,* **107**:239–252.

Cousens R. and Mortimer M. (1995). Dynamics of weed populations. Cambridge University Press, Cambridge.

Coutts A.D.M. and Forrest B.M. (2005). *Evaluation of eradication tools for the clubbed tunicate* Styela clava. Report No. 1110, Cawthron Institute, Nelson, New Zealand.

Coutts A.D.M. and Forrest B.M. (2007). Development and application of tools for incursion response: lessons learned from the management of a potential marine pest. *Journal of Experimental Marine Biology and Ecology* **342**:154–162.

Coutts A.D.M. and Taylor M.D. (2004). A preliminary investigation of biosecurity risks associated with biofouling on merchant vessels in New Zealand. *New Zealand Journal of Marine and Freshwater Research* **38**:215–229.

Coutts A.D.M. (2002). *A biosecurity investigation of a barge in the Marlborough Sounds.* Report No. 744. Cawthron Institute, Nelson, New Zealand.

Coutts A.D.M. and Dodgshun T.J. (2007). The nature and extent of organisms in vessel sea-chests: a protected mechanism for marine bioinvasions. *Marine Pollution Bulletin* **54**:875–886.

Coutts A.D.M., Moore K.M., and Hewitt C.L. (2003). Ships' sea-chests: an overlooked transfer mechanism for non-indigenous marine species? *Marine Pollution Bulletin* **46**:1510–1513.

Cowan P. (2000). Biological control of possums: Prospects for the future. In Montague T.L. (ed.) *The brushtail possum. Biology, impact and management of an introduced marsupial,* pp. 262–270. Manaaki Whenua Press, Lincoln, New Zealand.

Cowan P.E. (2005). Brushtail possum. In King C.M. (ed.) *The Handbook of New Zealand Mammals,* pp. 56–80. Oxford University Press, Melbourne, Australia.

Cowie R.H. (2005). Alien non-marine mollusks in the islands of the tropical and subtropical Pacific: a review. *American Malacological Bulletin* **20**:95–103.

Crall A.W., Meyerson L., Stohlgren T.J., Jarnevich C.S., Newman G.J., and Graham J. (2006). Show me the numbers: What data currently exist for non-native species in the U.S.? *Frontiers in Ecology and the Environment* **4**: 414–418.

Crawley M.J. (1987). What makes a community invasible? In Gray A.J., Crawley M.J., and Edwards P.J. (eds.) *Colonization, succession and stability,* pp. 429–453. Blackwell Press, Oxford.

Creagh C. (1991/92). A marauding weed in check. *Ecos* **70**:26–29.

Creese R.G., Davis A.R., and Glasby T.M. (2004). Eradicating and preventing the spread of the invasive alga *Caulerpa taxifolia* in NSW. Final Report to the Natural Heritage Trust's Coasts and Clean Seas Introduced Marine Pests Program, Project No. 35593. NSW Fisheries.

Critchley A.T. and Dijkema R. (1984). On the presence of the introduced brown alga *Sargassum muticum* attached to commercially imported *Ostrea edulis* in SW Netherlands. Botanica Marina **27**:211–216.

Critchley A.T., Farnham W.F., and Morrell S.L. (1986). An account of the attempted control of an introduced marine alga *Sargssum muticum* in Southern England UK. *Biological Conservation* **35**:313–332.

Cromarty P.L., Broome K.G., Cox A., Empson R.A., Hutchinson W.M., and McFadden I. (2002). Eradication planning for invasive alien animal species on islands—the approach developed by the New Zealand Department of Conservation. In Veitch C.R. and Clout M.N. (eds.) *Turning the tide: the eradication of invasive species,* pp. 85–91. IUCN SSC Invasive Species Specialist Group: IUCN, Gland, Switzerland and Cambridge, UK.

Crosier C.S. (2004). Synergistic methods to generate predictive models at large spatial extents and fine resolution. Colorado State University, Fort Collins.

Crosier C.S. and Stohlgren T.J. (2004). Improving biodiversity knowledge through dataset synergy: a case study of non-native vascular plants in Colorado. *Weed Technology* **18**:1441–1444.

Cruttwell McFadyen R.E. (2000). Successes in biological control of weeds. In Spenser, N.R. (ed.) *X International Symposium on Biological Control of Weeds,* Bozeman, MT, pp. 3–14. Advanced Litho Printing, Great Falls, MT.

Cruz F., Donlan C.J., Campbell K., and Carrion V. (2005). Conservation action in the Galàpagos: feral pig (*Sus scrofa*) eradication from Santiago Island. *Biological Conservation* **121**:473–478.

Cullen J.M. and Hassan S. (1988). Pathogens for the control of weeds. *Philosophical Transactions of the Royal Society of London* **318**:213–224.

Culver K.L. and Kuris A.M. (2000). The apparent eradication of a locally established introduced marine pest. *Biological Invasions* **2**:245–253.

Cunningham D.C., Woldendorp G., Burgess M.B., and Barry S.C. (2003). Prioritising sleeper weeds for eradication: selection of species based on potential impacts on agriculture and feasibility of eradication. Bureau of Resource Sciences, Canberra, Australia.

Daehler C.C. (2006). Invasibility of tropical islands by introduced plants: partitioning the influence of isolation and propagule pressure. *Preslia* **78**:389–404.

Dafforn K.A., Glasby T.M., and Johnston E.L. (2008). Differential effects of tributyltin and copper anti-foulants on recruitment of non-indigenous species. *Biofouling* **24**:23–33.

D'Antonio C.M. and Vitousek P.M. (1992). Biological invasions by exotic grasses, the grass fire cycle, and global change. *Annual Review of Ecology and Systematics* **23**:63–87.

Davidson D.W. (1998). Resource discovery versus resource domination in ants: a functional mechanism for breaking the tradeoff. *Ecological Entomology* **23**:484–490.

Davis M.A., Grime J.P., and Thompson K. (2000). Fluctuating resources in plant communities: a general theory of invasibility. *Journal of Ecology* **88**:528–534.

Davis N.E., O'Dowd D.J., Green P.T., and MacNally R.N. (2008). Effects of an alien ant invasion on the abundance, behavior and reproduction of endemic island birds. *Conservation Biology* **22**:1165–1176.

Davis P.R., Beggs J., Wylie R., and Dress B.M. (2004). Scientific review of the Australian red imported fire ant (*Solenopsis invicta*) eradication program. Report to the National Red Imported Fire Ant Committee.

Dawson H.A., Reinhardt U.G., and Savino J.F. (2006). Use of electrical or bubble barriers to limit the movement of Eurasian ruffe (*Gymnocephalus cernuus*). *Journal of Great Lakes Research* **32**:40–49.

Day R.K., Kairo M.T.K., Abraham Y.J., Kfir R., Murphy S.T., Mutitu K.E., and Chilma C.Z. (2003). Biological control of Homopteran pests of conifers in Africa. *In:* Neuenschwander P., Borgemeister C., and Langewald J. (eds.) *Biological Control in IPM Systems in Africa,* CABI Publishing, Wallingford, UK. 101–112.

Day T.D., Matthews L.R., and Waas J.R. (2003). Repellents to deter New Zealand's North Island robin *Petroica australis longiceps* from pest control baits. *Biological Conservation* **114**:309–316.

Dayan F.E. and Netherland M.D. (2005). Hydrilla, the perfect aquatic weed, becomes more noxious than ever. *Pesticide Outlook* **16**:277–282.

De Nicola A.J., Kesler D.J., and Swihart R.K. (1996). Ballistics of a biobullet delivery system. *Wildlife Society Bulletin* **24**:301–305.

De Poorter M. (2006). Nationale strategie en actieplannen voor invasieve exoten: ervaringen in andere landen [The need for a strategic national approach on invasive alien species (in Dutch)] In Branquart E., Baus E., Pieret N., Vanderhoeven S., and Desmet P. (eds.) *SOS Invasions ! Les espèces exotiques invasives en Belgique; Uitheemse invasieve Soorten in België (Brussels, 9 & 10 March 2006).* Belgian Biodiversity Platform.

De Poorter M. and Browne M. (2005). The Global Invasive Species Database (GISD) and international information exchange: using global expertise to help in the fights against invasive alien species In *Introduction and Spread of Invasive Species*, pp. 49–54. Proceedings of IPPC, Berlin, Germany, July 2005.

De Poorter M. and Clout M.N. (2005). Biodiversity conservation as part of Plant Protection: the opportunities and challenges of risk analysis. In *Introduction and Spread of Invasive Species*, Proceedings of IPPC, Berlin, Germany, July 2005, 55–60.

De Poorter M., Browne M., Lowe S., and Clout M.N. (2005). The ISSG Global Invasive Species Database and other aspects of an early warning system. In Mooney H.A., Mack R.N., Mcneely J.A., Neville L.E., Schei P.J. and J.K. Waage (eds.).*Invasive Alien Species—A New Synthesis*, pp. 59–83. Island Press, Washington, DC.

Deagle B.E., Bax N., Hewitt C.L., and Patil J.G. (2003). Development and evaluation of a PCR-based test for detection of *Asterias* (Echinodermata: Asteroidea) larvae in Australian plankton samples from ballast water. *Marine and Freshwater Research* **54**:709–719.

DeBach P. (1974). *Biological Control by Natural Enemies*. Cambridge University Press, Cambridge.

DeFerrari C.M. and Naiman R.J. (1994). A multi-scale assessment of the occurrence of exotic plants on the Olympic Peninsula, Washington. *Journal of Vegetation Science* **5**:247–258.

Dehnen-Schmutz K., Touza J., Perrings C., and Williamson M. (2007). The horticultural trade and ornamental plant invasions in Britain. *Conservation Biology,* **21**:224–231.

DeLoach C.J. (1997). Biological control of weeds in the United States and Canada. In Luken J.O. and Thieret J.W. (eds.) *Assessment and Management of Plant Invasions,* pp. 172–194. Springer-Verlag, New York.

Demeter A. (2006). International legislation and obligations on IAS-related issues; In Branquart E., Baus E., Pieret N., Vanderhoeven S. and Desmet P. (eds.) *SOS Invasions! Les espèces exotiques invasives en Belgique; Uitheemse invasieve Soorten in België* (Brussels, 9 & 10 March 2006). Belgian Biodiversity Platform. 63–65.

Denny C.M. (2007). *In situ* plastic encapsulation of the NZHMS Canterbury frigate: A trial of a response tool for marine fouling pests. Report No. 1271, Cawthron Institute, Nelson, New Zealand.

Denny C.M., Morley C.G., Chadderton W.L., and Hero J.M. (2005). *Demonstration project to eradicate invasive cane toads and mammals from Viwa Island, Fiji.* http://www.issg.org/cii/PII/Demonstration%20Projects/viwa.htm

Denslow J.S. (2003). Weeds in paradise: thoughts on the invasibility of tropical islands. *Annals of the Missouri Botanical Gardens* **90**: 119–127.

Department of Primary Industries (2008). *Carpsim 2.0.5 software.* http://www.dpi.vic.gov.au/DPI/Vro/vrosite.nsf/pages/pest_animals_carpsim (accessed September 2008).

Dewey S.A. and Anderson K.A. (2004). Distinct roles of surveys, inventories, and monitoring in adaptive weed management. *Weed Technology* **18**:1449–1452.

Diederich S., Nehls G., van Beusekom J.E., and Reise K. (2005). Introduced Pacific oysters (*Crassostrea gigas*) in the northern Wadden Sea: invasion accelerated by warm summers? *Helgolander Marine Research* **59**:97–106.

DiTomaso J.M. (2000). Invasive weeds in rangelands: Species, impacts, and management. *Weed Science* **48**:255–265.

DiTomaso J.M. and Johnson D.W. (eds.) (2006). *The Use of Fire as a Tool for Controlling Invasive Plants*, pp. 1–49.Cal-IPC Publication 2006–01. California Invasive Plant Council, Berkeley, CA.

DiTomaso J.M., Brooks M.L., Allen E.B., Minnich R., Rice P.M., and Kyser G.B. (2006). Control of invasive weeds with prescribed burning. *Weed Technology* **20**:535–548.

Dodd A. P. (1940). *The biological campaign against Prickly-Pear*. Commonwealth Prickly Pear Board Bulletin, Brisbane Australia.

Dodd J. (2004). Kochia (*Bassia scoparia* (L.) A.J. Scott) eradication in Western Australia: a review. In Sindel B.M. and Johnson S.B. (eds.) *Weed Management: Balancing People, Planet and Profit*. Proceedings of the 14th Australian Weeds Conference. Weed Society of New South Wales, Sydney, pp.496–500.

Dodd, J. & Moore, R.H. (1993) Introduction and status of *Kochia scoparia* in Western Australia. *Proceedings of the Tenth Australian and Fourteenth Asian-Pacific Weeds Conference* (ed. by J.T. Swarbrick), Volume 1, pp. 496–500. Weed Society of Queensland, Brisbane, Australia.

Donlan C.J., Tershy B.R., and Croll D.A. (2002). Islands and introduced herbivores: conservation action as ecosystem experimentation. *Journal of Applied Ecology* **39**:235–246.

Drescher J., Bluthgen N., and Feldhaar H. (2007). Population structure and intraspecific aggression in the invasive ant species *Anoplolepis gracilipes* in Malaysian Borneo. *Molecular Ecology* **16**: 1453–1465.

Duke D., O'Quinn P., and Sutton D.L. (2000). Control of hygrophila and other aquatic weeds in the Old Plantation Water Control District. *Aquatics* **22**:4–8.

Dukes J.S. and Mooney H.A. (1999). Does global change increase the success of biological invaders. *Trends in Ecology and Evolution* **14**:135–139.

Dumbauld B.R., Armstrong D.A., and Skalski J. (1997). Efficacy of the pesticide Carbaryl for thalassinid shrimp control in Washington State oyster (*Crassostrea gigas*, Thunberg, 1793) aquaculture. *Journal of Shellfish Research* **16**:503–518.

Dwyer W.P., Fredenberg W., and Erdahl D.A. (1993). Influence of electroshock and mechanical shock on survival of trout eggs. *North American Journal of Fisheries Management* 13:839–843.

Eason C.T. and Hickling G.J. (1992). Evaluation of a bio-dynamic technique for possum pest control. *New Zealand Journal of Ecology* **16**:141–144.

Eason C.T. and Wickstrom M. (2001). *Vertebrate pesticide toxicology manual (poisons)*. Department of Conservation, Wellington, New Zealand.

Ebbert S. (2000). Successful eradication of introduced foxes from large Aleutian islands. In Salmon T.P. and Crabb A.C. (eds.) *Proceedings of the 9th Vertebrate Pest Conference*, pp. 127–132. San Diego, California.

Eckhardt R.C. (1972). Introduced plants and animals in the Galapagos Islands. *BioScience* **22**:587–590.

Edwards D. and Musil C.J. (1975). *Eichhornia crassipes* in South Africa—a general review. *Journal of the Limnological Society of Southern Africa* **1**:23–27.

El-Ghareeb R. (1991). Vegetation and soil changes induced by *Mesembryanthemum crystallinum* L. in a Mediterranean desert ecosystem. *Journal of Arid Environments* **20**:321–330.

Elton C.S. (1958). *The Ecology of Invasions by Animals and Plants*. Methuen, London.

Engeman R.M. and Linnell M.A. (1998). Trapping strategies for deterring the spread of brown tree snakes from Guam. *Pacific Conservation Biology* **4**:548–553.

Environmental Risk Management Authority (2007). *The reassessment of 1080*. Wellington, Environmental Risk Management Authority.

Eplee R.E. (2001). Co-ordination of witchweed eradication in the USA. In Wittenberg R. and Cock M.J.W. (eds.) *Invasive alien species. A toolkit of best prevention and management practices.* CAB International, Wallingford, Oxford, UK.

EU (2008). Commission decision of 10 April 2008 concerning the non-inclusion of rotenone, extract from equisetum and chinin-hydrochlorid in Annex I to Council Directive 91/414/EEC and the withdrawal of authorisations for plant protection products containing these substances. *Official Journal of the European Union*, **L108**:30–32.

Evans H.C. (2000). Evaluating plant pathogens for biological control of weeds: an alternative view of pest risk assessment. *Australian Plant Pathology* **29**:1–14.

Evans H.C., Greaves M.P., and Watson A.K. (2001). Fungal biocontrol agents of weeds. *In:* Butto T.M., Jackson C., and Magan N. (eds.) *Fungi as Biocontrol Agents: Progress, Problems and Potential*, pp. 169–192. CAB International Publishing, Wallingford, UK.

Evans S.M., Birchenough A.C., and Brancato M.S. (2000). The TBT Ban: Out of the Frying Pan into the Fire? *Marine Pollution Bulletin* **40**:204–211.

Fagerstone K.A., Coffey M.A., Curtis P.D., Dolbeer R.A., Killian G.J., Miller L.A. *et al.* (2002). Fertility control. *The Wildlife Society Technical Review* 02–2.

Fajt J.R. (1996). *Toxicity of rotenone to common carp and grass carp: respiratory effects, oral toxicity, and evaluation of a poison bait.* Unpubl. PhD thesis, Auburn University, Alabama.

FAO (1996). *Code of Conduct for the Import and Release of Exotic Biological Control Agents*. ISPM No 3. FAO, Rome.

FAO (1997). *Export certification*. ISPM No. 7. FAO, Rome. https://www.ippc.int/id/13724?language=en

FAO (2004a). *Pest risk analysis for quarantine pests including analysis for environmental risks and living modified organisms*. ISPM No. 11. FAO, Rome. https://www.ippc.int/id/34163?language=en

FAO (2004b). *Guidelines for a phytosanitary import regulatory system*. ISPM No. 20. FAO, Rome. https://www.ippc.int/id/36755?language=en

FAO (2004c). *Pest risk analysis for regulated non quarantine pests*. ISPM No. 21. FAO, Rome. https://www.ippc.int/id/36757?language=ePMn

FAO (2005a). *Guidelines for inspection*. ISPM No. 23. FAO, Rome. https://www.ippc.int/id/76471?language=en

FAO (2005b). *Guidelines for the Export, Shipment, Import and Release of Biological Control Agents and Other Beneficial Organisms*. ISPM No 3. FAO, Rome.

FAO (2006). *FISH INFOnetwork Market Report: Nile perch industry has problems.* http://www.eurofish.dk/indexSub.php?id=3276 (accessed August 2008).

FAO (2007a). *Glossary of phytosanitary terms*. ISPM No. 5. FAO, Rome. https://www.ippc.int/id/184195?language=en

FAO (2007b). *Frame work for pest risk analysis*. ISPM No. 2. FAO, Rome. https://www.ippc.int/id/184204?language=en

FAO Inland Water Resources and Aquaculture Service (2003). *Review of the state of world aquaculture*. FAO Fisheries Circular No. 886, Rev. 2. FAO, Rome.

Feare C. (1999). Ants take over from rats on Bird Island, Seychelles. *Bird Conservation International* **9**:95–96.

Fenner F. and Myers K. (1978). Myxoma viruses and myxomatosis in retrospect: the first quarter century of a disease. In Kurstak E. and Maramorosch K. (eds.) *Viruses and the Environment*, pp. 539–570. Academic Press, New York, NY.

Fenner F. and Ratcliffe F.N. (1965). *Myxamatosis*. Cambridge University Press, London.

Fenner F., Henderson D.A., Arita I., Jeuek Z., and Lagyni I.D. (1988). *Smallpox and its eradication*. World Health Organization, Geneva.

Ferguson R. (2000). *The Effectiveness of Australia's Response to the Black Striped Mussel Incursion in Darwin, Australia*. A Report of the Marine Pest Incursion Management Workshop, Darwin.

Finlayson B.J., Schnick R.A., Cailteux R.L., DeMong L., Horton W.D., McClay W., et al. (2000). *Rotenone use in fisheries management: administrative and technical guidelines*. American Fisheries Society, Bethesda, Maryland.

Finlayson B.J., Schnick R.A., Cailteux R.L., DeMong L., Horton W.D., McClay W., et al. (2002). Assessment of antimycin A use in fisheries and its potential for reregistration. *Fisheries* **27**:10–18.

Floerl O., Pool T.K., and Inglis G.J. (2004). Positive interactions between nonindigenous species facilitate transport by human vectors. *Ecological Applications* **14**:1724–1736.

Floerl O., Inglis G.J., and Hayden B. (2005a). A risk-based predictive tool to prevent accidental introductions of non-indigenous marine species. *Environmental Management* **35**:765–778.

Floerl O., Inglis G.J., and Marsh H.M. (2005b). Selectivity in vector management: an investigation of the effectiveness of measures used to prevent transport of non-indigenous species. *Biological Invasions* **7**:459–475.

Formicki K., Tanski A., Sadwoski M., and Winnicki A. (2004). Effects of magnetic fields on fyke net performance. *Journal of Applied Ichthyology* **20**:402–406.

Forrest B.M. (2007). *Managing risks from invasive marine species: is post-border management feasible?* PhD thesis, Victoria University of Wellington, Wellington, New Zealand.

Forrest B.M. and Blakemore K.A. (2006). Evaluation of treatments to reduce the spread of a marine plant pest with aquaculture transfers. *Aquaculture* **257**:333–345.

Forrest B.M. and Gardner J.P.A. (2009). Internal borders for managing invasive marine species. *Journal of Applied Ecology* **46**:46–54.

Forrest B.M., Taylor M.D. and Sinner J. (2006). Setting priorities for the management of marine pests using a risk-based decision support framework. In Allen R.B. and Lee W.G. (eds.) *Biological invasions in New Zealand*, Ecological Studies **146**: pp. 299–405. Springer-Verlag, Berlin, Heidelberg.

Forrest B.M., Hopkins G.A., Dodgshun T.J., and Gardner J.P.A. (2007). Efficacy of acetic acid treatments in the management of marine biofouling. *Aquaculture* **262**:319–332.

Forsyth D.M. and Duncan R.P. (2001). Propagule size and the relative success of exotic ungulate and bird introductions to New Zealand. *The American Naturalist* **157**:583–595.

Foster B.A. and Willan R.C. (1979). Foreign barnacles transported to New Zealand on an oil platform. *New Zealand Journal of Marine and Freshwater Research* **13**:143–149.

Fowler S.V. (2004). Biological control of an exotic scale, *Orthezia insignis* Browne (Homoptera: Ortheziidae), saves the endemic gumwood tree, *Commidendrum robustrum* (Roxb.). DC. (Asteraceae) on the island of St. Helena. *Biological Control* **29**:367–374.

Fox A.M., Haller W.T., and Shilling D.G. (1996). Hydrilla control with split treatments of fluridone in Lake Harris, Florida. *Hydrobiologia* **340**:235–239.

Fox M.D. and Fox B.J. (1986). The susceptibility of natural communities to invasion. In Groves R.H. and Burdon J.J. (eds.) *Ecology of biological invasions*, pp. 57–66. Cambridge University Press, Cambridge.

Frank H. and Cave R. (2005). *Metamasius callizona* is destroying Florida's native bromeliads. In: *International Symposium on Biological Control of Arthropods*, September 12–16, 2005, Davos, Switzerland. USDA Forest Service Publication FHTET-2005–08, Morgantown, West Virginia, USA. 91–101.

Frank J.H. (1998). How risky is biological control? Comment. *Ecology* **79**:1829–1834.
Franks A.J. (2002). The ecological consequences of buffel grass *Cenchrus ciliaris* establishment within remnant vegetation of Queensland. *Pacific Conservation Biology*, **8**:99–107.
Freeman J., Stohlgren T.J., Hunter M., Omi P., Martinson E., Chong G.W., *et al.* (2007). Rapid Assessment of Postfire Plant Invasions in Coniferous Forests of the Western United States. *Ecological Applications* **17**:1656–1665.
Freeman T.E. and Charudattan R. (1985). Conflicts in the use of plant pathogens as biocontrol agents for weeds. In Delfosse E.S. (ed.) *Proceedings of the VI International Symposium on the Biological Control of Weeds, Vancouver, Canada.* Agriculture Canada, Ottawa, Canada. 351–357.
Freifeld H.B. (1999). Habitat relationships of forest birds on Tutuila Island, American Samoa. *Journal of Biogeography* **26**:1191–1213.
Fretwell S.D. (1987). Food chain dynamics, the central theory of ecology. *Oikos* **50**:291–301.
Fritts T.H. and Rodda G.H. (1998). Role of introduced species in the degradation of island ecosystems: A case history of Guam. *Annual Review of Ecology and Systematics* **29**:113–140.
Frölich K., Olgierd E.J., Kujawski G., Rudolph M., Ronsholt L., and Speck S. (2003). European brown hare syndrome virus in free-ranging European brown hares from Argentina. *Journal of Wildlife Diseases* **39**:121–124.
Fukami T., Wardle D.A., Bellingham P., Mulder C.P.H., Towns D.R., Yeates G.W., *et al.* (2006). Above- and below-ground impacts of introduced predators in seabird-dominated island ecosystems. *Ecology Letters* **9:** 12, 1299–1307.
Fuller P. (2007). Problems with the release of exotic fish. US Department of the Interior Geological Survey, Non-indigenous Aquatic Species website: http://nas.er.usgs.gov/taxgroup/fish/docs/dont_rel.asp
Galil B. (2000). A sea under siege—alien species in the Mediterranean. *Biological Invasions* **2**:177–186.
Garcia-Berthou E., Alcaraz C., Pou-Rovira Q., Zamora L., Coenders G., and Feo C. (2005). Introduction pathways and establishment rates of invasive aquatic species in Europe. *Canadian Journal of Fisheries and Aquatic Sciences* **62**:453–463.
Gavach C., Sandeaux R., Sandeaux J., Souard R., Uchimura M., and Escoubet P. (1996). Technique de destruction de *Caulerpa taxifolia* par emission d'ions cuivrique au moyen d'un panneau a membrane permselective. In Ribera M. A., Ballesteros E. Boudouresque C. F., Gomez A., and Gravez V. (eds.) Second International workshop on *Caulerpa Taxifolia*, pp. 139–144. Publications Universitat Barcelona.
Gavach C., Bonnal L., Uchimura M., Sandeaux R., Sandeaux J., Souard R., *et al.* (1999). Developpement pre-industriel de la destrcutyion de *Caulerpa taxifolia* par la technique de la cuverture a ions cuivrique. In Abstracts of the 4th International workshop on *Caulerpa taxifolia* in the Mediterranean Sea, February 1999.
Gehrke P.C. (2003). Preliminary assessment of oral rotenone baits for carp control in New South Wales. In *Managing invasive freshwater fish in New Zealand*, pp. 149–154. Proceedings of a workshop hosted by Department of Conservation, 10–12 May 2001, Hamilton, New Zealand.
Genovesi P. (2005). Eradications of invasive alien species in Europe: a review. *Biological Invasions* **7**:127–133.
Genovesi P. and Bertolino S. (2001). Human dimension aspects in invasive alien species issues: the case of the failure of the grey squirrel eradication project in Italy. In McNeely J.A. (ed.) 2001. *The Great Reshuffling: Human Dimensions in Invasive Alien Species*. IUCN, Gland, Switzerland and Cambridge, UK. 113–120.
Genovesi P. and Shine C. (2004). European strategy on invasive alien species—Final Version. Council of Europe, Strasbourg, T-PVS/Inf (2004).

Gertzen E., Familiar O., and Leung B. (2008). Quantifying invasion pathways: fish introductions from the aquarium trade. *Canadian Journal of Fisheries and Aquatic Sciences* **65**:1265–1273.

Gibb J.A. and Williams J.M. (1994). The rabbit in New Zealand. In Thompson H.V. and King C.M. (eds.). *The European rabbit: the history and biology of a successful colonizer*, pp. 158–204. Oxford University Press, Oxford.

Gilad O., Yun S., Adkison M.A., Way K., Willits N.H., Bercovier H., *et al*. (2003). Molecular comparison of isolates of an emerging fish pathogen, koi herpes virus, and the effects of water temperature on mortality of experimentally infected koi. *Journal of General Virology* **84**:2661–2668.

Gilligan D. and Faulks L. (2005). Assessing carp biology in NSW: the first step in a successful implementation strategy for daughterless carp, pp. 57–58. Proceedings of the 13th Australasian Vertebrate Pest Conference, Wellington, New Zealand.

Glasby T., Connell S., Holloway M., and Hewitt C. (2007). Nonindigenous biota on artificial structures: could habitat creation facilitate biological invasions? *Marine Biology* **151**:887–895.

Glasby T.M., Creese R.G., and Gibson P.T. (2005). Experimental use of salt to control the invasive marine alga *Caulerpa taxifolia* in New South Wales, Australia. *Biological Conservation* **122**:573–580.

Global Ballast Water Management Programme. (2004). Global Project Task Force (GPTF), Fifth Meeting, London, United Kingdom, 3–6 February 2004: Proceedings. IMO London.

Godwin L.S. (2003). Hull Fouling of Maritime Vessels as a Pathway for Marine Species Invasions to the Hawaiian Islands. *Biofouling* **19**(Suppl.):123–131.

Goka K., Okabe K., Yoneda M. (2006). Worldwide migration of parasitic mites as a result of bumblebee commercialization. *Population Ecology* **48**: 285–291.

Gollasch S. (2002). The importance of ship hull fouling as a vector of species introductions into the North Sea. *Biofouling* **18**:105–121.

Gollasch S. and Leppäkoski E. (1999). *Initial Risk Assessment of Alien Species in Nordic Coastal Waters*. Nordic Council of Ministers, Copenhagen.

Goodwin B.J., McAllister A.J., and Fahrig L. (1999). Predicting invasiveness of plant species based on biological information. *Conservation Biology* **13**:422–426.

Gopal B. (1987). *Water Hyacinth*. Elsevier, Amsterdam.

Goren M. and Galil B.S. (2005). A review of changes in the fish assemblages of Levantine inland and marine ecosystems following the introduction of non-native fishes. *Journal of Applied Ichthyology* **21**:364–370.

Gosling L.M. and Baker S.J. (1989). The eradication of muskrats and coypus from Britain. *Biological Journal of the Linnean Society* **38**:39–51.

Graham J., Newman G., Jarnevich C. S., Shory R., and Stohlgren T.J. (2007). A global organism detection and monitoring system for non-native species. *Ecological Informatics* **2**:177–183.

Greathead D.J. and Greathead A. (1992). Biological control of insect pests by insect parasitoids and predators: the BIOCAT database. *Biocontrol News and Information* **13**:61N–68N.

Green P.T., O'Dowd D.J., and Lake P.S. (1997). Control of seedling recruitment by land crabs in rain forest on a remote oceanic island. *Ecology* **78**:2474–2486.

Green P.T., Lake P.S., and O'Dowd D.J. (1999). Monopolization of litter processing by a dominant land crab on a tropical oceanic island. *Oecologia* **119**:435–444.

Green P.T., Slip D., and Comport S. (2002). *Environmental assessment*. http://www.environment.gov.au/cgi-bin/epbc/epbc_ap.pl?name=referral_detail&proposal_id=722).

Green P.T., O'Dowd D.J., and Lake P.S. (2008). Recruitment dynamics in a rainforest seedling community: context-dependent impact of a keystone consumer. *Oecologia* **156**:373–385.

Gregg M.D. and Hallegraeff G.M. (2007). Efficacy of three commercially available ballast water biocides against vegetative microalgae, dinoflagellate cysts and bacteria. *Harmful Algae* **6**:567–584.

Grice A.C. (1998). Ecology in the management of invasive rangeland shrubs: a case study of Indian jujube (*Ziziphus mauritiana*). *Weeds Science,* **46**:467–474.

Grice A.C. (2000). Weed management in Australian Rangelands. In Sindel B.M. (ed.) *Australian Weed Management Systems*. Richardson R.G. and F.J., Melbourne. 429–458.

Grice A.C. (2006). Commercially valuable weeds: can we eat our cake without choking on it? *Ecological Management and Restoration* 7: 40–44.

Grice A.C. and Ainsworth N. (2002). Sleeper weeds—a useful concept? *Plant Protection Quarterly,* **18**:35–39.

Grice A.C. and Campbell S.D. (2000). Weeds in pasture ecosystems—symptom or disease? *Tropical Grasslands,* **34**:264–270.

Grice A.C., Field A.R., and McFadyen R.E.C. (2004). Quantifying the effects of weeds on biodiversity: beyond Blind Freddy's test. In Sindel B.M. and Johnson S.B. (eds.) *Weed Management, Balancing People, Planet, Profit*. Proceedings of the 14th Australian Weeds Conference. Weed Society of New South Wales, Sydney. 464–468.

Griffin G.F. (1993). The spread of buffel grass in inland Australia: land use conflicts. *Proceedings of the 10th Australian Weeds Conference and 14th Asian Pacific Weed Science Society Conference*. The Weed Society of Queensland, Brisbane. 501–504.

Grime J.P. (1974). Vegetation classification by reference to strategies. Nature **250**:26–30.

Grosholz E.D. (2005). Recent biological invasion may hasten invasional meltdown by accelerating historical introductions. *Proceedings of the Academy of Natural Sciences* **102**:1088–1091.

Gross L. (2006). Assessing Ecosystem Services to Identify Conservation Priorities. *PLoS Biol* **4**(11): e392.

Gross M.R. (1998). One species with two biologies: Atlantic salmon (*Salmo salar*) in the wild and in aquaculture. *Canadian Journal of Fisheries and Aquatic Sciences* **55**(Suppl.1):131–144.

Grotkopp E. and Rejmánek M. (2007). High seedling relative growth rate and specific leaf area are traits of invasive species: phylogenetically independent contrasts of woody angiosperms. *American Journal of Botany* **94**:1–8.

Groves R.H. (1999). Sleeper weeds. In Bishop A.C., Boersma M. and Barnes C.D. (eds.) *Weed Management into the 21st Century: do we know where we are going?* Proceedings of the 12th Australian Weeds Conference. Tasmanian Weeds Society, Hobart. 632–636.

Groves R.H. (2006). Are some weeds sleeping? Some concepts and reasons. *Euphytica,* **148**: 111–120.

Groves R.H., Panetta F.D., and Virtue J.G. (eds.) (2001). *Weed Risk Assessment*, CSIRO Plant Industry Publishing, Collingwood, Victoria, Australia.

Gunasekera L. (1999). Alligator weed—an aquatic weed present in Australian backyards. *Plant Protection Quarterly* **14**:77–78.

Gunasekera L. and Bonila J. (2001). Alligator weed: tasty vegetable in Australian backyards? *Journal of Aquatic Plant Management* **39**:17–20.

Gurr G.M., Barlow N.D., Memmott J., Wratten S.D., and Greathead D.J. (2000). A history of methodical, theoretical and empirical approaches to biological control. In Gurr G. and Wratten S.D. (eds.) *Biological Control: Measures of Success*, pp. 3–37. Kluwer Academic Publishers, Dordrecht, The Netherlands.

Gust N. and Inglis G.J. (2006). Adaptive multi-scale sampling to determine an invasive crab's habitat usage and range in New Zealand. *Biological Invasions* **8**:339–353.

Haines I. and Haines J. (1978). Pest status of the crazy ant, *Anoplolepis longipes* (Jerdon) Hymenoptera: Formicidae), in the Seychelles. *Bulletin of Entomological Research* **68**:627–638.

Hairston N.G., Smith F.E., and Slobodkin L.B. (1969). Community structure, population control, and competition. *American Naturalist* **94**:421–425.

Haley K.B. and Stone L.D. (1979). *Search theory and applications*. NATO Scientific Affairs Division, Plenum Press, New York, NY.

Hamel J.F. and Mercier A. (1996). Evidence of chemical communication during the gametogenesis of holothuroids. *Ecology* **77**:1600–1616.

Hamer A.J., Lane S.J., and Mahony M.J. (2002). Management of freshwater wetlands for the endangered green and golden bell frog (*Litorea aurea*): roles for habitat determinants and space. *Biological Conservation* **106**:413–424.

Hamley J.M. (1975). Review of gillnet selectivity. *Journal of the Fisheries Research Board of Canada* **32**: 1943–1969.

Hansen L.P., Windsor M.L., and Youngson A.F. (1997). Interactions between salmon culture and wild stocks of Atlantic salmon: the scientific and management issues. *ICES Journal of Marine Science* **54**:963–964.

Harley K.L.S. (1990). The role of biological control in the management of water hyacinth, *Eichhornia crassipes*. *Biocontrol News and Information* **11**:11–22.

Harley K.L.S, Julien M.H., and Wright A.D. (1996). Water hyacinth: a tropical worldwide problem and methods for its control. In Brown H., Cussans G.W., Devine M.D., Duke S.O., Fernandez-Quintanilla C., Helweg A., *et al.* (eds.) *Proceedings of the Second International Weed Control Congress Volume II*, pp. 639–644. Copenhagen, 1996.

Harper J.L. (1977). *Population Biology of Plants*, pp. 83–110. Academic Press, New York, NY.

Harris J.H. (1997). Environmental rehabilitation and carp control. In Roberts J. and Tilzey R. (eds.) *Controlling carp: exploring the options for Australia, pp 21–36*. Proceedings of a Workshop 22–24 October 1996, Albury. CSIRO, Griffith.

Harris L.G. and Tyrrell M.C. (2001). Changing community states in the Gulf of Maine: synergism between invaders, overfishing and climate change. *Biological Invasions* **3**:9–21.

Harrison S. (1991). Local extinction in a metapopulation context: an empirical evaluation. In Gilpin M. and Hanski I. (eds.) *Metapopulation Dynamics*, pp.73–88. Academic Press, London.

Harrison S. and Wilcox C. (1995). Evidence that predator satiation may restrict the spatial spread of a tussock moth (*Orgyia vetusta*) outbreak. *Oecologia* **101**:309–316.

Hastings, A., Cuddington K., Davies K.F., Dugaw C.J., Elmendorf S., Freestone A., *et al.* (2005). The spatial spread of invasions: new developments in theory and evidence. *Ecology Letters* **8**: 91–101.

Hawksworth D. and Kalin-Arroyo M. (1995). Magnitude and distribution of biodiversity. In Heywood V. (ed.) *Global Biodiversity Assessment (UNEP)*, pp. 107–192. Cambridge University Press, Cambridge.

Hay C.H. and Luckens P.A. (1987). The Asian kelp *Undaria pinnatifida* (Phaeophyta: Laminariales) found in a New Zealand harbour. *New Zealand Journal of Botany* **25**:364–366.

Hayes K.R. and Hewitt C.L. (1998). *Risk Assessment Framework for Ballast Water Introductions*. Centre for Research on Introduced Pests, CSIRO Division of Marine Research, Hobart, Tasmania.

Hayes K.R. and Hewitt C.L. (2000). *Risk Assessment Framework for Ballast Water Introductions—Volume 2*. Centre for Research on Introduced Pests, CSIRO Division of Marine Research, Hobart, Tasmania.

Hayes K.R. and Sliwa C. (2003). Identifying Potential Marine Pests—A Deductive Approach Applied to Australia. *Marine Pollution Bulletin* **46**:91–98.

Hayes K.R., Cannon R., Neil K., and Inglis G.J. (2005). Sensitivity and cost considerations for the detection and eradication of marine pests in ports. *Marine Pollution Bulletin* **50**: 823–834.

Hazardous Substances and New Organisms Act, 1996, New Zealand. http://www.legislation.govt.nz/

Heinrich J.W., Mullett K.M., Hansen M.J. Adams J.V., Klar G.T., Johnson D.A., et al. (2003). Sea lamprey abundance and management in Lake Superior, 1957 to 1999. *Journal of Great Lakes Research* **29**(Suppl.1):566–583.

Herr J.C. (1999). Non-target impact of Rhinocyllus conicus (Coleoptera: Curculionidae) on rare native Californian Cirsium spp. thistles. In: *Abstracts, Tenth International Symposium on Biological Control of Weeds*. University of Montana, Bozeman, USA.

Herwig R.P., Cordell J.R., Perrins J.C., Dinnel P.A., Gensemer R.W., Stubblefield W.A., et al. (2006). Ozone treatment of ballast water on the oil tanker S/T Tonsina: chemistry, biology and toxicity. *Marine Ecology-Progress Series* **324**:37–55.

Hewitt C.L. and Hayes K. (2002). Risk assessment of marine biological invasions. In Leppäkoski E., Gollasch S., and Olenin S. (eds.) *Invasive aquatic species of Europe*, p.583. Kluwer Academic Publishers, Dordrecht, The Netherlands.

Hewitt C.L. and Martin R.B. (2001). *Revised Protocols for Baseline Port Surveys for Introduced Marine Species: Survey Design, Sampling Protocols and Specimen Handling*. Centre for Research on Introduced Pests, CSIRO Division of Marine Research, Hobart, Tasmania.

Hewitt C.L., Willing J., Bauckham A., Cassidy A.M., Cox C.M.S., Jones L., et al. (2004). New Zealand marine biosecurity: delivering outcomes in a fluid environment. *New Zealand Journal of Marine and Freshwater Research* **38**:429–438.

Hewitt C.L., Campbell M.L., McEnnulty F. Moore K.M., Murfet N.B., Robertson B., et al. (2005). Efficacy of physical removal of a marine pest, the intertidal kelp *Undaria pinnatifida* in a Tasmanian marine reserve. *Biological Invasions* **7**:251–263.

Hewitt C.L., Campbell M.L. and Gollasch S. (2006). Alien species in aquaculture. Considerations for responsible use. IUCN, Gland, Switzerland and Cambridge, U.K.

Hiebert R.D. (1997). Prioritizing invasive plants and planning for management. In Luken J.O. and Thieret J.W. (eds.) *Assessment and Management of Plant* Invasions, 195–212. Springer-Verlag, New York, NY.

Hight S.D., Carpenter K.A., Bloem K.A., Bloem S., Pemberton R.W., and Stiling P. (2002). Expanding geographical range of *Cactoblastis cactorum* (Lepidoptera: Pyralidae) in north America. *Florida Entomologist*. **84**:527–529.

Hill G. and Greathead D.J. (2000). Economic evaluation in classical biological control. In Perrings C., Williamson M., and Dalmazzone S. (eds.) *The Economics of Biological Invasions*. Edward Elgar, Cheltenham, UK. 208–223.

Hill M., Holm K., Vel T., Shah N.J., and Matyot P. (2003). Impact of the introduced yellow crazy ant *Anoplolepis gracilipes* on Bird Island, Seychelles. *Biodiversity and Conservation* **12**:1969–1984.

Hill M.P. (1999). Biological control of red water fern, *Azolla filiculoides* Lamarck (Pteridophyta: Azollaceae), in South Africa. *African Entomology Memoir* **1**:119–124.

Hill M.P. (2003). The impact and control of alien aquatic vegetation in South African aquatic ecosystems. *African Journal of Aquatic Science*, **28**, 19–24.

Hilliard R. (1999). *Best practice for the management of introduced marine pests: a review*. The Global Invasive Species Programme, Cape Town, South Africa.

Hinds L.A. and Pech R.P. (1996). Immuno-contraceptive control for carp. In Roberts J. and Tilzey R. (eds.) *Controlling carp: exploring the options for Australia*, pp. 108–118. Proceedings of a workshop. 22–24 October 1996, Albury. CSIRO, Griffith.

Hobbs, R.J. and Huenneke, L.F. 1992. Disturbance, diversity and invasion: implications for conservation. Conservation Biology **6**: 324–337.

Hobbs R.J. and Humphries S.E. (1995). An integrated approach to the ecology and management of plant invasions. *Conservation Biology* **9**:761–770.

Hochberg M.E. (2002). The decline of the regulation paradigm in classical biological control and the possible role of rare species. In *International Symposium on Biological Control of Arthropods*, January 14–18, 2002, Honolulu, Hawaii, USA. USDA Forest Service Publication FHTET-03-05, Morgantown, West Virginia, USA. 114–117.

Hockley J. (1974) . . . . and alligator weed spreads in Australia. *Nature* **250**:704.

Hoddle M.S. (2002). Classical Biological Control of Arthropods in the 21st Century. In *International Symposium on Biological Control of Arthropods*, pp. 3–16. January 14–18, 2002, Honolulu, Hawaii, USA. USDA Forest Service Publication FHTET-03–05, Morgantown, West Virginia, USA.

Hoffmann J.H. (1995). Biological control of weeds: the way forward, a South African perspective. In Stirton C.H., Chair. *Weeds in a Changing World*, pp. 77–89. British Crop Protection Council Proceedings, Farnham, Surrey, UK.

Hoffmann B.D. and O'Connor S. (2004). Eradication of two exotic ants from Kakadu National Park. *Ecological Management and Restoration* **5**:98–105.

Holdgate M.W. and Wace N.M. (1961). The influence of man on the floras and faunas of Southern Islands. *Polar Record* **10**:475–493.

Holdsworth A.R., Frelich L.E., and Riech P.B. (2007). Regional extent of an ecosystem engineer: earthworm invasion in northern hardwood forests. *Ecological Applications* **17**:1667–1677.

Holt J., Black R., and Abdallah R. (2006). A rigorus yet simple quantitative risk assessment method for quarantine pests and non-native organisms. *Annals of Applied Biology* **149**:167–173.

Holt J.S. (2004). Principles of weed management in agroecosystems and wildlands. *Weed Technology* **18**:1559–1562.

Holway D.A., Lach L.S., Suarez A., Tsutsui N., and Case T.J. (2002). The causes and consequences of ant invasions. *Annual Review of Ecology, Evolution and Systematics* **33**:181–233.

Hopkins G.A. and Forrest B.M. (2008). Management options for vessel hull fouling: an overview of risks posed by in-water cleaning. *ICES Journal of Marine Science* **65**:811–815

Hopper K.R. (1996). Making biological control introductions more effective. *In:* Waage J.K., Chair. *Biological Control Introductions—Opportunities for Improved Crop Production*, pp. 59–76. British Crop Protection Council Proceedings No 67. Farnham, Surrey, UK.

Horowitz M., Regev Y., and Herzlinger G. (1983). Solarization for weed control. *Weed Science* **31**:170–179.

Hoshovsky M.C. and Randall J.M. (2000). Management of invasive plant species. In Bossard C.C., Randall J.M., and Hoshovsky M.C. (eds.) *Invasive Plants of California's Wildlands*, pp. 19–27. University of California Press, Berkeley, CA.

Howld G., Donlan C.J., Galván J.P. Russell J., Parkes J., Samaniego A., et al. (2007). Invasive rodent eradication on islands. *Conservation Biology* **21**: 1258–1268.

Howald G.R., Faulkner K.R., Tershy B., Keitt B., Gellerman H., Creel E.M., et al. (2003). Eradication of black rats from Anacapa Island: biological and social considerations, pp. 299–312. In Garcelon D.K., Schwemm C.A. (eds.) Proceedings of the Sixth California Islands Symposium, Ventura, California, USA.

Howard-Williams C. (1993). Processes of aquatic weed invasions: the New Zealand example. *Journal of Aquatic Plant Management* **31**:17–23.

Howard-Williams C., and Davies J. (1988). The invasion of Lake Taupo by the submerged water weed *Lagarosiphon major* and its impact on the native flora. *New Zealand Journal of Ecology* **11**:13–19.

Howard-Williams C., Schwarz A.M., and Reid V. (1996). Patterns of aquatic weed regrowth following mechanical harvesting in New Zealand hydro-lakes. *Hydrobiologia* **340**:229–234.

Howath F.G. (1991). Environmental impact of classical biological control. *Annual Review of Entomology* **36**:485–509.

Huffaker C.B. and Kennett C.E. (1959). A ten-year study of vegetational changes associated with the biological control of Klamath weed. *Journal of Range Management* **12**:69–82.

Hughey K.F.D. and Parkes J.P. (1996). Thar management—planning and consultation under the Wild Animal Control Act. *The Royal Society of New Zealand Miscellaneous Series* **31**:85–90.

Huhta A.P., Rautio P., Tuomi J., and Laine K. (2001). Restorative mowing on an abandoned semi-natural meadow: Short-term and predicted long-term effects. *Journal of Vegetation Science* **12**:677–686.

Hulme P.E. (2006). Beyond control: Wider implications for the management of biological invasions. *Journal of Applied Ecology* **43**: 835–847.

Hulme P.E, Bacher S., Kenis M., Klotz S., Kühn I., Minchin D., *et al.* (2008). Grasping at the routes of biological invasions: a framework for integrating pathways into policy. *Journal of Applied Ecology* **45**:403–414.

Hummel M. and Kiviat E. (2004). Review of world literature on water chestnut with implications for management in North America. *Journal of Aquatic Plant Management* **42**:17–28.

Humphreys L.R. (1967). Buffel grass (*Cenchrus ciliaris*) in Australia. *Tropical Grasslands*, **1**:123–134.

Humphries S.E., Groves R.H., and Mitchell D.S. (1991). Plants invasions of Australian ecosystems: a status review and management directions. *Kowari*, **2**:1–116.

ICES (2005). ICES Code of Practice on the introductions and transfers of marine organisms, *International Council for the Exploration of the Seas*, Copenhagen, Denmark.

Ielmini M. and Ramos G. (2003). *A national early detection and rapid response system for invasive plants in the United States*. Federal Interagency Committee for the Management of Noxious and Exotic Weeds, Washington DC.

IFS (2004). *Carp Management Program Report: Lakes Crescent and Sorrell 1995—June 2004*. Inland Fisheries Service, Tasmania.

Ikeda S. (2006). Risk analysis, the precautionary approach and stakeholder participation in decision making in the context of emerging risks from invasive alien species. In Koike F., Clout M.N., Kawamichi M., De Poorter M., and Iwatsuki K. (eds.). *Assessment and control of biological invasion risks*, pp. 15–26. Shoukadoh Books Sellers, Kyoto and IUCN, Gland, Switzerland.

IMO (1997). *Guidelines for the control and management of ships' ballast water to minimise the transfer of harmful aquatic organisms and pathogens. Resolution A.868(20)*. International Maritime Organisation (IMO), London.

IMO (2001). International Conference on the Control of Harmful Anti-fouling Systems for Ships, Adoption of the Final Act of the Conference and Any Instruments, Recommendations and Resolutions Resulting From the Work of the Conference. International Maritime Organization, IMO Headquarters, London.

IMO Global Ballast Water Management Programme. http://globallast.imo.org/index.asp?page=resolution.htm&menu=true

Ingvarsdóttir A., Birkett M.A., Duce I., Genna R.L., Mordue W., Pickett J.A., *et al.* (2002). Semiochemical strategies for sea louse control: host location cues. *Pest Management Science* **58**:537–545.

Innes J., Nugent G., Prime K., and Spurr E.B. (2004). Reponses of kukupa (*Hemiphaga novaeseelandiae*) and other birds to mammal pest control at Motatau, Northland. *New Zealand Journal of Ecology*, **28**: 73–81.

International Organization for Standardization (1999). *International standard ISO 10990–4. Animal (mammal) traps—Part 4: methods for testing killing-trap systems used on land or underwater*. International Organization for Standardization, Geneva.

IPPC (1996). *Code of Conduct for the Import and Release of Exotic Biological Control Agents*. ISPM 3. FAO, Rome, Italy.

IPPC (1997). International Plant Protection Convention. https://www.ippc.int/id/13742?language=en

Ishii N., Hashimoto T., and Suzuki T. (2006). Population estimation and control of the introduced mongoose on Amami-oshima Island. In Koike F., Clout M.N., Kawamichi M., De Poorter M., Iwatsuki K. (eds.). *Assessment and control of biological invasion* risks, pp. 165–167. Shoukadoh Books Sellers, Kyoto and IUCN, Gland, Switzerland.

Ito H. (1991. Successful HOTAC methods for developing scallop sowing culture in the Nemuro district of East Hokkaido, Northern Japan. In Svrjeek R. S. (ed.) *Marine ranching: proceedings of the seventeenth US-Japan meeting on aquaculture, Ise, Mie Prefecture, Japan, October 16–18, 1988*. NOAA Technical Report. NMFS.

IUCN (2000). *ICUN Guidelines for the Prevention of Biodiversity Loss Caused by Alien Invasive Species Approved by 51st Meeting of the IUCN Council, February 2000*. http://www.issg.org/infpaper_invasive.pdf

Jackson J. (2005). Is there a relationship between herbaceous species richness and buffel grass (*Cenchrus ciliaris*)? *Austral Ecology,* **30**:505–517.

Jackson S., Storrs M., and Morrison J. (2005). Recognition of Aboriginal rights, interests and values in river research and management: Perspectives from northern Australia. *Ecological Management and Restoration*, **6**:105–110.

Jaffe M. (1994). *And No Birds Sing: The Story of an Ecological Disaster in a Tropical Paradise*. Simon & Schuster, New York, NY.

Jenkins J.A. and Thomas R.G. (2007). Use of eyeballs for establishing ploidy of Asian carp. *North American Journal of Fisheries Management* **27**:1195–1202.

Jervis M. and Kidd N. (1996). *Insect Natural Enemies*. Chapman and Hall, London, UK.

Jobling S., Nolan M., Tyler C.R., Brighty G., and Sumpter J.P. (1998). Widespread sexual disruption in wild fish. *Environmental Science and Technology* **32**:2498–2506.

Johns K., Chappell R., Masibalavu V., and Seniloli E. (2006). *Protecting the internationally important seabird colony of Vatu-I-Ra Island, Fiji: feasibility study*. http://www.issg.org/cii/PII/Demonstration%20Projects/Vatu-I-Ra.htm

Johnson L.E., Ricciardi A., and Carlton J.T. (2001). Overland dispersal of aquatic invasive species: a risk assessment of transient recreational boating. *Ecological Applications* **11**:1789–1799.

Johnstone, I.M. (1986). Plant invasion windows: a time-based classification of invasion potential. *Biological Review* **61**:369–394.

Johnstone I.M., Coffey B.T., and Howard-Williams C. (1985). The role of recreational boat traffic in interlake dispersal of macrophytes: a New Zealand case study. *Journal of Environmental Management* **20**:263–279.

Jousson O., Pawlowski J., Zaninetti L., Meinesz A., and Boudouresque C.F. (1998). Molecular evidence for the aquarium origin of the green alga *Caulerpa taxifolia* introduced to the Mediterranean Sea. *Marine Ecology Progress Series* **172**:275–280.

Julien M. and White J. (1997). *Biological Control of Weeds: Theory and Practical Application*. ACIAR, Canberra, Australia.

Julien M.H. (2001). Biological control of water hyacinth with Arthropods: a review to 2000. In Julien M.H., Hill M.P., Center T.D., and Ding J. (eds.) *Biological and Integrated Control of Waterhyacinth,* Eichhornia crassipes. *Proceedings of the 2nd Meeting of the Global Working Group for the Biological and Integrated Control of Waterhyacinth, Beijing, China*, pp. 8–20. Australian Centre for International Agricultural Research, Canberra, Australia, 9–12 October 2000.

Julien M.H. and Bourne A.S. (1988). Alligator weed is spreading in Australia. *Plant Protection Quarterly* **3**:91–96.

Julien M.H. and Griffiths M.W. (1998). *Biological Control of Weeds: A World Catalogue of Agents and their Target Weeds,* 4th edition. CABI Publishing, Wallingford, UK.

Julien M.H., Harley K.L.S., Wright A.D., Cilliers C.J., Hill M.P., Center T.D., *et al.* (1996). International co-operation and linkages in the management of water hyacinth with emphasis on biological control. In Moran V.C. and Hoffman J.H. (eds.) *Proceedings of the IX International Symposium on Biological Control of Weeds,* pp. 273–282, 21–26 January 1996, Stellenbosch, South Africa. University of Cape Town, Cape Town, South Africa.

Julien M.H., Griffiths M.W., and Stanley J.N. (2001). *Biological control of water hyacinth. The moths* Niphograpta albiguttalis *and* Xubida infusellus: *biologies, host ranges, and rearing, releasing and monitoring techniques for biological control of* Eichhornia crassipes. Monograph No. 79. Australian Centre for International Agricultural Research (ACIAR), Canberra.

Kairo M.T.K. (2005). Hunger, poverty, and protection of biodiversity: opportunities and challenges for biological control. In: *International Symposium on Biological Control of Arthropods*, September 12–16, 2005, Davos, Switzerland. USDA Forest Service Publication FHTET-2005–08, Morgantown, West Virginia, USA. 228–236.

Kairo M.T.K., Cock M.J.W., and Quinlan M.M. (2003). An assessment of the use of the Code of Conduct the import and release of exotic biological control agents (ISPM No 3) since its endorsement as an international standard. *Biocontrol News and Information* **24**:15N–27N.

Katahira L.K., Finnegan P., and Stone C.P. (1993). Eradicating feral pigs in montane mesic habitat at Hawaii Volcanoes National Park. *Wildlife Society Bulletin* **21**: 269–274.

Kay S.H. and Hoyle S.T. (2001). Mail order, the internet, and invasive aquatic weeds. *Journal of Aquatic Plant Management* **39**:88–91.

Keane R.M. and Crawley M.J. (2002). Exotic plant invasions and the enemy release hypothesis. *Trends in Ecology and Evolution* **17**:164–170.

Keeley J.E. and McGinnis T.W. (2007). Impact of prescribed fire and other factors on cheatgrass persistence in a Sierra Nevada ponderosa pine forest. *International Journal of Wildland Fire* **16**:96–106.

Keller R.P., Lodge D.M., and Finnoff D.C. (2007). Risk assessment for invasive species produces net bioeconomic benefits. *Proceedings of the National Academy of Sciences* **104**:203–207.

Kenis M., Tomov R., Svatos A., Schlinsog P., Vaamonde C.L., Heitland W., Grabenweger G., Girardoz S., Freise J., and Avtzis N. (2005). The horse-chestnut leaf miner in Europe. In *International Symposium on Biological Control of Arthropods*, September 12–16, 2005, Davos, Switzerland. USDA Forest Service Publication FHTET-2005–08, Morgantown, West Virginia, USA. 77–90.

Kessler C.C. (2002). Eradication of feral goats and pigs and consequences for other biota on Sarigan Island, Commonwealth of the Northern Mariana Islands. In Veitch C.R. and Clout M.N. (eds.) *Turning the tide: eradication of invasive species,* pp. 132–140. IUCN SSC Invasive Species Specialist Group, IUCN, Gland, Switzerland and Cambridge, UK.

Kilpatrick A.M., Daszak P., Goodman S.J., *et al.* (2006). Predicting pathogen introduction: West Nile Virus spread to Galápagos. *Conservation Biology* **20**: 1224–1231

King A. (1995). *The effects of carp on aquatic ecosystems: a literature review.* Unpublished report prepared for the Environment Protection Authority, New South Wales, Murray Region.

King C.M. and Murphy E.C. (2005). Stoat. In King C.M. (eds.). *The Handbook of New Zealand Mammals,* pp. 261–287. Oxford University Press, Melbourne, Australia.

Kludge R.L., Zimmerman H.G., Cilliers C.J., and Harding G.B. (1986). Integrated control of invasive alien weeds. In Macdonald I.A.W., Kruger F.J. and Ferrar A.A. (eds.) *Ecology and Management of Biological Invasions in Southern Africa,* pp. 294–302. Oxford University Press, Cape Town, South Africa.

Kogan M. (ed.) (1986). *Ecological Theory and Integrated Pest Management Practice.* John Wiley and Sons, New York, NY.

Kogan M. (1998). Integrated pest management: Historical perspectives and contemporary developments. *Annual Review of Entomology* **43**:243–270.

Kolar C.S. and Lodge D.M. (2001). Progress in invasion biology: predicting invaders. *Trends in Ecology and Evolution* **16**:199–204.

Kompas T. and Che N. (2001). *An economic assessment of the potential costs of red imported fire ants in Australia*. Report for Department of Primary Industries, Queensland, Canberra, August 2001. Canberra: Australian Bureau of Agriculture and Resource Economics.

Kriticos D.J., Alexander N.S., and Kolometz S.M. (2006). Predicting the potential geographic distribution of weeds in 2080. In Preston C., Watts J.H. and Crossman N.D. (eds.) *Managing Weeds in a Changing Climate*. Proceedings of the 15th Australian Weeds Conference. Weed Management Society of South Australia, Adelaide. 27–34.

Kriwoken L.K. and Hedge P. (2000). Exotic species in estuaries: managing *Spartina anglica* in Tasmania, Australia. *Ocean and Coastal Management* **43**:573–584.

KrKosek M., Lewis M.A., and Volpe J. P. (2005). Transmission dynamics of parasitic sea lice from farm to wild salmon. *Proceedings of the Royal Society B:* **272**:689–696.

Krueger-Mangold J.M., Sheley R.L., and Svejcar T.J. (2006). Toward ecologically-based invasive plant management on rangeland. *Weed Science* **54**:597–605.

Krutz L.J., Senseman S.A., Zablotowitz R.M., and Matocha M.A. (2005). Reducing herbicide runoff from agricultural fields with vegetative filter strips: a review. *Weed Science,* **53**:353–367.

Kuris A.M. (2003). Eradication of introduced marine pests. In Rapport D. J., Lasley B. L., Rolston D. E., Ole Nielsen N., Qualset C. O., and Damania A. B. (eds.) *Managing for healthy ecosystems*, pp. 543–550. CRC Press, Boca Raton, FL.

Kuris A.M., Lafferty K.D. and Torchin M.E. (2005). Biological control of the European Green Crab, *Carcinus maenas*: natural enemy evaluation and analysis of host specificity. In *International Symposium on Biological Control of Arthropods*, September 12–16, 2005, Davos, Switzerland. USDA Forest Service Publication FHTET-2005–08, Morgantown, West Virginia, USA. 102–115.

Lacey L.A., Frotos R., Kaya H.H., and Vail P. (2001). Insect pathogens as biological control agents: do they have a future? *Biological Control* **21**:230–248.

Langdon J.S. and Humphrey J.D. (1987). Epizootic haematopoietic necrosis, a new viral disease in redfin perch, *Perca fluviatilis* L., in Australia. *Journal of Fish Diseases* **10**:289–297.

Langeland K.A. (1996). *Hydrilla verticillata* (L.F.) Royle (Hydrocharitaceae), "The perfect aquatic weed". *Castanea* **61**:293–304.

Lavery S., Moritz C., and Fielder D.R. (1996). Indo-Pacific population structure and evolutionary history of the coconut crab *Birgus latro*. *Molecular Ecology* **5**:557–570.

Lavoie C., Cruz F., Carion G.V., Campbell K., Donlan C.J., Harcourt S., *et al.* (2007). The thematic atlas of project Isabela. Charles Darwin Foundation and Galapagos National Park, Puerto Ayora, Ecuador.

Lawes R.A., Murphy H.T., and Grice, A.C. (2006) Comparing agglomerative clustering and three weed classification frameworks to assess the invasiveness of alien species across spatial scale. *Diversity and Distributions* **12**:633–644.

Lee M. (2001). Non-native plant invasions in Rocky Mountain National Park: linking species traits and habitat characteristics. M.S. Thesis, Colorado State University, Fort Collins, CO.

Lennon R.E., Hunn J.B., Schnick R.A., and Burress R.M. (1971). Reclamation of ponds, lakes and streams with fish toxicants: a review. *FAO Fisheries Technical Papers* **100**.

Leung B., Bossenbroek J.M., and Lodge D.M. (2006). Boats, pathways, and aquatic biological invasions: estimating dispersal potential with gravity models. *Biological Invasions* **8**:241–254.

Lewis J.A. (2001). *Ship anti-foulants—tributyltin and substitutes*. National Shipping Industry Conference 2001, Sydney, New South Wales.

Li W., Siefkes M.J., Scott A.P., and Tetter J.H. (2003). Sex pheromone communication in the sea lamprey: implications for integrated management. *Journal of Great Lakes Research* **29**:85–94.

Li W., Twohey M., Jones M., and Wagner M. (2007). Research to guide the use of pheromones to control sea lamprey. *Journal of Great Lakes Research* **33** (special issue 2):70–86.

Liebhold A. and Bascompte J. (2003). The Allee effect, stochastic dynamics and the eradication of alien species. *Ecology letters* **6**, 133–140.

Liebman M. (2001). Managing weeds with insects and pathogens. In Liebman M., Mohler C.M. and Staver C.P. (eds.) *Ecological Management of Agricultural Weeds*, pp. 375–408. Cambridge University Press, New York, NY.

Lincoln R., Boxshall G., and Clark P. (1998). *A Dictionary of Ecology, Evolution and Systematics*. Cambridge University Press, Cambridge.

Ling N. (2003). Rotenone—a review of its toxicity and use in fisheries management. *Science for Conservation* **211**. Department of Conservation, Wellington, New Zealand.

Ling N. (2004). Gambusia in New Zealand: really bad or just misunderstood? *New Zealand Journal of Marine and Freshwater Research* **38**:473–480.

Lintermans M. and Raadik T. (2003). Local eradication of trout from streams using rotenone: the Australian experience. in *Managing invasive freshwater fish in New Zealand*, pp. 95–111. Proceedings of a workshop hosted by Department of Conservation, 10–12 May 2001, Hamilton, New Zealand.

Lipton J., Galbraith H., Burger J., and Wartenberg D. (1993). A paradigm for risk assessment. *Environmental Management* **17**:1–5.

Lockwood J.L., Cassey P., and Blackburn T. (2005). The role of propagule pressure in explaining species invasions. *Trends in Ecology and Evolution,* **20**:223–228.

Lodge D.M. (1993). Biological invasions: lessons for ecology. *Trends in Ecology and Evolution* **8**:133–137.

Lodge, D.M., Williams S., MacIsaac H., Hayes K., Leung B., Loope L., *et al.* (2006). Biological invasions: recommendations for policy and management [Position Paper for the Ecological Society of America]. *Ecological Applications* **16**:2035–2054.

Lonsdale W.M. (1994). Inviting trouble: Introduced pasture species in northern Australia. *Australian Journal of Ecology* **19**:345–354.

Lonsdale W.M. (1999). Global patterns of plant invasions and the concept of invasibility. *Ecology* **80**: 1522–1536.

Loosanoff V.L. (1961). Biology and methods on controlling the starfish, *Asterias forbesi* (Desor) Fisheries Leaflet, *US Fish and Wildlife Service* **520**:1–11.

Lorvelec O. and Pascal M. (2005). French attempts to eradicate non-indigenous mammals and their consequences for native biota. *Biological Invasions* **7**:135–140.

Louda S.M., Arnett A.E., Rand T.A., and Russell F.L. (2003). Invasiveness of some biological control insects and adequacy of their ecological risk assessment and regulation. *Conservation Biology* **17**:73–82.

Lovett G.M., Canham C.D., Arthur M.A., Weathers K.C., and Fitzhugh R.D. (2006). Forest ecosystem responses to exotic pests and pathogens in Eastern North America. *Bioscience* **56**:395–405.

Low T. (1997). Tropical pasture plants as weeds. *Tropical Grasslands* **31**:337–343.

Lowe S., Browne M., Boudjelas S., and De Poorter M. (2000). *100 of the world's worst invasive alien species: A selection from the Global Invasive Species Database*. Invasive Species Specialist Group (ISSG) of the World Conservation Union (IUCN), Auckland, New Zealand.

Lucking R.S. and Ayerton V.J. (1994). Seychelles magpie robin recovery plan: November 1992–November 1994. Birdlife International Report.

Luken J.O. (1997). Management of plant invasions: Implicating ecological succession. In Luken J.O. and Thieret J.W. (eds.) *Assessment and Management of Plant Invasions*, pp. 133–144. Springer-Verlag, New York, NY.

Lumpkin T.A. and Plucknett D.L. (1982). *Azolla as a green manure: use and management in crop production*. Westview Tropical Agriculture Series No. 5. Westview Press, Boulder, CO.

Lynch L.D. and Thomas M.B. (2000). Nontarget effects in the biocontrol of insects with insects, nematodes and microbial agents: the evidence. *Biocontrol News and Information* **21**:117N–129N.

MacArthur R.H. and Wilson E.O. (1967). *The theory of island biogeography*. Princeton University Press.

Macdonald D.W. and Harrington L.A. (2003). The American mink: the triumph and tragedy of adaptation out of context. *New Zealand Journal of Zoology* **30**:421–441.

Mack R.N. and Foster S.K. (2004). Eradication or control? Combating plants through a lump sum payment or on the instalment plan. In Sindel B.M. and Johnson S.B. (eds.) *Weed Management: Balancing People, Planet and Profit*. Proceedings of the 14th Australian Weeds Conference. Weed Society of New South Wales, Sydney. 56–61.

Mack R.N. and Lonsdale W.M. (2001). Humans as global plant dispersers: Getting more than we bargained for. *Bioscience* **51**:95–102.

Mack R.N. and Lonsdale W.M. (2002). Eradicating invasive plants: hard-won lessons for islands. In Veitch C.R. and Clout M.N. (eds.) *Turning the tide: eradication of invasive species*, pp. 164–172. IUCN SSC Invasive Species Specialist Group, IUCN, Gland, Switzerland and Cambridge, UK.

Mack R.N., Simberloff D., Lonsdale W.M., Evans H., Clout M.N., and Bazzaz F.A. (2000). Biotic invasions: causes, epidemiology, global consequences, and control. *Ecological Applications* **10**:689–710.

Mackey A.P. (1998). *Acacia nilotica* ssp. *indica* (Benth.) Brenan. In Panetta F.D., Groves R.H., and Shepherd R.C.H. (eds.) *The Biology of Australian Weeds*. Richardson R.G. and F.J., Melbourne. 1–18.

MacNair N. and Smith M. (1998). *An investigation into the effects of lime and brine immersion treatments on Molgula sp. (sea grape) fouling oyster collectors on P.E.I.* Technical Report Series. P.E.I. Department of Fisheries and Aquaculture, Charlottetown, P.E.I., Canada.

Madeira P.T., Coetzee J.A., Center T.D., White E.E., and Tipping P.W. (2007). The origin of *Hydrilla verticillata* recently discovered at a South African dam. *Aquatic Botany* **87**:176–180.

MAF (2005). *Two biosecurity threats discovered in Auckland area*. MAF media release, 23 March 2005. http://www.maf.govt.nz/mafnet/press/230305marron.htm (accessed August 2008).

Manly B.F.J. (2000). Statistics for environmental science and management. Chapman & Hall/CRC, Boca Raton, FL.

Marking L.L. and Olsen L.E. (1975). Toxicity of the lampricide 3-trifluoromethyl-4-nitrophenol (TFM) to non-target fish in static tests. *Investigations in Fish Control* **60**. U.S. Department of the Interior, Fish and Wildlife Service, Washington.

Marks C.A. (2001). The 'Achilles Heel' principle. *Proceedings of the 12th Australasian Vertebrate Pest Conference*, pp. 330–335, Melbourne, Australia.

Marohasy J. (1996). Host shifts in biological weed control: real problems, semantic difficulties or poor science? *International Journal of Pest Management* **42**:71–75.

Marr R.M., O'Dowd D.J., and Green P.T. (2003). *Assessment of non-target impacts of Presto®01 ant bait on litter invertebrates in Christmas Island National Park, Indian Ocean*. Unpublished report to Parks Australia North, Darwin, NT, Australia.

Mascall L. (1590). *A booke of fishing with hook and line and other instruments there unto belonging*. John Wolfs, London.

Masifwa W.F., Twongo T., and Denny P. (2001). The impact of water hyacinth, *Eichhornia crassipes* (Mart) Solms on the abundance and diversity of aquatic macroinvertebrates along the shores of northern Lake Victoria, Uganda. *Hydrobiologia* **452**:79–88.

Matthews S. (2004). Tropical Asia invaded—the growing danger of invasive alien species. GISP, South African National Biodiversity Institute, Cape Town, South Africa.

Matthews S. (ed.) (2005). South America invaded—the growing danger of invasive alien species. GISP, South African National Biodiversity Institute, Cape Town, South Africa.

Matthews S. and Brand K. (2004). Africa invaded—the growing danger of invasive alien species. GISP, South African National Biodiversity Institute, Cape Town, South Africa.

Maxime V. (2008). The physiology of triploid fish: current knowledge and comparisons with diploid fish. *Fish and Fisheries* **9**:67–78.

McClay W. (2000). Rotenone use in North America (1988–1997). *Fisheries Management* **25**:15–21.

McClay W. (2002). Rotenone use in North America: an update. *Fisheries* **27**:19–20.

McColl K.A., Sunarto A., Williams L.M., and Crane M. St J. (2007). Koi herpes virus: dreaded pathogen or white knight? *Aquaculture Health International* **9**:4–6.

McDonald G. and Kolar C. (2006). *Research to guide the use of lampricides for controlling sea lamprey.* Great Lakes Fishery Commission, LaCrosse.

McDowall R.M. (1990). *New Zealand freshwater fishes: a natural history and guide.* Revised edition. Heinemann Reid, Auckland, New Zealand.

McDowall R.M. (2004). Shoot first, and then ask questions: a look at aquarium fish imports and invasiveness in New Zealand. *New Zealand Journal of Marine and Freshwater Research* **38**:503–510.

McDowall R.M. (2006). Crying wolf, crying foul or crying shame: alien salmonids and a biodiversity crisis in the southern cool-temperate galaxioid fishes. *Reviews in Fish Biology and Fisheries* **16**:233–422.

McEnnulty F.R., Bax N.J., Schaffelke B., and Campbell M.L. (2001). *A Review of Rapid Response Options for the Control of ABWMAC Listed Introduced Marine Pest Species and Related Taxa in Australian Waters.* Centre for Research on Introduced Marine Pests, CSIRO Marine Research, Hobart.

McFadyen R.E.C. (1998). Biological control of weeds. *Annual Review of Entomology* **43**:369–393.

McGlynn T.P. (1999). The worldwide transfer of ants: geographical distribution and ecological invasions. *Journal of Biogeography* **26**:535–548.

McKinney M.L. (2001). Effects of human population, area, and time on non-native plant and fish diversity in the United States. *Biological Conservation* **100**:243–252.

McLennan J. (2006). Strategies to reduce predation on bird populations In Allen R.B. and Lee W.G. (eds.) Biological Invasions in New Zealand. *Ecological Studies* **186**:371–387 Springer Berlin Heidelberg.

McLoughlin R. and Bax N. (1993). *Scientific discussions in Japan and Russia on the northern Pacific seastar.* CSIRO Division of Fisheries, Hobart.

McNeeley J.A. (2006). As the world gets smaller, the chances of invasion grow. *Euphytica* **48**:5–15.

McNeeley J.A., Mooney H.A., Neville L.E., Schei P.J., and Waage J.K. (2005). A global strategy on invasive alien species: synthesis and ten strategic elements. In Mooney H.A., Mack R.N., McNeeley J.A., Neville L.E., Schei P.J., and Waage J.K. (eds.) *Invasive Alien Species: a new synthesis*, pp. 332–345. Island Press, Washington, DC.

Meinesz A. (1999). *Killer Algae: the true tale of a biological invasion* (translated by Daniel Simberloff). University of Chicago Press.

Meinesz A., Cottalorda J.M., Chiaverini D., Thibaut T., and Vaugelas J. (2001). Evaluating and disseminating information concerning the spread of *Caulerpa taxifolia* along the French Mediterranean coasts. Second International Conference on Marine Bioinvasions, New Orleans, LA.

Meyerson L.A. and Mooney H.A. (2007). Invasive alien species in an era of globalization. *Frontiers in Ecology and Environment* **5**:199–208.

Meyerson L.A. and Reaser J.K. (2002). Biosecurity: moving towards a comprehensive approach. *BioScience* **52**:593–600.

Michel A., Scheffler B.E., Arias R.S., Duke S.O., Netherland M.D., and Dayan F.E. (2004). Somatic mutation-mediated evolution of herbicide resistance in the non-indigenous invasive plant hydrilla (*Hydrilla verticillata*). *Molecular Ecology* **13**:3229–3237.

Micol T. and Jouventin P. (2002). Eradication of rats and rabbits from Saint-Paul Island, French Southern Territories. In Veitch C.R. and Clout M.N. (eds.) *Turning the tide: eradication of invasive species*, pp. 199–205. IUCN SSC Invasive Species Specialist Group, IUCN, Gland, Switzerland and Cambridge, UK.

Midgley J.M., Hill M.P., and Villet M.H. (2006). The effect of water hyacinth, *Eichhornia crassipes* (Martius) Solms-Laubach (Pontederiaceae), on benthic biodiversity in two impoundments on the New Year's River, South Africa. *African Journal of Aquatic Science* **31**:25–30.

Miller A.W., Chang A.L., Cosentino-Manning N., and Ruiz G.M. (2004). A new record and eradication of the northern Atlantic alga Ascophyllum nodosum (Phaeophyceae) from San Francisco bay, California, USA. *Journal of Phycology* **40**:1028–1031.

Miller M.L. and Gunderson L.H. (2004). Biological and cultural camouflage: the challenges of seeing the harmful invasive species problem and doing something about it. In Miller M.L. and Fabian R.N. (eds.) *Harmful Invasive Species: Legal Responses*, pp. 1–22. Environmental Law Institute, Washington D.C.

Milon J.W., Yingling J., and Reynolds J.E. (1986). An economic analysis of the benefits of aquatic weed control in North-Central Florida. Economics Report No. 113. Food and Resource Economics, Agricultural Experiment Station, Institute of Food and Agricultural Sciences, University of Florida, Gainesville 32611.

Minchin D. (1996). Management of the introduction and transfer of marine molluscs. *Aquatic Conservation: Marine and Freshwater Ecosystems* **6**:229–244.

Minchin D. (2007). Aquaculture and transport in a changing environment: overlap and links in the pread of alien biota. *Biofouling* **55**:302–313.

Minchin D. and Gollasch S. (2003). Fouling and ships' hulls: How changing circumstances and spawning events may result in the spread of exotic species. *Biofouling* **19**:111–122.

Mineur F., Belsher T., Johnson M.P., Maggs C.A., and Verlaque M. (2007). Experimental assessment of oyster transfers as a vector for macroalgal introductions. *Biological Conservation* **137**:237–247.

Moller H. (1996). Lessons for invasion theory from social insects. *Biological Conservation* **78**:125–142.

Montgomery A.D. (2007). The eradication of snowflake coral on the island of Kauai: 18 months later. Fifth International Conference on Marine Bioinvasions, Cambridge, Massachusetts.

Moody M.E. and Mack R.N. (1988). Controlling the spread of plant invasions: the importance of nascent foci. *Journal of Applied Ecology*, **25**:1009–1021.

Mooney H.A. (2005). Invasive alien species: the nature of the problem. In Mooney H.A., Mack R.N., McNeeley J.A., Neville L.E., Schei P.J., and Waage J.K. (eds.) *Invasive Alien Species: A New Synthesis* pp. 1–15. Island Press, Washington, DC.

Moore G.M. (1999). *Crossing the Chasm: Marketing and selling high-tech products to mainstream customers*. Revised Edition. Harper Business.

Moorhouse T.M., Agaba P., and McNab T.J. (2001). Recent efforts in biological control of water hyacinth in the Kagera River headwaters of Rwanda. In Julien M.H., Hill M.P., Center T.D., and Ding Jianqing (eds.) *Biological and integrated control of water hyacinth*, Eichhornia crassipes, pp. 39–42. Proceedings of the second meeting of the Global Working Group for the

Biological and Integrated Control of Water Hyacinth, Beijing, China, 9–12 October 2000. ACIAR Proceedings No. 102.

Morgan D. and Hickling G. (2000). Techniques used for poisoning possums. In Montague TL (ed.) *The brushtail possum,* pp.143–153. Manaaki Whenua Press, Lincoln, New Zealand.

Morgan D.R. (1993). Multi-species control by aerial baiting: a realistic option? *New Zealand Journal of Zoology* **20**:367–372.

Morgan D.R., Nugent G., and Warburton B. (2006). Benefits and feasibility of local elimination of possum populations. *Wildlife Research* **33**:605–614.

Morisette J.T., Jarnevich C.S., Ullah A., Cai W.J., Pedelty J.A., Gentle J.E., et al. (2006). A tamarisk habitat suitability map for the continental United States. *Frontiers in Ecology and the Environment* **4**:11–17.

Morrell S.L. and Farnham W.F. (1982). Some effects of substratum on *Sargassum muticum. British Phycological Journal* **17**:236–237.

Morris K.D. (2002). The eradication of the black rat (*Rattus rattus*) on Barrow and adjacent islands off the north-west coast of Western Australia. In Veitch C.R. and Clout M.N. (eds.) *Turning the tide: the eradication of invasive species.* Occasional Paper of the IUCN Special Survival Commission No. 27, 219–225.

Morrison S.A., Macdonald N., Walker K., Lozier L., and Shaw M.R. (2007). Facing the dilemma at eradication's end: uncertainty of absence and the Lazarus effect. *Frontiers in Ecology and Evolution* **5**:271–276.

Morriss G.A., Nugent G., Lorigan R., and Speedy C. (2005). *Development and testing of a deer repellent cereal bait for possum control. Part III: Cereal bait field test.* Landcare Research Contract Report LC0506/44.

Morriss G.A., O'Connor C.E., Airey A.T., and Fisher P. (2008). Factors influencing palatability and efficacy of toxic baits in ship rats, Norway rats and house mice. *Science for Conservation* **282**, Department of Conservation, Wellington, New Zealand.

Mountfort D., Hay C.H., Taylor M., Buchanan S., and Gibbs W. (1999). Heat Treatment of Ships' Ballast Water: Development and Application of a Model Based on Laboratory Studies. *Journal of Marine Environmental Engineering* **5**:193–206.

Mountfort D., Rhodes L., Broom J., Gladstone M., and Tyrrell J. (2007). Fluorescent *in situ* hybridisation assay as a species-specific identifier of the northern Pacific seastar, *Asterias amurensis. New Zealand Journal of Marine and Freshwater Research* **41**:283–290.

Moyes A.B., Witter M.S., and Gamon J.A. (2005). Restoration of native perennials in a California annual grassland after prescribed spring burning and solarization. *Restoration Ecology* **13**:659–666.

Muirhead J.R. and MacIsaac H.J. (2005). Development of inland lakes as hubs in an invasion network. *Journal of Applied Ecology* **42**:80–90.

Mulongoy K.J., Webbe J., Ferreira M., and Mittermeier C. (2006). The wealth of islands—a global call for conservation. Special issue of the CBD Technical series, Montreal, Canada.

Murphy C.A., Stacey N.E., and Corkum L.D (2001). Putative steroidal pheromones in the round goby, Neogobius melanostomus: olfactory and behavioural responses. *Journal of Chemical Ecology* **27**:443–470.

Murphy E. and Bradfield P. (1992). Change in diet of stoats following poisoning of rats in a New Zealand forest. *New Zealand Journal of Ecology* **16**:137–140.

Murphy E. and Fechney L. (2003). What's happening with stoat research? Fifth report on the five-year stoat research programme. Department of Conservation, Wellington, New Zealand.

Murphy E., Clapperton B.K., Bradfield P.M.F., and Speed H.J. (1998a). Brodifacoum residues in target and non-target animals following large-scale poison operations in New Zealand. *New Zealand Journal of Zoology* **25**:307–314.

Murphy E.C., Clapperton B.K., Bradfield P.M.F., and Speed H.J. (1998b). Effects of rat-poisoning operations on abundance and diet of mustelids in New Zealand podocarp forests. *New Zealand Journal of Zoology* **25**:315–328.

Muyt A. (2001). *Bush Invaders of Southeast Australia*. R.G. and F.J. Richardson. P.O. Box 42, Meredith, Victoria, Australia.

Myers J.H., Simberloff D., Kuris A.M., and Carey J.R. (2000). Eradication revisited: dealing with exotic species. *Trends in Ecology and Evolution* **15**:316–320.

National Academy of Sciences (1983). Risk assessment in the federal government: Managing the process. National Research Council. National Academy Press. Washington DC.

Navarro L. and Phiri G. (2000). Water hyacinth in Africa and the Middle East. A survey of problems and solutions. International Development Research Centre, Ottawa, Canada.

Naylor R.L., Williams S.L., and Strong D.R. (2001). Aquaculture: a gateway for exotic species. *Science* **294**:1655–1656.

Nehls G., Diederich S., and Thieltges D.W. (2006). Wadden Sea mussel beds invaded by oysters and slipper limpets: competition or climate control? *Helgolander Marine Research* **60**:135–143.

Nel R., Coetzee P.S., and Vanniekerk G. (1996). The evaluation of two treatments to reduce mud worm (*Polydora hoplura claparede*) infestation in commercially reared oysters (*Crassostrea gigas* Thunberg). Aquaculture **141**:31–39.

Newman D.G. (1994). Effects of a mouse, *Mus musculus,* eradication program and habitat change on lizard population on Mana Island, New Zealand, with special reference to McGregor skink, *Cyclodina macgregori. New Zealand Journal of Zoology,* **21**:443–456.

Newsome A.E. and Noble I.R. (1986). Ecological and physiological characters of invading species. In Groves R.H. and Burdon J.J. (eds.) *Ecology of biological invasions*, pp. 1–15. Cambridge University Press, Cambridge.

NIMPIS (2002a). National Introduced Marine Pest Information System. Web publication. In Hewitt C.L., Martin R.B., Sliwa C., McEnnulty F.R., Murphy N.E., Jones T., *et al.* (eds.) http://crimp.marine.csiro.au/nimpis Accessed 12/09/2007.

NIMPIS (2002b). Asterias amurensis species summary. National Introduced Marine Pest Information System. Web publication. In Hewitt C.L., Martin R.B., Sliwa C., McEnnulty F.R., Murphy N.E., Jones T., *et al.* (eds.) http://crimp.marine.csiro.au/nimpis Accessed 12/09/2007.

Noble J.C. (1997). *The Delicate and Noxious Scrub*. CSIRO, Melbourne.

Nogales M., Martin A., Tershy B.R. Donlan C.J., Witch D., Puerta N., *et al.* (2004). A review of feral cat eradication on islands. *Conservation Biology* **18**:310–319.

Nordstrom M., Hogmander J., Laine J., Nummelin M., Laanetu N., and Korinaki E. (2002). Effect of feral mink removal on seabirds, waders and passerines on small islands in the Baltic Sea. *Biological Conservation* **109**:359–368.

Nowell D.C. and Maynard G.V. (2005). International guidelines for the export, shipment, import and release of biological control agents and other beneficial organisms (International Standard for Phytosanitary Measures No 3). In *International Symposium on Biological Control of Arthropods,* September 12–16, 2005, Davos, Switzerland. USDA Forest Service Publication FHTET-2005–08, Morgantown, West Virginia, USA. 726–734.

Nugent G. and Fraser K.W. (1993). Pest or valued resource: conflicts in game management. *New Zealand Journal of Zoology* **20**:361–366.

Nugent G. and Fraser K.W. (2005). Red deer. In King C.M. (ed.). *6 e Handbook of New Zealand Mammals,* pp. 401–420. Oxford University Press, Melbourne, Australia.

Nugent G., Morgan D.R., Sweetapple P., and Warburton B. (2007). Developing strategy and tools for the local elimination of multiple pest species. In *Managing vertebrate invasive species*. National Wildlife Research Centre, Fort Collins, USA, 7–9 August, 2007.

O'Dowd D.J., Green P.T., and Lake P.S. (1999). *Status, impact, and recommendations for research and management of exotic invasive ants in Christmas Island National Park*. Environment Australia, Darwin, Northern Territory. http://www.issg.org/database/species/reference_files/Christmas_Island_Report.pdf (accessed February 2007).

O'Dowd D.J., Green P.T., and Lake P.S. (2003). Invasional 'meltdown' on an oceanic island. *Ecology Letters* **6**:812–817.

O'Dowd D.J. and Green P.T. (2002). *After the aerial campaign: monitoring, surveillance and follow-up action*. Unpublished Discussion Paper for Parks Australia North, 29 November.

O'Dowd D.J. and Green P.T. (2003). Potential for indirect biological control of the yellow crazy ant (*Anoplolepis gracilipes*) on Christmas Island, Indian Ocean. Discussion paper for the Crazy Ant Steering Committee.

O'Dowd D.J. and Green P.T. (2009). Invasional meltdown: do invasive ants facilitate secondary invasions? In Lach L., Parr C., and Abbott K. (eds.) *Ant Ecology*. Oxford University Press, Oxford, in press.

O'Keefe S. (2005). Investing in conjecture: eradicating the red-eared slider in Queensland. *Proceedings of the 13th Australasian Vertebrate Pest Conference*, pp. 169–176. Wellington, New Zealand.

Ogutu-Ohwayo R. (1990). The decline of the native fishes of Lakes Victoria and Kyoga (East Africa) and the impact of introduced species, especially the Nile perch, *Lates niloticus*, and the Nile tilapia, *Oreochromis niloticus*. *Environmental Biology of Fishes* **27**:81–96.

OIE World Organisation for Animal Health (OIE). http://www.oie.int/eng/en_index.htm

Orchard M., Comport S., and Green P.T. (2002). *Control of the invasive yellow crazy ant* (Anoplolepis gracilipes) *on Christmas Island, Indian Ocean; progress, problems, and future scenarios*. Unpublished report to the Crazy Ant Steering Committee. Parks Australia Christmas Island.

Padilla D.K. and Williams S.L. (2004). Beyond ballast water: aquarium and ornamental trades as sources of invasive species in aquatic ecosystems. *Frontiers in Ecology and the Environment* **2**:131–138.

Page A.R. and Lacey K.L. (2006). *Economic impact assessment of Australian weed biological control*. CRC for Australian Weed Management, Technical Series No 10, Adelaide.

Pancho J.V. and Soerjani M. (1978). *Aquatic Weeds of Southeast Asia*. National Publication Cooperative, Quezon City, Philippines.

Panetta F.D. (2007). Evaluation of the performance of weed eradication programs: containment and extirpation. *Diversity and Distributions* **11**:33–41.

Panetta F.D. and Lawes R. (2005). Evaluation of the performance of weed eradication programs: the delimitation of extent. *Diversity and Distributions* **11**:435–442.

Panetta F.D. and Lawes R. (2007). Evaluation of the Australian branched broomrape (*Orobanche ramosa*) eradication program. *Weed Science* **55**:644–651.

Panetta F.D. and Mitchell N.D. (1991). Homoclime analysis and the prediction of weediness. *Weed Research* **31**:273–284.

Panetta F.D. and Timmins S.M. (2004). Evaluating the feasibility of eradication for terrestrial weed invasions. *Plant Protection Quarterly* **19**:5–11.

Pannell A. and Coutts A.D.M. (2007). *Treatment methods used to manage Didemnum vexillum in New Zealand*. New Zealand Marine Farming Association. Report prepared for Biosecurity New Zealand.

Parkes J.P. (1990a). Feral goat control in New Zealand. *Biological Conservation* **54**:335–348.

Parkes J.P. (1990b). Eradication of feral goats on islands and habitat islands. *Journal of the Royal Society of New Zealand* **20**:297–304.

Parkes J.P. (1993). The ecological dynamics of pest-resource-people systems. *New Zealand Journal of Zoology* **20**:223–230.

Parkes J.P. (2006a). Does commercial harvesting of introduced wild mammals contribute to their management as conservation pests? In Allen R.B., Lee W.G. (eds.) Biological invasions in New Zealand. *Ecological Studies* **186**:407–420.

Parkes J. (2006b). *Feasibility plan to eradicate common mynas (*Acridotheres tristis*) from Mangaia Island, Cook Islands.* Landcare Research Contract Report LC0506/184. http://www.issg.org/CII/PII/Mangaia.htm

Parkes J. (2006c). Eradication of vertebrate pests: are there any general lessons? In Feare C.J., Cowan D.P. (eds.) *Advances in Vertebrate Pest Management IV*, pp. 91–110. Filander Verlag.

Parkes J. and Murphy E. (2003). Management of introduced mammals in New Zealand. *New Zealand Journal of Zoology* **30**, 335–359.

Parkes J.P., Paulson J., Donlan C.J., and Campbell K. (2008a). Control of North American beavers in Tierra del Fuego: feasibility of eradication and alternative management options. Landcare Research Contract Report LC0708.

Parkes J.P., Glentworth B. and Sullivan G. (2008b). Changes in immunity to rabbit haemorrhagic disease virus, and in abundance and rates of increase of wild rabbits, Mackenzie Basin, New Zealand. *Wildlife Research* **35**: 775–779.

Parkes J.P., Morrison S., Ramsey D., and Macdonald (2008c). Eradication of feral pigs (*Sus scrofa*) from Santa Cruz Island, California. *Proceedings of the 14th Australasian Vertebrate Pest Conference, Darwin, Australia.*

Pascal M., Siorat F., Lorvelec O., Yésou P. and Simberloff D. (2005). A pleasing consequence of Norway rat eradication: two shrew species recover. Diversity and Distributions **11**: 193–198.

Passera L. (1994). Characteristics of tramp species. In Williams D.F. (ed.). *Exotic ants: biology, impact, and control of introduced species*, pp. 23–43. Westview Press, Boulder, Colorado.

Pemberton R.W. and Cordo H. (2001). Potential and risk of biological control of *Cactoblastis cactorum* (Lepidoptera: Pyralidae) in North America. *Florida Entomologist* **84**: 513–526.

Petch T. (1925). Entomophagous fungi and their use in controlling insect pests. *Bulletin of the Department of Agriculture, Ceylon* **71**:1–40.

Petts J. and Leach B. (2000). *Evaluating methods for public participation: literature review. R&D Technical Report: E135.* Environment Agency, UK.

Pheloung P., Williams P.A., and Halloy S. (1999). A weed risk assessment model for use as a biosecurity tool evaluating plant introductions. *Journal of Environmental Management*, **57**:239–251.

Pimentel D. (2002). *Environmental and economic costs of alien plant, animal and microbe invasions.* CRC Press, Boca Raton, FL.

Pimentel D., Wilson C., McCullum C., Huang R., Dwen P., Flack J., *et al.* (1997). Economic and environmental benefits of biodiversity. *BioScience* **47**:747–757.

Pimentel D., Lach L., Zuniga R., and Morrison D. (1999). Environmental and economic costs of nonindigenous species in the United States. *BioScience*, **50**, 53–65.

Pimentel D., Zuniga R., and Morrison D. (2005). Update on the environmental and economic costs associated with alien-invasive species in the United States. *Ecological Economics* **52**:273–288.

Piola R.F. and Johnston E.L. (2006a). Differential resistance to extended copper exposure in four introduced bryozoans. *Marine Ecology and Progress Series* **311**:103–114.

Piola R.F. and Johnston E.L. (2006b). Differential tolerance to metals among populations of the introduced bryozoan *Bugula neritina*. *Marine Biology* **148**:997–1010.

Planty-Tabacchi, A,-M., Tabacchi E., Naiman R.J., DeFerrari C., and Decamps H. (1996). Invasibility of species-rich communities in riparian zones. *Conservation Biology* **10**: 598–607

Plowes R.M., Dunn J.G., and Gilbert, E. (2007). The urban fire ant paradox: native fire ants persist in an urban refuge while invasive fire ants dominate natural habitats. *Biological Invasions*, **9**:825–836.

Polis G.A. (1999). Why are parts of the world green? Multiple factors control productivity and the distribution of biomass? *Oikos*, **86**:3–15.

Polis G.A. and Strong D.R. (1996). Food web complexity and community dynamics. *American Naturalist*, **147**:813–846.

Ponniah A.G. and Husin N.M. (2005). Current knowledge of aquatic alien invasive species in ASEAN and their management of the context of aquaculture development. In NACA. 2005. *The Way Forward: Building capacity to combat impacts of aquatic invasive alien species and associated trans-boundary pathogens in ASEAN countries*, pp. 97–114. Final report of the regional workshop, hosted by the Department of Fisheries, Government of Malaysia, on 12th–16th July 2004. Network of Aquaculture Centres in Asia-Pacific, Bangkok, Thailand.

Powlesland R.G., Knegtmans J.W., and Marshall I.S.J. (1999). Costs and benefits of aerial 1080 possum control operations using carrot baits to North island robins (*Petroica australis longiceps*), Pureora Forest Park. *New Zealand Journal of Ecology* **23**:149–159.

Pringle R.M. (2005). The origins of the Nile perch in Lake Victoria. *Bioscience* **55**:780–787.

Proulx G. (1999). Review of current mammal trap technology in North America. In Proulx G. (ed.) *Mammal trapping*, pp. 1–46. Alpha Wildlife Research & Management, Canada.

Pullar D., Kingston M., and Panetta F.D. (2006). Weed surveillance strategies using models of weed dispersal in landscapes. In Sindel B.M., Johnson S.B. (eds.) *Proceedings of the 14th Australian Weed Conference*, pp. 711–714. Wagga Wagga, NSW, Australia.

Pycha R.L. (1980). Changes in mortality of lake trout (*Salvelinus namaycush*) in Michigan waters of Lake Superior in relation to sea lamprey (*Petromyzon marinus*) predation 1968–78. *Canadian Journal of Fisheries and Aquatic Sciences* **37**:2063–2073.

QNRME (Queensland Department of Natural Resources, Mines and Energy) (2004a). *Prickly Acacia: National Case Studies Manual. Approaches to the management of prickly acacia (Acacia nilotica subsp. indica) in Australia*. State of Queensland (Department of Natural Resources, Mines and Energy), Cloncurry, Australia.

QNRME (Queensland Department of Natural Resources, Mines and Energy) (2004b). *Rubber Vine Management: Control Methods and Case Studies*. State of Queensland (Department of Natural Resources, Mines and Energy), Rockhampton, Australia.

Queensland Government (2004). *Policy to reduce the weed threat of Leucaena*. LPG/2005/1910– November 2004. Queensland Government (Natural Resources and Mines, Environmental Protection Agency, Department of Primary Industries), Brisbane, Australia. http://www.nrm.qld.gov.au/policy/documents/1910/pdfs/lpg_2005_1910.pdf

Radosevich S.R., Holt J.S. and Ghersa C.M. (2007). *Ecology of Weeds and Invasive Plants: Relationship to Agriculture and Natural Resource Management*, 3rd edition, John Wiley and Sons, New York, NY.

Raghu S., Wilson J.R., and Dhileepan K. (2006). Refining the process of agent selection through understanding plan demography and plant response to herbivory. *Australian Journal of Entomology* **45**:308–316.

Rahel F.J. (2000). Homogenization of fish faunas across the United States. *Science* **288**:854–856.

Rainbolt R.E. and Coblentz B.E. (1997). A different perspective on eradication of vertebrate pests. *Wildlife Society Bulletin* **25**, 189–191.

Ramsey D. (2005). Population dynamics of brushtail possums subject to fertility control. *Journal of Applied Ecology* **42**:348–360.

Ramsey D.S.L., Parkes J. and Morrison S.A. (2009). Quantifying eradication success: the removal of feral pigs from Santa Cruz Island, California. *Conservation Biology* **23**: 449–459.

Ratcliff D.A. (1967). Decreases in eggshell weight in certain birds of prey. *Nature* **215**:208–210.

Rattray M.R. (1995). The relationship between P, Fe and Mn uptakes by submersed rooted angiosperms. *Hydrobiologia* **308**:117–120.

Rayner M.J., Hauber M.E., Imber M.J., Stamp R.K., and Clout M.N. (2007). Spatial heterogeneity of mesopredator release within an oceanic island system. *PNAS*, **104**:20862–20865.

Rayner T.S. and Creese R.G. (2006). A review of rotenone use for the control of non-indigenous fish in Australian fresh waters, and an attempted eradication of the noxious fish, *Phalloceros caudimaculatus*. *New Zealand Journal of Marine and Freshwater Research* **40**:477–486.

Reaser J.K., Meyerson L.A., Cronk, Q., DePoorter, M., Eldridge, L., Green, E., et al. (2007). Ecological and socioeconomic impacts of invasive alien species in island ecosystems. *Environmental Conservation* **34**:98–111.

Reever M.J.K. and Seastedt T.R. (1999). Effects of soil nitrogen reduction on nonnative plants in restored grasslands. *Restoration Ecology* **7**:51–55.

Regan T.J., McCarthy M.A., Baxter P.W.J., Panetta F.D., and Possingham H.P. (2006). Optimal eradication: when to stop looking for an invasive plant. *Ecology Letters* **9**:759–766.

Reich R.M. and Bravo V.A. (1998). Integrating spatial statistics with GIS and remote sensing in designing multiresource inventories. In North America Science Symposium: Toward a unified framework for inventorying and monitoring forest ecosystem resources, pp. 202–207. USDA Rocky Mountain Research Station Proceedings, RMRS-P-12, Guadalajara, Jalisco, Mexico.

Reich R.M., Lundquist J.E., and Bravo V.A. (2004). Spatial models for estimating fuel loads in the Black Hills, South Dakota, USA. *International Journal of Wildland Fire* **13**:119–129.

Reichard S.E. (1997). Prevention of invasive plant introductions on national and local levels. In Luken J.O. and Thieret J.W. (eds.) *Assessment and Management of Plant Invasions*, pp. 215–227. Springer-Verlag, New York, NY.

Reichard S.H. and Hamilton C.W. (1997). Predicting invasions of woody plants introduced into North America. *Conservation Biology* **11**:193–203.

Rejmánek M. (1996). Invasive plant species and invasible ecosystems. In Sandlund O.T., Schei P.J., and Viken A. (eds.) *Invasive species and biodiversity management*, pp. 79–102. Kluwer Academic Publishers, Dordrecht.

Rejmánek M. and Pitcairn M.J. (2002). When is eradication of exotic pest plants a realistic goal? In Veitch C.R. and Clout M.N. (eds.) *Turning the tide: the eradication of invasive species*, pp. 249–253. IUCN SSC Invasive Species Specialist Group. IUCN, Gland, Switzerland and Cambridge, U.K.

Rejmanek M. and Richardson D.M. (1996). What attributes make some plant species more invasive? *Ecology* **77**:1655–1661.

Rejmánek M., Richardson D.M., Higgin S.I., Pitcairn M.J., and Grotkopp E. (2005). Ecology of invasive plants: State of the Art. In Mooney H.A., Mack R.N., McNeeley J.A., Neville L.E., Schei P.J., and Waage J.K. (eds.) *Invasive Alien Species: A New Synthesis*, pp. 104–161. Island Press, Washington, DC.

Ricciardi A. and Rasmussen J.B. (1998). Predicting the identity and impact of future biological invaders: a priority for aquatic resource management. *Canadian Journal of Fisheries and Aquatic Sciences* **55**:1759–1765.

Rice P.M. (2004). *Fire as a tool for controlling nonnative invasive plants: A review of current literature*. Center for Invasive Plant Management, Montana State University, Bozeman, MO.

Richardson D.M., Pyek P., Rejmanek M., Barbour M.G., Panetta F.D., and West C.J. (2000). Naturalization and invasion of alien plants: concepts and definitions. *Diversity and Distributions,* **6**:93–107.

Risbey D.A., Calver M.C., Short J., Bradley J.S., and Wright I.W. (2000). The impact of cats and foxes on the small vertebrate fauna of Heirisson Prong, Western Australia. II. A field experiment. *Wildlife Research,* **27**:223–225.

Rixon C.A.M., Duggan I.C., Bergeron N.M.C., Ricciardi A., and Macisaac H.J. (2005). Invasion risks posed by the aquarium trade and live fish markets on the Laurentian Great Lakes. *Biodiversity and Conservation* **14**:1365–1381.

Robertson B.C. and Gemmell N.J. (2004). Defining eradication units to control invasive pests. *Journal of Applied Ecology* **41**:1042–1048.

Robertson H.A., Hay J.R., Saul E.K., and McCormack G.V. (1994). Recovery of the kakerori—an endangered forest bird of the Cook Islands. *Conservation Biology* **8**:1078–1086.

Robinson G.R., Quinn J.F., and Stanton M.L. (1995). Invasibility of experimental habitat in California winter annual grassland. *Ecology* **79**: 786–794.

Rodda G.H., Fritts T.H., Campbell E.W., Dean-Bradley K., Perry G., and Qualls C.P. (2002). Practical concerns in the eradication of island snakes. In Veitch C.R. and Clout M.N. (eds.) *Turning the tide: the eradication of invasive species.* Occasional Paper of the IUCN Species Survival Commission No. 27, 260–265.

Roemer G.W., Coonan T.J., Garcelon D.K., Bascompte J., and Laughrin L. (2001). Feral pigs facilitate hyperpredation by golden eagles and indirectly cause the decline of the island fox. *Animal Conservation* **4**:307–318.

Rocmer G.W., Donlan C.J., and Courchamp F. (2002). Golden eagles, feral pigs, and insular carnivores: how exotic species turn native predators into prey. *PNAS* **99**:791–796.

Rolls E.C. (1969). *They All Ran Wild.* Angus and Robertson, Sydney, Australia.

Roni P., Beechie T.J., Bilby R.E., Leonetti F.E., Pollock M.M., and Pess G.R. (2002). A review of stream restoration techniques and a hierarchical strategy for prioritizing restoration in Pacific Northwest watersheds. *North American Journal of Fisheries Management* **22**:1–20.

Ross D.J., Johnson C.R., and Hewitt C.L. (2003). Assessing the ecological impacts of an introduced seastar: the importance of multiple methods. *Biological Invasions* **5**:3–21.

Ross M.A. and Lembi C.A. (1999). *Applied Weed Science,* 2nd edition, pp. 48–75. Burgess Publishing, Minneapolis, MN.

Rossiter N.A., Setterfield S.A., Douglas M.M., and Hutley L.B. (2003). Testing the grass-fire cycle: alien grass invasion in the tropical savannas of northern Australia. *Diversity and Distributions,* **9**:169–176.

Rouget M., Richardson D.M., Nel J.L., Van Wilgen B.W. (2002). Commercially important trees as invasive aliens – towards spatially explict risk assessment at a national scale. *Biological Invasions* **4**: 397–412.

Rouland P. (1985). Les castors canadiens de la Puisaye. *Bulletin Mensuel de l'Office National de las Chasse* **91**, 35–40.

Roy J. (1990). In search of the characteristics of plant invaders. In Di Castri F., Hansen A., and Debussche M. (eds.) *Biological invasions in Europe and the Mediterranean Basin,* pp. 335–352. Kluwer Academic Publishers, Dordrecht, Netherlands.

Rubec P.J., Cruz F., Pratt V., Oellers R., McCullough B., and Lallo F. (2001). Cyanide-free net-caught fish for the marine aquarium trade. *Aquarium Sciences and Conservation* **3**:37–51.

Ruiz G.M., Fofonoff P.W., and Anson A.H. (1999). Non-indigenous species as stressors in estuarine and marine communities: assessing invasion impacts and interactions. *Limnology and Oceanography* **43**:950–972.

Ruiz G.M., Fofonoff P.W., Carlton J.T., Wonhom M.J., and Hines A.H. (2000). Invasion of coastal marine communities in North America: apparent patterns, processes and biases. *Annual Review of Ecology and Systematics* **31**:481–531.

Ruscoe W. (2001). Advances in New Zealand mammalogy 1990–2000: House mouse. *Journal of the Royal Society of New Zealand* **31**:127–134.

Russell A.W. and Sparrow R. (2008). The case for regulating intragenic GMOs. *Journal of Agricultural and Environmental Ethics* **21**:153–181.

Russell J.C., Towns D.R., and Clout M.N. (2008). Preventing rat invasion of islands. *Science for Conservation 286*. Department of Conservation, Wellington, New Zealand.

Sainty G., McCorkelle G., and Julien M. (1998). Control and spread of Alligator Weed *Alternanthera philoxeroides* (Mart.) Griseb., in Australia: lessons for other regions. *Wetlands Ecology and Management* **5**:195–201.

Sakai A.K., Allendorf F.W., Holt J.S., Lodge D.M., Molofsky J., With K.A., *et al.* (2001). The population biology of invasive species. *Annual Review of Ecology and Systematics* **32**:305–332.

Samson R.A., Evans H.C., and Latge J.P. (1988). *Atlas of Entomopathogenic Fungi*. Springer-Velag, Berlin, Germany.

Sanderson J.C. (1990). A preliminary survey of the distribution of the introduced macroalga, *Undaria pinnatifida* (Harvey) Suringar on the east coast of Tasmania. *Botanica Marina* **33**:153–157.

Sarty M., Abbott K.L., and Lester P.J. (2007). Community level impacts of an ant invader and food mediated coexistence. *Insectes Sociaux* **54**:166–173.

Sarty M., Senior Adviser, Post Border, MAF Biosecurity New Zealand, personal communication.

Saunders D.L., Meeuwig J.J., and Vincent A.C.J. (2002). Freshwater protected areas: strategies for conservation. *Conservation Biology* **16**:30–41.

Saunders G., Lane C., Harris S., and Dickman C. (2006). *Foxes in Tasmania: a report on an incursion of an invasive species*. Invasive Animals Cooperative Research Centre, Canberra, Australia. http://www.invasiveanimals.com

Savarie P.J., Pan P., Hayes D.J., Roberts J.D., Dasch G.L., Felton R. *et al.* (1983). Comparative acute oral toxicity of para-aminopropiophenone. *Bulletin of Environmental Contamination and Toxicology* **30**:122–126.

Scheltema R.S. (1971). Larval dispersal as a means of genetic exchange between geographically separated populations of shallow-water benthic marine gastropods. *Biological Bulletin* **140**:284–322.

Schiller C.B. (1988). *Spawning and Larval Recruitment in the Coconut Crab (*Birgus Latro*) on Christmas Island, Indian Ocean*. Unpublished report for the Australian National Parks and Wildlife Service, Canberra, ACT, Australia.

Schmitz D.C., Nelson B.V., Nall L.E., and Schardt J.D. (1991). Exotic aquatic plants in Florida: a historical perspective and review of present aquatic plant regulation program. In Center T.D., Doren R.F., Hofstetter R.L., Myers R.L., and Whiteaker L.D. (eds.) *Proceedings of a Symposium on Exotic Pest Plants*, pp. 303–336. November 2–4, 1988, Miami, Florida. United States Department of the Interior, National Park Service, Washington, DC.

Schnase J.L., Lane M.A., Bowker G.C., Star S.L., and Silberschatz A. (2000). Building the next-generation biological-information infrastructure. In P.H. Raven (ed.), *Nature and Human Society: The Quest for a Sustainable World*, pp. 291–300. National Research Council, National Academy Press, Washington, DC.

Schnase J.L., Smith J.A., Stohlgren T.J., Quinn J.A., and Graves S. (2002a). Biological invasions: A challenge in ecological forecasting. In Proceedings of the International Geoscience and Remote Sensing Symposium, 2002 (IGARSS '02, Toronto, June). *IEEE 2002 International* **1**:154–156.

Schnase J.L., Stohlgren T.J., and Smith J.A. (2002b). The national invasive species forecasting system: A strategic NASA/USGS partnership to manage biological invasions. *Earth Observation Magazine* **11**: 46–49.

Schoonbee H.J. (1991). Biological control of fennel-leaved pondweed, *Potamogeton pectinatus* (Potamogetonaceae), in South Africa. *Agriculture, Ecosystems and Environment* **37**:231–237.

Schrader G. (2003). New working programme of the European and Mediterranean plant protection organisation on invasive alien species *Aliens* **18**:14–15.

Secord D. (2003). Biological control of marine invasive species: cautionary tales and land-based lessons. *Biological Invasions* **5**:117–131.

Seier M. K. and Evans H.C. (2007). Fungal pathogens associated with *Heracleum mantegazzianum* in its native and invaded distribution range. In Pyšek P., Cock M.J.W., Nentwig W., Ravn H.P., and Wade W. (eds.) *Ecology and Management of Giant Hogweed*, pp. 189–208. CABI Publishing, Wallingford, UK.

Semmens B.X., Buhle E.R., Salomon A.K., and Pattengill-Semmens C.V. (2004). A hotspot of non-native marine fishes: evidence for the aquarium trade as an invasion pathway. *Marine Ecology Progress Series* **266**:239–244.

Seymour A., Varnham K., Roy S., Harris S., Bhageerutty L., Church S., *et al.* (2005). Mechanisms underlying the failure of an attempt to eradicate the invasive Asian musk shrew *Suncus murinus* from an island nature reserve. *Biological Conservation* **125**:23–35.

Shafii B., Price W.J., Prather T.S., Lass L.W., and Thill D.C. (2003). Predicting the likelihood of yellow starthistle (*Centaurea solstitialis*) occurrence using landscape characteristics. *Weed Science* **51**:748–751.

Shafroth P.B., Friedman J.M., and Ischinger L.S. (1995). Effects of salinity on establishment of *Populus fremontii* (cottonwood) and *Tamarix ramosissma* (salt cedar) in southwestern United States. *Great Basin Naturalist*, **55**:58–65.

Shearer L.W. and MacKenzie C.L. (1961). The effects of salt solutions of different strengths on oyster enemies. *Proceedings of the National Shellfisheries Association,* **50**:97–103.

Sheley R., Petroff J., and Borman M. (1999). Introduction to biology and management of noxious rangeland weeds. In Sheley R. and Petroff J. (eds.) *Biology and Management of Noxious Rangeland Weeds*, pp. 1–3 Oregon State University Press, Corvallis, OR.

Sheppard A.W. and Raghu S. (2005). Working at the interface of art and science; how best to select an agent for classical biological control? *Biological Control* **34**:233–235.

Sheppard A.W., Hill R., DeClerck-Floate A., McClay A., Olckers T., Quimby Jr. P.C., and Zimmermann H.G. (2003). A global review of risk-benefit-cost analysis for the introduction of classical biological control agents against weeds: a crisis in the making. *Biocontrol News and Information* **24**:91N–108N.

Sherley G. (ed.) (2000). *Invasive species in the Pacific: A technical review and draft regional strategy.* South Pacific Regional Environment Programme: Apia, Samoa.

Shine C., Williams N., and Gündling L. (2000). *A guide to designing legal and institutional frameworks on alien invasive species.* IUCN, Gland, Switzerland Cambridge and Bonn. xvi +138 pp.

Shine C., Williams N., and Burhenne-Guilmin F. (2005). Legal and institutional Frameworks for Invasive Alien Species. In Mooney H.A., Mack R.N., McNeely J.A., Neville L.E., Schei P.J., and Waage J.K. (eds.).*Invasive Alien Species – A New Synthesis*, pp. 233–284. Island Press, Washington, DC.

Silvertown J. and Charlesworth D. (2001). *Introduction to plant population biology*, 4th edition. pp. 290–294. Blackwell Science, Oxford, UK.

Simberloff D. (1996). Impacts of Introduced Species in the United States. *Consequences* **2**:13.

Simberloff D. (2002). Today Tiritiri Matangi, tomorrow the world! Are we aiming too low in invasives control? In Veitch C.R. and Clout M.N. (eds.) *Turning the tide: the eradication of invasive species,* pp. 4–13. IUCN SSC Invasive Species Specialist Group. IUCN, Gland, Switzerland and Cambridge, U.K.

Simberloff D. (2003a) Confronting introduced species: a form of xenophobia? *Biological Invasions* **5**:179–192.

Simberloff D. (2003b). How much information on population biology is needed to manage introduced species? *Conservation Biology* **17**: 83–92.

Simberloff D. (2005). The politics for assessing risk for biological invasions: the USA as a case study. *Trends in Ecology and Evolution* **20**:216–222.

Simberloff D. and Gibbons, L. (2004). Now you see them, now you don't—population crashes of established introduced species. *Biological Invasions* **6**:161–172.

Simberloff D. and Stiling P. (1996a). Risk of species introduced for biological control. *Biological Conservation* **78**:185–192.

Simberloff D. and Stiling P. (1996b). How risky is biological control? *Ecology* **77**:1965–1974.

Simberloff D. and Stiling P. (1998). How risky is biological control? Reply. *Ecology* **79**:1834–1836.

Simberloff D., Parker I.M., and Windle P.N. (2005). Introduced species policy, management, and future research needs. *Frontiers in Ecology and the Environment* **3**:12–20.

Simmonds F.J., Franz J.M., and Sailer R.I. (1976). History of biological control. *In:* Huffaker C.B. and Messenger P.S. (eds.) *Theory and Practice of Biological Control*, pp. 17–39. Academic Press, New York, NY.

Simon K.S. and Townsend C.R. (2003). Impacts of freshwater invaders at different levels of ecological organisation, with emphasis on salmonids and ecosystem consequences. *Freshwater Biology* **48**:982–994.

Simpson A., Sellers E., Grosse A., and Xie Y. (2006). Essential elements of online information networks on invasive alien species. *Biological Invasions* **8**:1579–1587.

Sinclair A.R.E. (1995). Serengeti past and present. In Sinclair A.R.E. and Arcese P. (eds.). *Serengeti II. Dynamics, management and conservation of an ecosystem*, pp. 3–30. University of Chicago Press.

Sinden J., Jones R., Hester S., Odom D., Kalisch C., James R., and Cacho O. (2003). The economic impact of weeds in Australia. *CRC for Australian Weed Management*, Glen Osmond, South Australia.

Sinner J., Forrest B.M., and Taylor M.D. (2000). *A strategy for managing the Asian kelp* Undaria: Cawthron Report 578, Cawthron Institute, Nelson, New Zealand.

Sisler S. (2005). Behavioural evidence for aggregation pheromones in goldfish (*Carassius auratus*) and common carp (*Cyprinus carpio*). Unpublished PhD thesis, University of Minnesota, MN

Sivakumar K. (2003). Introduced mammals in Andaman and Nicobar Islands (India): a conservation perspective. *Aliens* **17**:11.

Skerman T.M. (1960). Ship-fouling in New Zealand waters: a survey of marine fouling organisms from vessels of the coastal and overseas trades. *New Zealand Journal of Science* **3**:620–648.

Smith B.R. and Tibbles J.J. (1980). Sea lamprey (*Petromyzon marinus*) in Lakes Huron, Michigan, and Superior: history of invasion and control, 1936–78. *Canadian Journal of Fisheries and Aquatic Sciences* **37**:1780–1801.

Smith R.G., Maxwell B.D., Menalled F.D., and Rew L.J. (2006). Lessons from agriculture may improve the management of invasive plants in wildland systems. *Frontiers in Ecology and Environment* **4**:428–434.

Smith V.R., Avenant N.L., and Chown S.L. (2002). The diet and impact of house mice on a sub-Antarctic island. *Polar Biology* **25**:703–715.

Society of Environmental Toxicology and Chemistry (1987). *Workshop Report.* Research Priorities in Environmental Risk Assessment. August 16–21, 1987.

Solow A., Seymour A., Beet A., and Harris S. (2008). The untamed shrew: on the termination of an eradication programme for an introduced species. *Journal of Applied Ecology*, **45**:424–427.

Sorensen P.W. and Hoye T.R. (2007). A critical review of the discovery and application of a migratory pheromone in an invasive fish, the sea lamprey *Petromyzon marinus* L. *Journal of Fish Biology* **71**(Suppl.D):100–114.

Sorensen P.W. and Stacey N.E. (2004). Brief review of fish pheromones and discussion of their possible uses in the control of non-indigenous teleost fishes. *New Zealand Journal of Marine and Freshwater Research* **38**:399–417.

Stanaway M.A., Zalucki M.P., Gillespie P.S., Rodriguez C.M., and Maynard G.V. (2001). Pest risk assessment of insects in sea cargo containers. *Australian Journal of Entomology* **40**:180–192.

Stead D.H. (1971a). *A preliminary survey of mussel stocks in Pelorus Sound.* No. 16. New Zealand Fisheries Technical Report.

Stead D.H. (1971b). *Pebrus Sound mussel survey.* December, 1969. New Zealand Fisheries Technical Report.

Steinhaus E.A. (1949). *Principles of Insect Pathology.* McGraw-Hill, New York, NY.

Sterner J.D. and Barrett R.H. (1991). Removing feral pigs from Santa Cruz Island, California. *Transactions of the Western Section of the Wildlife Society* **27**:47–53.

Stöck M., Lamatsch D.K., Steinlein C., Epplen J.T., Grosse W., Hock R., et al. (2002). A bisexually reproducing all-triploid vertebrate. *Nature Genetics* **30**:325–328.

Stohlgren T.J. and Schnase J.L. (2006). Risk analysis for biological hazards: What we need to know about invasive species. *Risk Analysis* **26**:163–173.

Stohlgren T.J., Quinn J.F., Ruggiero M., and Waggoner G. (1995). Status of biotic inventories in U.S. National Parks. *Biological Conservation* **71**: 97–106.

Stohlgren T.J., Chong G.W., Kalkhan M.A., and Schell L.D. (1997). Rapid assessment of plant diversity patterns: a methodology for landscapes. *Environmental Monitoring and Assessment* **48**:25–43.

Stohlgren T.J., Bull K.A., Otsuki Y., Villa C.A., and Lee L. (1998). Riparian zones as havens for exotic plant species in the central grasslands. *Plant Ecology* **138**:113–125.

Stohlgren T.J., Binkley D., Chong G.W., Kalkhan M.A., Schell L.D., Bull K.A., et al. (1999a). Exotic plant species invade hot spots of native plant diversity. *Ecological Monographs* **69**:25–46.

Stohlgren T.J., Schell L.D., and Vanden Heuvel B. (1999b). How grazing and soil quality affect native and exotic plant diversity in Rocky Mountain grasslands. *Ecological Applications* **9**:45–64.

Stohlgren T.J., Chong G.W., Schell L.D., Rimar K.A., Otsuki Y., Lee M., et al. (2002). Assessing vulnerability to invasion by nonnative plant species at multiple scales. *Environmental Management* **29**:566–577.

Stohlgren T.J., Guenther D., Evangelista P., and Alley N. (2005). Patterns of plant rarity, endemism, and uniqueness in an arid landscape. *Ecological Applications* **15**:715–725.

Stohlgren T.J., Barnett D., Flather C., Fuller P., Peterjohn B., Kartesz J., et al. (2006). Species richness and patterns of invasion in plants, birds, and fishes in the United States. *Biological Invasions* **8**:427–457.

Stoner D.S. (1992). Vertical distribution of a colonial ascidian on a coral reef: The roles of larval dispersal and life-history variation. *American Naturalist* **139**:802–824.

Stork N., Kitching R., Cermack M., Davis N., and McNeil K. (2003). *The impact of aerial baiting for control of the crazy ant, Anoplolepis gracilipes, on the vertebrates and canopy-dwelling arthropods on Christmas Island.* Unpublished report to Parks Australia North, Darwin NT Australia.

Strong D.R. (1997). Fear no weevil? *Science* **277**:1058–1059.

Stuart I.G., Williams A., McKenzie J., and Holt T. (2006). Managing a migratory pest species: a selective trap for common carp. *North American Journal of Fisheries Management* **26**:888–893.

Stuart M. and Chadderton W.L. (1997). The attempted eradication of the adventive Asian seaweed, *Undaria pinnatifida* from Big Glory Bay Stewart Island. A Progress Report of the DOC Eradication Programme. Unpublished DOC internal report.

Stuart M.D. (2002). *Incursion response tools for undesirable exotic marine organisms*. Ministry of Fisheries.

Suarez A.V., Holway D.A., and Ward P.S. (2005). The role of opportunity in the unintentional introduction of nonnative ants. *Proceeding of the National Academy of Sciences of the United States of America* **102**:17032–17035.

Sutton C.A. and Hewitt C.L. (2004). Detection kits for community-based monitoring of introduced marine pests. Revised Final Report to National Heritage Trust/Coasts and Clean Seas. NHT 21247.

Swales A., Ovenden R., MacDonald I.T., Lohrer A., and Burt K.L. (2005). Sediment remobilisation from decomposing cordgrass (*Spartina Anglica*) patches on a wave-exposed tidal flat. *New Zealand Journal of Marine and Freshwater Research* **39**:1305–1319.

Swanton C.J. and Booth B.D. (2004). Management of weed seedbanks in the context of populations and communities. *Weed Technology* **18**:1496–1502.

Syrett P., Briese D.T., and Hoffmann J.H (2000). Success in biological control of terrestrial weeds by arthropods. In Gurr G. and Wratten S.D. (eds.) *Biological Control: Measures of Success*, pp. 189–230. Kluwer Academic Publishers, Dordrecht, The Netherlands.

Tanaka H. and Larson B. (2006). The role of the International Plant Protection Convention in the prevention and management of invasive alien species. In Fumito Koike F., Clout M.N., Kawamichi M., De Poorter M., and Iwatsuki K. (eds.) *Assessment and Control of Biological Invasion Risks*. Houkadoh Book Sellers, Kyoto, Japan and IUCN, Gland, Switzerland. 56–62.

Tang Z.J., Butkus M.A., and Xie Y.F.F. (2006). Crumb rubber filtration: A potential technology for ballast water treatment. *Marine Environmental Research* **61**:410–423.

Tassin J., Riviére J.-N., Cazanova M., and Bruzzese E. (2006). Ranking of invasive woody species for management on Réunion Island. *Weed Research* **46**:388–403.

Taylor C.M. and Hastings A. (2004). Finding optimum control strategies for invasive species: a density-structured model for *Spartina alterniflora*. *Journal of Applied Ecology,* **41**:1049–1057.

Taylor M.D., Hay C.H., and Forrest B.M. (1999). Patterns of marine bioinvasion in New Zealand and mechanisms for internal quarantine. In Pederson J. (ed.) *Marine bioinvasions*. Proc 1st National (US) Conf Marine Bioinvasions. MIT Sea Grant College Program, Massachusetts Institute of Technology, Cambridge, MA.

Taylor M.D., MacKenzie L.M., Dodgshun T.J., Hunt C.D., and de Zwart E.J. (2007). Trans-Pacific shipboard trials on planktonic communities as indicators of open ocean ballast water exchange. *Marine Ecology and Progress Series* **350**:41–54.

Taylor R.H. (1968). Introduced mammals and islands: priorities for conservation and research. *Proceedings of the New Zealand Ecological Society* **15**:61–67.

Taylor R.H. (1979). How the Macquarie Island parakeet became extinct. *New Zealand Journal of Ecology* **2**:42–45.

Thomas M. (2005). *Non-target impact study of Fipronil based ant bait on Robber Crabs (Birgus latro): the effectiveness of chicken pellet lures*. Unpublished report to Parks Australia North, Christmas Island, Indian Ocean.

Thompson, K., Hodgson J.G., and Rich T.C.G. (1995). Native and alien invasive plants: more of the same? *Ecography* **18**:390–402.

Thorp J.R. and Lynch R. (2000). *The Determination of Weeds of National Significance*. National Weeds Strategy Executive Committee, Launceston, Tasmania.

Thresher R. (2007). Genetic options for the control of invasive vertebrate pests: prospects and constraints. USDA National Wildlife Research Centre Symposia: managing vertebrate invasive species. University of Nebraska, Lincoln. http://digitalcommons.unl.edu/nwrcinvasive/52 (accessed August 2008)

Thresher R.E. and Kuris A.M. (2004). Options for managing invasive species. *Biological Invasions* **6**:295–300.

Tisdale E.W. (1976). Vegetational responses following biological control of *Hypericum perforatum* in Idaho. *Northwest Science* **60**:61–75.

Tiwary B.K., Kirubagaran R., and Ray A.K. (2004). The biology of triploid fish. *Reviews in Fish Biology and Fisheries* **14**:391–402.

Tlusty M. (2002). The benefits and risks of aquacultural production for the aquarium trade. *Aquaculture* **205**:203–219.

Toft J.D., Simenstad C.A., Cordell J.R., and Grimaldo L.F. (2003). The effects of introduced water hyacinth on habitat structure, invertebrate assemblages, and fish diets. *Estuaries* **26**:746–758.

Tomley A.J. (1998). *Cryptostegia grandiflora* Roxb. ex R.Br. In Panetta, F.D., Groves R.H., and Shepherd R.C.H. (eds.) *The Biology of Australian Weeds*, pp. 63–76. R.G. and F.J Richardson, Melbourne.

Tomley A.J. and Panetta F.D. (2002). Eradication of the exotic weeds *Helenium amarum* (Rafin) HL and *Eupatorium serotinum* Michx from south-eastern Queensland. In Spafford J.H., Dodd J., and Moore J.H. (eds.) Proceedings of the 13th Australian Weeds Conference, Perth, Australia, 293–296.

Torr N. (2002). Eradication of rabbits and mice from subantarctic Enderby and Rose Islands. In Veitch C.R. and Clout M.N. (eds.) *Turning the tide: eradication of invasive species*, pp. 319–328. IUCN SSC Invasive Species Specialist Group, IUCN, Gland, Switzerland and Cambridge, UK.

Towns D.R. and Broome K.G. (2003). From small Maria to massive Campbell: forty years of rat eradications from New Zealand islands. *New Zealand Journal of Zoology* **30**:377–398.

Towns D.R., Atkinson I.A.E., and Daugherty C.H. (2006). Have the harmful effects of introduced rats on islands been exaggerated? *Biological Invasions* **8**:863–891.

Tsutsui N.D. and Suarez A.V. (2003). The colony structure and population biology of invasive ants. *Conservation Biology* **17**:48–58.

Turner J.W., Liu I.K.M., Rutberg A.T., and Kirkpatrick J.P. (1996). Immunocontraception limits foal production in free-roaming feral horses in Nevada. *Journal of Wildlife Management* **61**:873–880.

Tuyttens F.A.M., Macdonald D.W., Delahay R., Rogers L.M., Mallinson P.J., Donnelly C.A., *et al.* (1999). Differences in trappability of European badgers *Meles meles* in three populations in England. *Journal of Applied Ecology* **36**:1051–1062.

Twigg L.E., Lowe T.J., Martin G.R., Wheeler A.G., Gray G.S., Griffin S.L., *et al.* (2000). Effects of surgically imposed sterility on free-ranging rabbit populations. *Journal of Applied Ecology* **37**:16–39.

Tyndale-Biscoe H. (1994). Virus-vectored immunocontraception of feral mammals. *Reproduction, Fertility and Development* **6**:9–16.

Tyser R.W. (1992). Vegetation associated with two alien plant species in a fescue grassland in Glacier National Park, Montana. *Great Basin Naturalist* **52**:189–193.

Uchimura M., Rival A., Nato A., Sandeaux R., Sandeaux J., and Baccou J.C. (2000). Potential use of Cu^{2+}, K^+ and Na^+ for the destruction of *Caulerpa taxifolia*: Differential effects on photosynthesis parameters. *Journal of Applied Phycology* **12**:15–23.

UNEP (2001). Status, impacts and trends of alien species that threaten ecosystems, habitats and species. *Note by the Executive Secretary, Subsidiary Body on Scientific, Technical and Technological Advice, Convention on Biological Diversity*, Sixth meeting, Montreal, 12–16 March 2001; UNEP/CBD/SBSTTA/6/INF/11.

References

UNEP (2005a). Implications of the findings of the Millennium Ecosystem Assessment for the future work of the Convention—*Addendum—Summary for decision makers of the biodiversity synthesis report*. UNEP/CBD/SBSTTA/11/7/Add.1 (31 August 2005).

UNEP (2005b) Millennium ecosystem assessment. *Ecosystems and Human Well-being; Biodiversity Synthesis*. World Resources Institute, Washington DC.

UNEP-MAP-RAC/SPA (2005). Action plan concerning species introductions and invasive species in the Mediterranean Sea. Ed. RAC/SPA, Tunis.

Unger J.G. (2003). The IPPC Standard on Environmental Risks of Plant Pests. *Aliens* **18**:13–14.

USGS (2004). *Facts about bighead and silver carp*. Columbia Environmental Research Center, USGS, Columbia.

Usher M.B. (1988). Biological invasions of nature reserves: a search for generalizations. *Biological Conservation* **44**:119–135.

Valentine E.W. and Walker A.K. (1991). *Annotated catalogue of New Zealand Hymenoptera*, pp. 1–84. DSIR Plant Protection report No. 4. General Printing Services Ltd, Nelson, New Zealand.

Van Driesche R. and Bellows Jr. T.S. (1996). *Biological Control*. Chapman and Hall, New York, NY.

Van Driesche R.G. and Ferro D.N. (1987). Will the benefits of classical biological control be lost in the biotechnology stampede? *American Journal of Alternative Agriculture* **2**:50–56.

van Schagen J.J., Davis P.R., and Widmer M.A. (1994). Ant pests of Western Australia, with particular reference to the Argentine ant, (*Linepithema humile*). In *Exotic Ants: Biology, Impact and Control of Introduced Species*. Westview Press, Boulder, CO, 174–180.

Van T.K., Wheeler G.S., and Center T.D. (1999). Competition between *Hydrilla verticillata* and *Vallisneria americana* as influenced by soil fertility. *AquaticBotany* **62**:225–233.

Van Vuren D. and Coblentz B.E. (1987). Some ecological effects of feral sheep on Santa Cruz Island, California, USA. *Biological Conservation* **41**:253–268.

Vanderwoude C., Elson-Harris M., Hargreaves J.R., Harris E.J., and Plowman K.P. (2003). An overview of the red imported fire ant (*Solenopsis invicta* Buren) eradication plan for Australia. *Records of the South Australian Museum Monograph Series* **7**:11–16.

Varnham K. (2006). Non-native species in UK Overseas Territories: a review. *Joint Nature Conservation Committee Report* No. 372.

Vazquez-Dominguez E., Ceballos G., and Cruzado, J. (2004). Extirpation of an insular subspecies by a single introduced cat: the case of the endemic deer mouse *Peromyscus guardia* on Estanque Island, Mexico. *Oryx* **38**:347–350.

Veitch C.R. and Bell B.D. (1990). Eradication of introduced animals from the islands of New Zealand. In Towns D.R., Daugherty C.H., and Atkinson I.A.E. (eds.) Ecological restoration of New Zealand islands, Conservation Sciences Publication No. 2, 137–146.

Veitch C.R. and Clout M.N. (2002). *Turning the tide: the eradication of invasive species*. Proceedings of the International Conference on Eradication of Island Invasives. Occasional Paper of the IUCN Species Survival Commission No. 27. IUCN, Gland, Switzerland and Cambridge, UK.

Veltman C.J., Nee S., and Crawley M.J. (1996). Correlates of introduction success in exotic New Zealand birds. *6 e American Naturalist* **147**: 542–547.

Vencill W.K. (ed.) (2002). *Herbicide Handbook*, 8th edition. Weed Science Society of America, Lawrence, KS.

Venevski S, and Veneskaia I. (2003). Large-scale energetic and landscape factors of vegetation diversity. *Ecology Letters* **6**:1004–1016.

Verling E., Ruiz G.M., Smith L.D., Galil B., Miller A.W., and Murphy K.R. (2005). Supply-side invasion ecology: characterizing propagule pressure in coastal ecosystems. *Proceedings of the Royal Society B* **272**:1249–1257.

Vertebrate Pest Committee (2002). List of exotic vertebrate animals in Australia. VPC, Canberra, Australia.
Vetemaa M., Eschbaum R., Albert A., and Saat T. (2005). Distribution, sex-ratio and growth of *Carrasius gibelio* (Bloch) in coastal and inland waters of Estonia (north-eastern Baltic Sea). *Journal of Applied Ichthyology* **21**:287–291.
Vigueras A.L. and Portillo L. (2002). Uses of Opuntia species and the potential impact of *Cactoblastis cactorum* (Lepidoptera: Pyralidae) in Mexico. *Florida Entomologist.* **84**:493–498.
Vitousek P.M., Mooney H.A., Lubchenco J., and Melillo J.M. (1997). Human Domination of Earth's Ecosystems. *Science* **277**:494–499.
Vivrette N.J. and Muller C.H. (1977). Mechanisms of invasion and dominance of coastal grassland by *Mesembryanthemum crystallinum*. *Ecological Monographs* **47**:301–318.
Vogler W. and Lindsay A. (2002). The impact of the rust fungus *Maravalia cryptostegiae* on three rubber vine (*Cryptostegia grandiflora*) populations in tropical Queensland. In Spafford Jacob H., Dodd J., and Moore J.H. (eds.) *Weeds: Threats Now and Forever*. Proceedings of the 13th Australian Weeds Conference. Plant Protection Society of Western Australia, Perth, Australia, 180–182.
Wainger L.A. and King D.M. (2001). Priorities for weed risk assessment: using a landscape context to assess indicators of functions, services, and values. In Groves R.H., Panetta F.D., and Virtue J.G. (eds.) *Weed Risk Assessment*, pp. 34–51. CSIRO Publishing, Australia.
Walker B. and Weston E.J. (1990). Pasture development in Queensland – a success story. *Tropical Grasslands,* **24**:257–268.
Walsh C., Morrison M.A., and Middleton C. (2003). Invasion of the Asian goby, *Acentrogobius pflaumii*, into New Zealand, with new locality records of the introduced bridled goby, *Arenigobius bifrenatus*. *New Zealand Journal of Marine and Freshwater Research* **37**:105–112.
Walters L.J., Brown K.R., Stam W.T., and Olsen J.L. (2006). E-commerce and *Caulerpa*: unregulated dispersal of invasive species. *Frontiers in Ecology and Environment* **4**:75–79.
Walton C. (2003). The biology of Australian weeds. 42. *Leucaena leucocephala* (Lamark) de Wit. *Plant Protection Quarterly,* **18**:90–98.
Wanless R.M., Angel A., Cuthbert R.J., Hilton G.M., and Ryan P.G. (2007). Can predation by invasive mice drive seabird extinctions? *Biology Letters* **3**:241–244.
Wapshere A.J. (1975). A protocol for programmes for biological control of weeds. *PANS* **21**:295–303.
Warburton B., Poutu N., Peters D., and Waddington P. (2008). Traps for killing stoats (*Mustela erminea*): improving welfare performance. *Animal Welfare* **17**:111–116.
Ward D.F., Beggs J.R., Clout M.N., Harris R.J., and O'Connor S. (2006). The diversity and origin of exotic ants arriving in New Zealand via human-mediated dispersal. *Diversity and Distributions* **12**:601–609.
Warwick H., Morris P., and Walker D. (2006). Survival and weight changes of hedgehogs (*Erinaceus europaeus*) translocated from the Hebrides to mainland Scotland. *Lutra* **49**:89–102.
Wasphere A.J. (1989). A testing sequence for reducing rejection of potential biological control agents for weeds. *Annals of Applied Biology* **114**:515–526.
Wasson K., Zabin C.J., Bedinger L., Diaz M.C., and Pearse J.S. (2001). Biological invasions of estuaries without international shipping: the importance of intraregional transport. *Biological Conservation* **102**:143–153.
Waters J.M. and Roy M.S. (2004). Out of Africa: the slow train to Australasia. *Systematic Biology* **53**:18–24.
Watson G.J., Bentley M.G., Gaudron S.M., and Hardege J.D. (2003). The role of chemical signals in the spawning induction of polychaete worms and other marine invertebrates. *Journal of Experimental Marine Biology and Ecology* **294**:169–187.

Weedbusters http://www.weedbusters.info/

Weigle S.M., Smith L.D., Carlton J.T., and Pederson J. (2005). Assessing the risk of introducing exotic species via the live marine species trade. *Conservation Biology* **19**:213–223.

Wells R.D.S., Winton M.D., and Clayton J.S. (1997). Successive macrophyte invasions within the submerged flora of Lake Tarawera, Central North Island, New Zealand. *New Zealand Journal of Marine and Freshwater Research* **31**:449–459.

Werksman J. (2004). Invasive alien species and the multilateral trading system. In Miller M.L. and Fabian R.N. (eds.) *Harmful Invasive Species: Legal Responses*, pp. 203–218. Environmental Law Institute, Washington DC, 203–214.

Westbrooks R. (2003). *A national early detection and rapid response system for invasive plants in the United States: Conceptual Design*. Federal Interagency Committee for the Management of Noxious and Exotic Weeds (FICMNEW).

Westphal M.I., Browne M., MacKinnon K., and Noble I. (2008). The link between international trade and global distribution of invasive alien species. *Biological Invasions* **10**:391–398.

Whisenant S.G. (1999). *Repairing Damaged Wildlands: A Process-Oriented, Landscape-Scale Approach*. Cambridge University Press, Cambridge, UK.

Whitfield P., Gardner T., Vives S.P., Gilligan M.R., Courtenay W.R., Carleton-Ray G., et al. (2002). Biological invasions of the Indo-Pacific lionfish (*Pterois volitans*) along the Atlantic coast North America. *Marine Ecology Progress Series* **235**: 289–297.

Wilcove D.S., Rothstein D., Bubow J., Phillips A., and Losos E. (1998). Quantifying threats to imperilled species in the United States. *BioScience* **48**(8):607–615.

Willan R.C. (1987). The mussel *Musculista senhousia* in Australasia; another aggressive alien highlights the need for quarantine at ports. *Bulletin of Marine Science* **41**:475–489.

Williams J.E., Sada D.W., and Williams C.D. (1988). American Fisheries Society guidelines for introductions of threatened and endangered fishes. *Fisheries* **13**: (5) 5–11.

Williams K., Parer I., Coman B., Burley J. and Braysher M. (1995). *Managing vertebrate pests: Rabbits*. Bureau of Resource Sciences and CSIRO Division of Wildlife Ecology, Canberra, Australia.

Williamson M.H. and Fitter A. (1996a). The characters of successful invaders. *Biological Conservation* **78**:163–170.

Williamson M.H. and Fitter A. (1996b). The varying success of invaders. *Ecology* **77**:1661–1666.

Wilson C.G. and McFadyen R.E.C. (2000). Biological control in the developing world: safety and legal issues. In Spencer N.R. (ed.) *Proceedings of the Tenth International Symposium on Biological Control Weeds*, pp. 505–511. Montana State University, Bozeman, MT.

Wilson E.O. (2005). Early ant plagues in the New World. *Nature* **433**:32.

Wilson E.O. and Taylor R.W. (1967). The ants of Polynesia (Hymenoptera: Formicidae). *Pacific Insects Monograph* **14**. Bernice P. Bishop Museum. Honolulu, Hawaii.

Wilson G., Dexter N., O'Brien P., and Bomford M. (1992). Pest animals in Australia: a survey of introduced wild mammals. Bureau of Rural Resources and Kangaroo Press, Canberra, Australia.

Wilson J.R.U., Richardson, D.M., Rouget, M., Procheş, Ş., Amis, M.A., Henderson, L. et al. (2007). Residence time and potential range: Crucial considerations in modelling plant invasions. *Diversity and Distributions* **13**:11–22.

Winston J.E., Gregory M.R., and Stevens L.M. (1996). Encrusters, epibionts, and other biota associated with pelagic plastics: a review of biogeographical, environmental, and conservation issues. In Coe J.M. and Rogers D.B. (eds.) *Marine debris: sources, impacts and solutions*, pp. 81–97. Springer, New York.

Wirf L. (2006). Using simulated herbivory to predict the efficacy of a biocontrol agent: the effect of manual defoliation and *Macaria pallidata* Warren (Lepidoptera: Geometridae) herbivory on *Mimosa pigra* seedlings. *Australian Journal of Entomology* **45**, 324–326.

Witkowski A. (1996). Introduced fish species into Poland: pros and cons. *Archives of Polish Fisheries* **4**:101–112.

Witmer G., Campbell E., and Boyd F. (1998). Rat management for endangered species protection in the U.S. Virgin Islands. *Proceedings of the Vertebrate Pest Conference* **18**:281–286.

Witte F., Maku B.S., Wanink J.H., Seehausen O., and Katunzi E.F.B. (2000). Recovery of cichlid species in Lake Victoria: An examination of factors leading to differential extinction. *Reviews in Fish Biology and Fisheries* **10**:233–241.

Wittenberg R. and Cock M.J.W. (2001). *Invasive Alien Species: A Toolkit of Best Prevention and Management Practices*. CAB International, Wallingford, Oxon, UK.

Wittenberg R. and Cock M.J.W. (2005). Best practices for the prevention and management of invasive alien species. In Mooney H.A., Mack R.N., McNeely J.A., Neville L.E., Schei P.J., and Waage J.K. (eds.) *Invasive Alien Species*, pp. 209–232. Island Press, Washington, DC.

Wittington R.J. and Chong R. (2007). Global trade in ornamental fish from an Australian perspective: the case for revised import risk analysis and management strategies. *Preventive Veterinary Medicine* **81**:92–116.

Woldendorp G. and Bomford M. (2004). Weed eradication: strategies, timeframes and costs. Bureau of Resource Sciences, Canberra, Australia.

Wonham M.J., Carlton J.T., Ruiz G.M., and Smith L.D. (2000). Fish and ships: relating dispersal frequency to success in biological invasions. *Marine Biology* **136**:1111–1121.

Wonhom M.J., O'Connor M., and Harley C.D.G. (2005). Positive effects of a dominant invader on introduced and native mudflat species. *Marine Ecology Progress Series* **289**:109–116.

Woodward K.N. (2005). Veterinary pharmacovigilance. Part 3. Adverse effects of veterinary medicinal products in animals and on the environment. *Journal of Veterinary Pharmacology and Therapeutics* **28**:171–184.

Wooton D.M. and Hewitt C.L. (2004). Marine biosecurity post-border management developing incursion response systems for New Zealand. *New Zealand Journal of Marine and Freshwater Research* **38**:553–559.

Worrall J. (2002). *Review of systems for early detection and rapid response*. U.S. Department of Agriculture, Forest Service, Forest Health Protection. Report for the National Invasive Species Council.

Wotton D.M. and Hewitt C.L. (2004). Marine biosecurity post-border management: developing incursion response systems for New Zealand. *New Zealand Journal of Marine and Freshwater Research* **38**:553–559.

Wotton D.M., O'Brien C., Stuart M.D., and Fergus D.J. (2004). Eradication success down under: heat treatment of a sunken trawler to kill the invasive seaweed *Undaria pinnatifida*. *Marine Pollution Bulletin* **49**:844–849.

Wright A.D. and Purcell M.F. (1995). *Eichhornia crassipes* (Mart.) Solms-Laubach. in Groves R.H. and H. Shepherd R.C.H. (eds.) *The Biology of Australian Weeds Vol.* **1**: 111–121

Wyatt A.S.J., Hewitt C.L., Walker D.I., and Ward T.J. (2005). Marine introductions in the Shark Bay World Heritage Property, Western Australia: a preliminary assessment. *Diversity and Distributions* **11**:33–44.

Xu H., Ding H., Li M., Qiang S., Guo J., Han Z., *et al.* (2006b). The distribution and economic losses of alien species invasion to China. *Biological Invasions* **8**:1495–1500.

Xu H., Qiang S.H., Han Z.H., Guo J.Y., Huang Z.G., Sun H.Y., *et al.* (2006a). The status and causes of alien invasion in China. *Biodiversity and Conservation* **15**:2893–2904.

References

Young, T.R. 2006. National and regional legislation for promotion and support to the prevention, control, and eradication fo invasive species. *Biodiversity series paper; no. 108. The World Bank Environment Department.* vii + 88 pp.

Zambrano L., Scheffer M., and Martinez-Ramos M. (2001). Catastrophic response of lakes to benthivorous fish introduction. *Oikos* **94**:344–350.

Zamora D.L., Thill D.C., and Eplee R.E. (1989). An eradication plan for plant invasions. *Weed Technology* **3**:2–12.

Zavaleta E.S. (2002). It's often better to eradicate, but can we eradicate better? In Veitch C.R. and Clout M.N. (eds.) *Turning the tide: eradication of invasive species,* pp. 393–403. IUCN SSC Invasive Species Specialist Group, IUCN, Gland, Switzerland and Cambridge, UK.

Zavaleta E.S., Hobbs R.J., and Mooney H.A. (2001). Viewing invasive species removal in a whole-ecosystem context. *Trends in Ecology & Evolution* **16**:454–459.

Zimmermann H.G., Moran V.C., and Hoffmann J.H. (2002). The renowned cactus moth, *Cactoblastis cactorum* (Lepidoptera: Pyralidae): Its natural history and threat to native *Opuntia* floras in Mexico and the United States of America. *Florida Entomologist* **84**:543–551.

Zink T.A. and Allen M.F. (1998). The effects of organic amendments on the restoration of a disturbed coastal sage scrub habitat. *Restoration Ecology* **6**:52–58.

Zulijevic A. and Antolic B. (1999a). *Partial eradication of Caulerpa taxifolia in Starigrad Bay.* In Abstracts of the 4th International workshop on *Caulerpa taxifolia* in the Mediterranean Sea, February 1999.

Zulijevic A. and Antolic B. (1999b). *Appearance and eradication of* Caulerpa taxifolia *in the Barbat Channel.* In Abstracts of the 4th International workshop on *Caulerpa taxifolia* in the Mediterranean Sea, February 1999.

Index

Note: page numbers in *italics* refer to Figures and Tables.

Aarhus Convention 94
abundance of invasive species,
 relationship to impact 66–7
abundance maps 68
Acacia nilotica (prickly acacia),
 introduction to Australia 71
accidental introductions 4, 5–7
acetic acid, in control of marine
 species 222, 225
Achatina fulica (giant African snail), biological
 control 82
Achilles heel approach, poisoning 180
Acridotheres species (mynas), control
 tools *183*, 184
 A. tristis, eradication 58
Adoption Model *106*
adventitious roots 127–8
adverse effects, eradication programmes 49
aerial baiting 55
 rabbits 56, 58
 yellow crazy ant *156*, 161–2, 167, 170–1
  assessment of success 164–5
  bait dispersal *156*, 163
  evaluation and lessons learned 167–70
  helicopter *156*, 162–3
  legislative approach 162
  non-target impacts 165–6
  trial 164
African big-headed ant, eradication 99
Agasicles hygrophila 150
Agreement on the Conservation of African
 Eurasian Migratory Waterbirds 114
agriculture, international legislative
 frameworks 3
alligator weed (*Alternanthera
  philoxeroides*) *142*
 biological control 150
 eradication cost 54
 mode of spread 145–6
American bullfrog (*Rana catesbeiana*) 114
Amsterdam Island, mesopredator release
 effect 241
Anacapa Island, rat eradication 49
Andaman Islands, legislation as impediment to
 IAS management 109
Andropogon gayanus (gamba grass),
 control 76
Andropogon virginicus, eradication cost 54
angiosperms, classification 126–7
annual plants 126

Anoplolepis gracilipes (yellow crazy ant) 155–6
 invasion of Christmas Island 156–7, *158*
 aerial control campaign 161–7, 170–1
 interim response 157–61
Anoplolepis longipes, mesopredator release
 effect 241
antifouling paints 209, 214, 230
antimycin-A (Fintrol®) 194, 196
ants 154–5
 African big-headed ant, eradication 99
 Pacific Ant Prevention Programme
  (PAPP) 123
 *see also Anoplolepis gracilipes; Anoplolepis
  longipes*
aquaculture 209–10, 228
 for aquarium trade 191
 chemical control 225
 for food 191–2
 heat treatment 224
 sea squirts, removal from mussel
  shells 215
aquarium trade 145, 189–91, 210
aquatic plants viii, 141
 importance of plant characteristics 141,
  144–5
 management
  biological control 148–*51*
  herbicidal control 148
  integrated control 151
  manual and mechanical control 146–7
  utilization 146
 modes of introduction and spread 145–6
 potential quarantine issues *17*
 prevention, early detection, and rapid
  response 152
aquatic species, eradication 55, 58
arthropods, biological control, safety of
 agents 90
artichoke thistle (*Cynara cardunculus*) *130*
Arundo donax (giant reed) *132*, *135*
Ascophyllum nodosum, physical removal 215
Asparagus asparagoides (bridal creeper),
 biological control 104
Asterias amurensis (northern Pacific
 sea-star) 209, 221
 chemical control 226
 detection 229
Asterias forbesi, control 223
Atlantic comb jelly (*Mnemiopsis leidyi*),
 biological control 82

Australia
  biological control of rubber vine, economics 84
  bridal creeper
    biological control 104
    public participation 104–6
  community-led initiatives 97–8
  containment of *Leucaena leucocephala* 74
  containment of rubber vine 73–4
  control of invasive pasture grasses 75–6
  eradication of trout 195
  mimosa trees 37–8
  plant introductions 8
  prevention of introductions 4–5
  Queensland fruit fly (*Bactrocera tryoni*) 5
avian malaria vi
Azolla filiculoides (red water fern) *142*, 144
  biological control 150

BACI (before-after control-impact) protocol 246
bacteria, potential quarantine issues *17*
baits 180
ballast water 189, *208*–9
  control and management 118–19, 211, 229–30
Ballast Water Convention, IMO 2005 94
ballast water exchange (BWE) 211
banded rail (*Rallus philippensis*) 235
Barbados, Programme of Action for the Sustainable Development of Small Island Developing States 121
barriers
  in exclusion of terrestrial vertebrates 176
  prevention of spread of invasive fish 192–3
Bassia scoparia (tumbleweed), eradication 53
  cost 54
Bayluscide® 197
Beauveria banana use in biological control 80
beavers (*Castor canadensis*)
  eradication 52, 55
  trapping 178
bench inspection equipment 13–14
Bern Convention (1979) 122
Beroe ovata (comb jelly) 82
biennial plants 126
binding instruments, international legislation 111
biodiversity, Convention on Biological Diversity 112–13
bioeconomic models 188
biofuels vi, 9
biological control viii, 7–8, 77, 92
  in aquaculture 191–2
  of aquatic plants 148–*51*
  classical
    early history and development 79–80
    Enemy Release Hypothesis 78
  projects in natural ecosystems 80–2
  successes, failures, and economics 83–4
  constraints 91–2
  of fish 197–9
  modern methods 86
    characteristics of efficacious agents 86–7
    risk assessment 87–90
  potential quarantine issues *17*
  of rubber vine 73–4
  of terrestrial plants *131*, 138
  of terrestrial vertebrates 181–2
biosecurity 1
  legislative frameworks 3–4
  prediction of invasiveness 2–3
Biosecurity Act 1993, New Zealand 94
Bird Island, mesopredator release effect 241
Birgus latro (robber crab), impact of aerial fipronil baiting 166
blackberries (*Rubus* species), eradication 57
black carp, use in biological control 191–2
black mussel (*Mytilopsis adamsi*), eradication 213, 225
black rat vi
black-striped mussel (*Mytilopsis sallei*), control 223
Boiga irregularis (brown tree snake) 175
  control tools *183*
  eradication 58
  introduction to Guam 20, 32
borders, preventive actions 11
bottom-up regulation 233, 237
branched broomrape (*Orobanche ramosa*), eradication campaign 53
bridal creeper (*Asparagus asparagoides*), biological control 104
broad-spectrum baits 180
bromine compounds, in control of marine species 222
Bromus tectorum (cheatgrass) 25–6, *28*, 33
brown tree snake (*Boiga irregularis*) 175
  control tools *183*
  eradication 58
  introduction to Guam 20, 32
brushtail possum (*Trichosurus vulpecula*) 51
  control tools *183*
  poisoning 180, *181*
bubble barriers 193
bud banks 128, *129*, 137
buffel grass (*Cenchrus ciliaris*), control 75–6
Bufo marinus (cane toad) 82
  control tools *183*
bull frog (*Rana catesbeina*), control tools *183*
Bush Friendly Garden, Floriade Canberra *105*

CAB International, Invasive Species Compendium 42
Cactoblastis cactorum (prickly pear moth) 8, 80
  risk of biological control 21

calicivirus, use in biological control 182
Cambomba caroliniana (fanwort) *143*
Canadian waterweed (*Elodea canadensis*) *143*
 control 147
canals, spread of marine fish 193
cane toad (*Bufo marinus*) 82
 control tools *183*
 Fiji 96–7
Capra hircus (goat)
 control tools *183*
 eradication *51*, 52
 helicopter culling 179
Carassius auratus (goldfish) 187
carbaryl, in control of marine species 226
carbohydrate storage, perennial plants *127*
Carcinus maenas (European green crab) 209
 biological control 82
 chemical control 226
Carijoa riisei (snow-flake coral) 222
Caring for the Country Unit, Northern Territory 98
carp 186
 effect of river restoration 198
 rotenone-impregnated baits 196
carp control programme, Tasmania 203, 204
carp herpes virus 198
Carpsim software 204
Castor canadensis (beaver)
 eradication 52, 55
 trapping 178
cats (*Felis catus*)
 control, New Zealand 74–5
 control tools *183*
 eradication *51*, 52
Caulerpa taxifolia 190, 192, 210, 213
 chemical control 226
 EDRA programme 40
 physical removal 215
 wrapping 223
Cenchrus ciliaris (buffel grass), control 75–6
Cenchrus echinatus, eradication cost *54*
Centaurea tricocephala, eradication cost *54*
Cervus elaphus (red deer), control tools *184*
CFT legumine 196
chaining 135
cheatgrass (*Bromus tectorum*) 25–6, *28*, 33
chemical control
 fish 193–7
 marine species 218–19, 224–7
 terrestrial plants *131*, 138–9
chemical hazards, risk assessment 19
Cherax tenuimanus (marron crayfish), eradication in New Zealand 192
chlorine, in control of marine species 222, 223
Chondrilla juncea, eradication cost *54*
Christmas Island, invasion by yellow crazy ant 156–7, *158*

aerial control campaign 161–7, 170–1
 evaluation and lessons learned 167–70
 interim response 157–61
Chrysemys picta (painted turtle) 114
Cinara cupressi (cypress aphid), biological control 81
Ciona intestinalis 215, 225, 226
CITES (Convention on International Trade in Endangered Species) 114
Citharexylum gentryi, eradication cost *54*
Cladophora spp., chemical control 225
classical biological control 77, 92
 early history and development 79–80
 Enemy Release Hypothesis 78
 projects in natural ecosystems 80–2
 successes, failures, and economics 83–4
 see also biological control
Cleome rutidosperma, eradication cost *54*
climate change v, vi
climate matching 26
cloning, terrestrial plants 128
Coccinella undecimpunctata (ladybird), use in biological control 79
Code of Conduct for Responsible Fisheries 115
Codium fragile spp. *tomentosoides*, chemical control 225
colonization, deliberate introductions 7
Columba livia (pigeon), eradication 52
commercial exploitation, conflicting interests 72
community-led initiatives 97, 107
 see also public participation
competition, role in weed control 137–8
competitor release effect 236, 242–3
 control strategies 245–6
compliance, value of public participation 93–5
conflicting interests 72
constraints 2–3
containment viii, 61–3, 76
 conflicting interests 72
 coordinated management of multiple species 69
 early action 69
 evaluation of impact 67
 examples
  Leucaena leucocephala 74
  rubber vine, northern Australia 73–4
 exploitation of natural barriers 70
 feasibility 63–4
 indications for use 63
 knowledge of species' biology and ecology 67–8
 long-term commitment 71–2
 monitoring 72–3
 principles, prioritization 68–9
 targeting of effort 70
 of terrestrial plants 130
containment potential, assessment 33

containment strategies 65
contingency planning, marine species 213–14
contingency response plans 14–15
control 62, 76, 213
   conflicting interests 72
   coordinated management of multiple species 69
   early action 69–70
   evaluation of impact 67
   examples
      grasses in Australia 75–6
      mammalian predators in New Zealand 74–5
   exploitation of times of low population 71
   indications for use 63, 65–7
   knowledge of species' biology and ecology 67–8
   long-term commitment 71–2
   monitoring 72–3
   principles, prioritization 68–9
   targeting of effort 70
   of terrestrial plants 130–1
   of terrestrial vertebrates 176, 183–4
      fertility control 182
      poisoning 179–81
      shooting 178–9
      snares and traps 176–8
Convention on Biological Diversity (CBD) 112–13, 121
Convention on Conservation of Migratory Species of Wild Animals (CMS) 114–15
Convention on International Trade in Endangered Species (CITES) 114
Convention on Wetlands (Ramsar 1971) 113–14
Cook's petrel (*Pterodroma cookii*), interactions 241
Cooperative Initiative on Invasive Alien Species on Islands (CII) 122–3
copper sulphate, in control of marine species 223, 225
coqui frog (*Eleutherdactylus coqui*), control tools 183
cost:benefit analysis, invasive fish 188
costs, estimation of 24, 32–3
coypu (*Myocaster coypus*) 52
   control tools 184
   trapping 178
crazy ant (*Anoplolepis longipes*), mesopredator release effect 241
Criocers species 104
Cryptochaeyum iceryae, use in biological control 79
Cryptomeri japonica (Japanese red cedar), functional role 234
Cryptostegia grandiflora (rubber vine) 85
   biological control, economics 84
   containment 73–4

cultural methods of weed control 131, 136
   competition 137–8
   prevention of invasion 136–7
Cyanoramphus novaezelandiae eryhtrotis, extinction 235
Cynara cardunculus (artichoke thistle) 130
cypress aphid (*Cinara cupressi*), biological control 81

Daktulosphaira vitifoliae 3
data, species reporting requirements 45, 46
databases 41–2, 124–5
   components and potential uses 42–5
data completeness 24, 30–1
data management 41
data-sharing 40, 41
date mussel (*Musculista senhousia*) 220–1
daughterless induction, fish 199
   daughterless carp project 58
DDT 19
deliberate introductions 4, 7–9, 22
   fish 185–6
delimitation 56–7
Delivering Alien Invasive Species Inventories for Europe (DAISIE) 42
dense waterweed (*Egeria densa*) 143
   control 147
desiccation, marine species 220–1
detection
   terrestrial vertebrates 175–6
   see also early detection
development aid 6–7
dewatering 199–200
   combination with piscicide use 194–5
dicots 126
Didemnum vexillum
   control by wrapping 222, 223
   removal from vessel hulls 215
direction of effort 70
disaster relief 6–7
disasters, role in introductions 5
distribution maps 38, 68
distribution modelling 43
dogs, use in hunting 179
donkeys (*Equus asinus*), eradication 52
dredging, in control of marine species 220–1
Dublin Core 45
Dutch elm disease 32

early containment action 69
early control 69–70
early detection 36, 37
   aquatic plants 152
early detection and rapid assessment (EDRA) programmes 38–40, 45–6
   data and information management 41–5
   guiding principles 40–1
   marine species 212–14, 227–8

novel solutions 229–30
species reporting requirements 45, *46*
early response, to invasive fish 192
Earth Summit (Rio de Janeiro 1992) 112, 120–1
ecological risks, biological control 87–90
economic analyses, biological control 83–4
economic costs of invasive species 20
ecosystem, effect of introduced species 20
ecosystem approaches 120, 232–3, 247
  species with functional roles 234
  species with long lasting effects 234
  see also interactions
ecosystem rules 233
effectiveness, public participation in IAS management 95–9
Egeria densa (dense waterweed) *143*
  control 147
Eichhornia crassipes (water hyacinth) vi, 108, *142*, 144, 146
  biological control *149*, 150
  eradication cost 54
  mechanical control *147*
  mode of spread 145
  use of herbicides 148
electrical barriers to fish 193
electrofishing 200, *202*
Eleutherdactylus coqui (coqui frog), control tools *183*
Elodea canadensis (Canadian waterweed) *143*
  control 147
emergency aid 6–7
emergency preventive actions 14–15
encapsulation techniques, marine species 218–19, 221–4, *222*, 230
enclosure experiments 244
Enemy Release Hypothesis (ERH) 78
entomogenous fungi, use in biological control 79–80
Environment Risk Management Authority (ERMA), New Zealand 94
Equus asinus (donkey), eradication 52
eradication viii, 47–8, 59–60, 61
  delimiting boundaries 56–7
  detection of survivors or immigrants 57
  feasibility 48–50
  institutional commitment 58–9
  local elimination 59
  on mainlands 55
  scale 56
  of terrestrial plants 130, *131*
  tricky species 58
  of vertebrate pests 50–2
  of weeds 52–3
    costs 53–*4*
    seed banks 57
Erinaceus europaeus (hedgehog), eradication programmes 49, 52
ethics, public participation 93

Euclasta whalleyi 74
Eupatorium serrotinum, eradication cost 54
European Brown Hare Syndrome virus 182
European green crab (*Carcinus maenas*) 209
  biological control 82
  chemical control 226
European perch (*Perca fluviatilis*)
  control by pathogens 198
  introduction to New Zealand 197
European Strategy 122
European Union policy on IAS 122
exclusion, terrestrial vertebrates 176
explosives, use in fish eradication 200
exposure estimation 24, 27, 29

fanwort (*Cambomba caroliniana*) 143
FAO, Code of Conduct for biological control agents 91–2
feasibility studies
  containment 63–4
  eradication programmes 48
Federal Noxious Weed Act 1975 (USA) 137
Felis catus (cat)
  control, New Zealand 74–5
  control tools *183*
  eradication 51, 52
fences, in exclusion of terrestrial vertebrates 176
feral horses, fertility control 182
fertility control
  fish 198–9
  terrestrial vertebrates 182
Fiji, eradication of Pacific rats 96–7, 100–1
Fijian ground frog (*Platymantis vitianus*) 96
Fintrol® (antimycin-A) 194, 196
fipronil 161, 163
  non-target impacts 165–6
fires
  effect of introduced species 20, 26
  as metaphor for invasion 37
  use in vegetation control 133
fish ix, 186–7, 203–4
  Code of Conduct for Responsible Fisheries 115
  costs of eradication or control 188
  early response 192
  human introductions 186–7
  redistribution of indigenous species 188–9
  response tools
    biocontrol 197–9
    chemical control 193–7
    physical removal 199–203
    prevention of spread 192–3
  risk assessment 187
  routes of introduction and spread
    aquaculture 191–2
    ballast water and vessel hull transport 189
    trade 189–91
  see also marine species

flowering plants 128
flowing waters, chemical eradication of fish 195
fluridone, in control of hydrilla 148
food-web interactions 233
fouling of maritime vessels 209
  management 211, 214–15, 223, 224, 230
founder populations, eradication 55
foxes, eradication campaign 57
freshwater, in control of marine
  species 220–1, 224
freshwater fish 188
  see also fish
functional roles of alien species 234
funding 167–8
  for biological control 92
  containment and control programmes 71–2
fungi
  potential quarantine issues 17
  use in biological control 79–80, 90

Galápagos Islands
  eradication of blackberries 57
  predictive model for West Nile virus
    introduction 21
gamba grass (*Andropogon gayanus*), control 76
Gambusia spp. (mosquitofish) 187
  G. holbrooki 198
Gecarcoidea natalis (red crab), impact of yellow
  crazy ant 157
geckos (*Hemidactylus frenatus*) 174
generic response frameworks 153, *154*, 170
genetic engineering, fertility control 182, 199
genetic manipulation 58
  of fish 198–9
geostatistical modelling 34
giant African snail (*Achatina fulica*), biological
  control project 82
giant hogweed (*Heracleum mantegazzianum*),
  biological control 88–9
giant reed (*Arundo donax*) *132*, *135*
gibel carp 199
gill nets 200, *202*
GIS (Geographic Information System)
  software 168
Global Invasive Species Database
  (GISD) 41, 124
Global Invasive Species Information Network
  (GISIN) 41, 125
Global Invasive Species Programme
  (GISP) 123–4
globalization 108
Global Organism Detection and Monitoring
  (GODM) system 42
Global Register of Invasive Species (GRIS) 41
goats (*Capra hircus*)
  control tools *183*
  eradication *51*, 52

helicopter culling 179
gobies, methods of spread 193
Gobio gobio (gudgeon), eradication in
  New Zealand 192
golden eagle (*Aquila chrysaetos*), interaction
  with pigs 237–8
goldfish (*Carassius auratus*) 187
GPS (Global Positioning System) 168
  use in aerial baiting 162
  use in supercolony mapping 163
Grand Staircase-Escalante National
  Monument, Utah, study of
  non-native species 27, *28*
grass carp, in control of aquatic plants 150
grasses, control in Australia 75–6
'gravity model', zebra mussel
  (*Dreissena polymorpha*) 21
grazing, role in weed control 138
Great Lakes
  sea lamprey control 197
  zebra mussel 'gravity model' 21
green mussel (*Perna viridis*), control 223
grey fox (*Urocyon littoralis*),
  hyperpredation 237–8
grey squirrels (*Sciurus carolinensis*) 55
  control tools *184*
Guadalupe Island, effects of eradication
  programme 49
Guam, brown tree snake introduction 20, 32
gudgeon (*Gobio gobio*), eradication in
  New Zealand 192

habitat modification, role in invasive
  fish control 198
habitat quality, effects of invasive species 32
habitats, matching to species traits 25–7
habitat-specific risk assessments 21, *26*
habitat suitability maps 31
Hakea sericea, control in South Africa 83
hand pulling of weeds 132–3
Hawaii
  effect of herbivore removal 240
  impact of avian malaria vi
Hazardous Substances and New Organisms
  (HSNO) Act, New Zealand 94, 187
hazardous waste disposal 12
heat treatment, marine species 220–1, 224
hedgehog (*Erinaceus europaeus*) eradication
  programmes 49, 52
Helenium amarum, eradication 58
  cost 54
helicopter culling 179
Hemidactylus frenatus (gecko) 174
Hemitragus jemlahicus (Himalayan thar),
  New Zealand 49
Heracleum mantegazzianum (giant hogweed),
  biological control 88–9

Index | 301

herbicides
  in control of aquatic plants 146, 148
  in control of marine species 226
  in control of terrestrial plants 139
herbivore removal 238–40
Herpestes javanicus (mongoose) 52
  control tools *184*
heuristic models 153, *154*, 170
Hieracium pilosella, eradication cost 54
high pressure spraying, role in defouling 215, *216–17*
Himalayan thar (*Henitragus jemlahicus*), New Zealand 49
homogenization of fish species 188–9
horse-chestnut leaf miner 81
host specificity, biological control 86, 87–90, 138
hull-fouling *209*
  management 211, 214–15, 223, 224, 230
human dimension in invasive species management 167
human health risks 32–3
humans, role in spread of invasive species v–vi, 99
  accidental introductions 5–7
  deliberate introductions 7–9
  fish 186–7
  marine species 207–10
hunting with dogs 179
Hydrilla verticillata 143, 144
  competitive release effect 242
  control 148, 150
  mode of introduction 145
Hydrocotyle ranunculoides (pennywort) *143*
Hygrophila polysperma, competitive release 242
Hyperaspis pantherina, use in biological control 81
Hypericum perforatum (St John's wort), biological control 80–1, 242
hyperpredation 235–8, *236*
  control strategies 245
Hypocharis radicata, eradication cost 54

IAS management *110*
  ecosystem approach 120
  precautionary principle 120–1
iceplant (*Mesembryanthemum crystallinum*), long lasting effect 234
Icerya purchasi (scale), biological control 79
immunocontraceptive control, fish 199
IMO, Ballast Water Convention 94
impacts of invasive species vi, 1–2, 20, 65–6
  assessment 32–3, 67
  fish 185
impedance, weed eradication 53
implementation, public participation 103
Import of Live Fish Act 1980 (UK) 190

India, Wildlife Protection Act 109
indigenous fish, redistribution 188–9
information
  addition to databases 42
  provision by public 102
  provision to rapid assessment teams *43*
information management 34, 41
insecticides, in control of marine species 226
inspection facilities and processes 11–12
  bench inspection equipment 13–14
integrated management
  aquatic plants 151
  terrestrial plants *131*, 139–40
interactions ix, 232
  competitor release effect *236*, 242–3
  control strategies 245–6
  herbivore removal 238–40
  hyperpredation 235–8, *236*
  Japanese red cedar and macaques 234
  mesopredator release effect *236*, 241–2
  mitigation 244
Inter-American Biodiversity information Network (IABIN) 42
intercontinental invasions 38, *39*
internal borders, marine species management 228
international action, necessity for 108–9
International Civil Aviation Organization (ICAO) 119
International Council for the Exploration of the Sea (ICES) Code of Practice 115–16
international legislation, binding instruments 111
international legislative frameworks 3
International Maritime Organization (IMO) 118–19
international phytosanitary standards (ISPMs) 3, 10
International Plant Protection Convention (IPPC) 3, 117
international programmes and organizations 123–5
international trade vi, 5–6
international travel 4, 108
International Union for Conservation of Nature (IUCN) 124
introduction, definitions 111
invasion, five stages 61, *62*
invasion fronts, containment effort 70
invasion process 205, *206*
'invasion windows' 27
invasive alien species (IAS), definition 108
invasiveness 2–3, 16
  species traits 23, *25*
Invasive Species Compendium, CAB International 42
Invasive Species Specialist Group (ISSG) 124

invertebrates viii, 153–4
  eradication 55
  potential quarantine issues 17
in-water defouling 214–15, *216–17*, 230
islands
  eradication of vertebrate pests 50–2
  vulnerability to invasion 154–5
island-wide survey, Christmas Island 159, *160*
iterative sampling approach 29, *30*

Jacaranda bug (*Orthezia insignis*), biological control 81
Japanese red cedar (*Cryptomeri japonica*), functional role 234
Jatropha curcas, eradication cost 54
joint decision making 102–3
jute matting, use in control of marine species 223

Kakadu National Park (KNP)
  eradication of invasive ants 99
  rapid response to mimosa trees 37–8
Kentucky bluegrass (*Poa pratensis*) 28
Kochia scoparia, introduction to Australia 71
koi carp, control in New Zealand 201

ladybirds, use in biological control 79, 81
Lagarosiphon major 143, 144
  control 147, 150
  mode of introduction 145
Lake Victoria, removal of Nile perch 201, 203, 204
lampreys (*Petromyzon marinus*), eradication 58
Lantana camara 9
  biological control 80
legal mandates 33
legislation 3–4, 94–5, 109, 125
  and aerial baiting 162
  biological control 138
  fish trade 187, 190
  global instruments for conservation of biological diversity
    Code of Conduct for Responsible Fisheries 115
    Convention on Biological Diversity (CBD) 112–13
    Convention on Conservation of Migratory Species of Wild Animals 114–15
    Convention on International Trade in Endangered Species 114
    Convention on Wetlands (Ramsar) 113–14
    UN Convention on the Law of the Sea 115
  instruments relating to transport operations 118–19
  instruments relating to phytosanitary and sanitary measures 117–18
  international instruments and approaches 120–1
  international programmes and organizations 123–5
  prevention of terrestrial vertebrate invasions 174
  prevention of weed problems 137
  redistribution of indigenous species 189
  regional agreements 116–17
  regional strategies and plans 122–3
  scope 110
  types of international instruments 110–11
  WTO Agreement on the Application of Sanitary and Phytosanitary Measures 119–20
Lepeoptheirus salmonis (sea lice), use of semiochemicals 230
Leucaena leucocephala, containment 74
life cycles, exploitation 71
lime, in control of marine species 226
live trapping 178
living maps 34
local elimination 59
local invasions 38, *39*
local relevance of IAS 95, 9607
location-focused priorities 68
long lasting effects of alien species 234
long-term strategies 71–2
lures 178
Lythrum salicaria (purple loosestrife), early detection programme 40

macaque (*Macaca fascicularis*)
  control tools *184*
  interaction with Japanese red cedar 234
Macquarie Island
  effects of rabbit removal 240
  hyperpredation 235
magnets, use in fish capture 200
mahua oil cake 196
mainlands, eradications 55
malaria, avian vi
mammalian predators, control in New Zealand 74–5
manageability potential 228
management of landscapes, effect on feasibility of containment 64
management options *206*
management tools *206*
manual control, aquatic invasive plants 146–7
mapping, ant supercolonies *160*, 163
Maravalia cryptostegiae 73–4
marine environment
  Code of Conduct for Responsible Fisheries 115
  International Council for the Exploration of the Sea (ICES) Code of Practice 115–16

International Maritime Organization
    (IMO) 118–19
  UN Convention on the Law of the Sea 115
  United Nations Environment Programme
    (UNEP) Regional Seas Programme 116
marine fish 188, 191
  methods of spread 193
Marine Protected Areas (MPAs),
    legislation 109
marine species ix, 16, 205–7, 227–9, 231
  early detection and rapid response 212–14
    novel solutions 229–30
  eradication 55
  human-mediated invasion
    pathways 207–10
  post-border management 211–12
  pre-border management 210–11
  response tools 214, 216–21
    chemicals 224–7
    desiccation 224
    freshwater baths 224
    heat treatment 224
    physical removal 214–15, 220–1
    wrapping and smothering 221–4
marron crayfish (*Cherax tenuimanus*),
    eradication in New Zealand 192
mechanical control, aquatic invasive
    plants 146–7
mechanical methods of invasive plant 134–5
Mediterranean, action plan concerning IAS
    introductions 116
Melaegris gallopavo (turkey), eradication 52
meristems 127
Mesembryanthemum crystallinum (iceplant),
    long lasting effect 234
mesopredator release effect *236*, 241–2
  control strategies 245
metadata 34, 42, 45
Metarhizium anisopliae, use in biological
    control 80
mice (*Mus musculus*) 174
  competitive release 242–3
  control tools *184*
  eradication *51*, 52
Middle Island, rat eradication 49
mimosa (*Mimosa pigra*)
  management programme 98
  rapid response 37–8
mink (*Mustela vison*) 52
  trapping 178
Mnemiopsis leidyi (Atlantic comb jelly),
    biological control 82
Molgula spp., chemical control 225
mongoose (*Herpestes javanicus*) 52
  control tools *184*
monitoring 37–8
  containment and control programmes 72–3
  post-eradication 246

monocots 126
mosquitofish (*Gambusia* spp.) 187
mowing, use in weed control 134–5
mulching 136
multiple species, coordinated
    management 69, 76, 232–3
Musculista senhousia (date mussel) 220–1
musk shrews (*Suncus murinus*), eradication 58
Mustela species
  control in New Zealand 75
  M. erminea
    control tools *184*
    trapping 177
  M. vison (mink) 52
    trapping 178
mynas (*Acridotheres* species) 174
  control tools *183*, 184
  eradication 58
Myocaster coypus (coypu) 52
  control tools *184*
  trapping 178
Myriophyllum aquaticum 142
Myriophyllum spicatum (spiked water
    milfoil) *143*, 144
  biological control 150
  control 146–7
Mytilopsis species
  eradication 55, 213, 225
  M. sallei (black-striped mussel),
    control 223
myxoma virus 82, 181–2

National Introduced Maritime Pest Information
    System (NIMPIS) database 214
national legislative frameworks 3–4
natural barriers, exploitation 70
naturalization 61
natural spread 4, 5
Neochetina species *149*, 150
neophobia 58
netting, nuisance fish 200, *202*
networking 169
New Zealand 60, 212
  control of mammalian predators 74–5
  European perch introduction 197
  fish control *201*, *202*
  fish introductions 186
    early responses 192
  Hazardous Substances and New Organisms
    Act 1996 187
  Himalayan thar (*Henitragus jemlahicus*) 49
  legislation 94
  local elimination of pests 59
  management of aquatic invasive plants 147
  public participation 95
  rabbits 48
  vertebrate pests *51*
niclosamide 194, 197

Nile perch, Lake Victoria 201, 203, 204
Niphograpta albiguttalis 149
Nonindigenous Species Database network (NISbase) 42
non-target populations
   impact of aerial baiting 165–6
   protection during eradication programmes 49, 60
   risks from poisoning 180
      piscicides 196
   safety of biological control 89–90
northern Pacific sea-star (*Asterias amurensis*) 209, 221
   detection 229
nutria (*Myocaster coypus*), control tools 184
nutrient availability, aquatic plants 144
Nymphaea mexicana (yellow waterlily) 142
   mode of introduction 145

'off label' usage, piscicides 194
oil platforms, transport of fish 189
Operculina ventricosa
   effect of herbivore removal 239
   Sarigan Island 49
Ophiostoma ulmi 32
opportunity costs 33
Opuntia species (prickly pears), biological control 8, 21, 80
ornamental plant introductions 8–9, 99, 145
Orobanche ramosa (branched broomrape), eradication campaign 53
Orthezia insignis (Jacaranda bug), biological control 81
Oryctolagus cuniculus (rabbit)
   biological control 82, 181–2
   containment 63
   control tools 184
   effect of calcivirus 71
   eradication 50, 52, 56, 58
Oxyura jamaicensis (ruddy duck) 114
   eradication 52
   shooting 179

Pacific Ant Prevention Programme (PAPP) 123
Pacific Invasives Initiative (PII) 122–3
Pacific rat eradication
   Vatu-i-Ra 100–1
   Viwa Island 96–7
painted turtle (*Chrysemys picta*) 114
partial containment 62
Paspalum unvillei, eradication cost 54
passive public participation 102
pasture grasses, control in Australia 75–6
'pathophobia' 89
pathways of introduction 4–5
   accidental introductions 5–7
   deliberate introductions 7–9

   natural spread 5
pennywort (*Hydrocotyle ranunculoides*) 143
Perca fluviatilis (European perch)
   control by pathogens 198
   introduction to New Zealand 197
perennial plants, management 126–7, 134, *135*
Perna viridis (Asian green mussel), control 223
Pest Management Strategies (PMS), New Zealand 94
Pest Risk Analysis, IPPC 117
pests, definitions 117
Petromyzon marinus (sea lamprey), eradication 58
pet trade 175
pheromone lures 58, 178
   in control of invasive fish 197–8
   in control of marine species 230
Phloeospora heraclei 88
Phragmites australis, chemical control 226
Phylloxera vastatrix (now *Daktulosphaira vitifoliae*) 3
Phylloxera vitifolia, biological control 79
physical control
   fish 199–200
      carp control programme, Tasmania 203
      Nile perch 201, 203
   terrestrial plants *131*, 132
      fire 133
      hand pulling and hoeing 132–3
      mechanical methods 134–5
      mulching and solarization 136
   marine species 214–15, *216–20*, 220–1
PIERS database 16
pigeons (*Columba livia*), eradication 52
pigs
   control tools 184
   eradication 56
   facilitation of golden eagles 238
   helicopter culling 179
   hunting with dogs 179
   trapping 178
piscicides 194–7
Pistia stratiotes (water lettuce) 142, 144
   biological control 150, *151*
plant introductions 8–9
Platymantis vitianus (Fijian ground frog) 96
Poa foliosa, effect of rabbit removal 240
Poa pratensis (Kentucky bluegrass) 28
poisoning
   fish 193–7
   terrestrial vertebrates 179–81, 184
pollinators, functional role 234
polymerase chain reaction (PCR), in ballast water management 229–30
pontoons, wrapping *222*
post-eradication monitoring 246

potential distribution models 31
pre-border preventive actions 10
precautionary principle 120–1
predatory fish, use in biological control 197
predictive spatial modelling *44*
pre-eradication studies 244
presence and abundance surveys *24*
Presto®01 bait *156*, 161, 163, 164
   knockdown effect 165
   non-target impacts 165–6
preventive actions 9–10, 16, 61
   against aquatic plants 152
   against marine species 207, 210–11, 227
   against terrestrial plants 129–30, *131*, 136–7
   against terrestrial vertebrates 174–5
   bench inspection equipment 13–14
   borders 11
   emergency actions 14–15
   inspection facilities and processes 11–12
   pre-entry 10
prey switching 241–2
prickly acacia (*Acacia nilotica*), introduction to Australia 71
prickly pear moth (*Cactoblastis cactorum*) 8
   risk of biological control 21
prickly pears (*Opuntia* species), biological control 80
prioritization 68–9
   species and sites 34–5
propagule pressure 27, 29, 63–4
Pterodroma cookii (Cook's petrel), interactions 241
public awareness 168
public participation 35, 93
   Adoption Model *106*
   challenges 107
   compliance 93–5
   effectiveness 95–9
   and ethics 93
   Participation Model *102*–3, 107
public support 98–9
   Vatu-i-Ra Pacific rat eradication 100–1
Puccinia myrsiphylli 104
Pueraria phaseoloides eradication cost *54*
purple loosestrife (*Lythrum salicaria*), early detection programme 40
Pycnonotus cafer (red-vented bulbul), control tools *183*

quarantine 1
   conflicts with trading practices 6
   legislative frameworks 3–4
   and traditional movements 6
quarantine facilities 11–12
   bench inspection equipment 13–14

quarantine systems 9–10
Queensland fruit fly (*Bactrocera tryoni*) 5

rabbit calicivirus, introduction to Australia 71
rabbit haemorrhagic disease 182
rabbits (*Oryctolagus cuniculus*)
   biological control 82, 181–2
   containment 63
   control tools *184*
   eradication 50, 52, 56, 58
   management in New Zealand 48
Rallus philippensis (banded rail), extinction 235
Ramsar 113–4
Rana catesbeiana (American bullfrog) 114
   control tools *183*
range of invasive species, influencing factors 61–2
ranking systems for invasive species 33
rapid assessment teams, information provision *43*
rapid response 37
   to aquatic plants 152
   to marine species 213–14
   see also early detection and rapid assessment (EDRA) programmes
rate of spread, estimation *24*
rats (*Rattus* species) 174, *247*
   control
     competitive release of mice 242–3
     New Zealand 74–5
   control tools *184*
   eradication 49, *51*, 52
     Vatu-i-Ra Island 100–1
     Viwa Island 96–7
record verification, databases 42
recruitment, management 56
red crab (*Gecarcoidea natalis*), impact of yellow crazy ant 157
red deer (*Cervus elaphus*), control tools *184*
red-eared slider (*Trachemys scripta*) 114
   control tools *183*
red fox (*Vulpes vulpes*), control tools *184*
red imported fire ant (RIFA) (*Solenopsis invicta*), New Zealand 95
red imported fire ant (*Solenopsis invicta*), eradication 55
red-vented bulbul (*Pycnonotus cafer*), control tools *183*
red water fern (*Azolla filiculoides*) *142*, 144
   biological control 150
regulation, biological control 91–2
reporting requirements 45, *46*
reproductive methods, aquatic plants 144–5
resource availability, relationship to invasion success 25–6
resource demands, competition 168–9

resources, value of public participation 95, 97–8
responsibility for invasive species management 168
  containment and control programmes 71
Réunion Islands, invasive species 21
Rhinocyllus conicus, use in biological control 87–8
rinderpest, biological control 181
riparian areas, public participation 103–4
risk assessment 19
  biological control 87–90
  components 23, *24*
    containment potential, costs, and opportunity costs 33
    estimating potential distribution and abundance 31
    estimating potential rate of spread 31–2
    estimation of impact and costs 32–3
    exposure estimation 27, 29
    information on species traits 23, *25*
    legal mandates and social considerations 33
    matching species traits to habitats 25–7
    surveys of current distribution and abundance 29–30
    understanding of data completeness 30–1
  current state 21–2
  fish 187
  information management 34
  need for formal approach 20–1
  pre-entry *10*
  selection of priority species and sites 34–5
  ultimate challenge 22
risk mitigation, inspection facilities 11
river restoration, role in invasive fish control 198
robber crabs (*Birgus latro*), impact of aerial fipronil baiting 166
Rodolia (Vedalia) cardinalis, use in biological control 79
roots 127
rotenone 194, 195–6
Round Island, herbivore removal 238
rubber vine (*Cryptostegia grandiflora*) 85
  biological control, economics 84
  containment 73–4
Rubus species (blackberries)
  eradication 57
  R. glaucus, eradication cost *54*
ruddy duck (*Oxyura jamaicensis*) 114
  eradication 52
  shooting 179

safety, biological control 87–90
St Helena, use of biological control 81
St John's wort (*Hypericum perforatum*), biological control 80–1, 242

Saint Paul Island, competitive release effect 242
salmon farming 192
salmonids 188
  eradication in Australia 195
  introduction to New Zealand 186
salt, in control of marine species 226
Salvinia molesta 142, 144
  biological control 150
  eradication cost *54*
sampling, iterative approach 29, *30*
San Beito Islands, effect of herbivore removal 240
San Cristobal Island, herbivore removal 239
Santa Cruz Island
  eradication of feral pigs 56
  hyperpredation *237*–8
Santiago Island, eradication of feral pigs 56
saponins 196
Sargassum muticum
  burial 223
  chemical control 225, 226
  physical removal 215
Sarigan Island
  eradication of feral goats and pigs 49
  herbivore removal *239*
saturation 61
scale (*Icerya purchasi*), biological control 79
schools, involvement in invasive species management 104
science-management interface 168
scientific understanding, role in invasive species management 167
Sciurus carolinensis (grey squirrel) 55
  control tools *184*
sea-chests, transport of fish 189
sea lamprey
  control in Great Lakes 197
  pheromone traps 197
sea lice (*Lepeoptheirus salmonis*), use of semiochemicals 230
search theory 57
sea squirts, removal from mussel shells 215
secondary poisoning 180
seed banks 57, 128, *129*, *130*, 137
seed dispersers, functional role 234
seines 200
semiochemicals
  in control of marine species 230
  see also pheromone lures
sensitivity to piscicides 195
ships' hulls, transport of fish 189
shooting 178–9
simultaneous control, poisoning 180
smothering, marine species *218–19*, 221–4
smuggling 9
snares 176–8
snow-flake coral (*Carijoa riisei*) 222
social considerations 33

socio-economic considerations, containment and control 71–2, 76
sodium hypochlorite, in control of marine species 225
solarization 136
Solenopsis invicta (red fire ant), eradication 55
South Africa, plant introductions 8
Southern California Caulerpa Action Team 40
South Pacific Regional Environment Programme (SPREP) 122
Spartina angelica
   chemical control 226
   eradication cost 54
spatial modelling 45
   predictive 44
species reporting requirements 46
species-specific risk assessments 21
species traits
   matching to habitats 24
   relationship to invasiveness 23, 24, 25
spiked water milfoil (*Myriophyllum spicatum*) 143, 144
   control 146–7, 150
spread, fire as metaphor 37
spread models 31–2
standardization, traps 177
starling (*Sternus vulgaris*), control tools 183
steering committees 161, 169
stem fragmentation, aquatic plants 144
stoats (*Mustela erminea*)
   control in New Zealand 59, 75
   control tools 184
   trapping 177
strategic goals 61
stratified-random sampling 30
Stratiotes aloides (water soldier) 142
Strawberry Reservoir, rotenone application 194
stressors to ecosystems 20
Striga asiatica, eradication campaign 53
Styela clava, control 215, 222, 225, 226
suction devices, value in control of marine species 215, 216–17
Suez Canal, spread of marine fish 193
Suncus murinus (musk shrew), eradication 58
supercolonies, yellow crazy ant 155, 157, 160
   mapping 163
superpredators 241
surveillance programmes, marine species 212–13, 228
surveys, in risk assessment 29–30
sustainable development programmes 121

Tamarix species 38
   long lasting effect 234
target species 20
Tasmania
   carp control programme 203, 204
   red foxes, eradication attempts 57

teaseed cake 196
Terebrasabella heterouncinata, heat treatment 224
terrestrial plants viii
   abundance, relationship to impact 66–7
   classification 126–7
   containment, feasibility 64
   eradication 52–3, 58
      costs 53–4
      seed banks 57
   management 126, 128
      biological control 138
      chemical control 138–9
      cultural methods 136–8
      integrated weed management 139–40
      physical methods of control 132–6
      principles of prevention, eradication, containment, and control 128–31
      relationship to plant characteristics 127–8
   potential quarantine issues 16–17
   species' biology 67
terrestrial vertebrates viii–ix, 173–4
   control tools 176, 183–4
      biocontrol 181–2
      fertility control 182
      poisoning 179–81
      shooting 178–9
      snares and traps 176–8
   detection tools 175–6
   eradication 50–2
   exclusion 176
   potential quarantine issues 17
   prevention of invasion 174–5
TFM (3-trifluoromethyl-4-nitrophenol) 194, 197
thistles, biological control 87–8
top-down regulation 233
traceability 12
Trachemys scripta (red-eared slider) 114
   control tools 183
trade, as pathway of introduction 5–6, 208
traditional movements, role in introductions 6
Tragapogon dubius (western goatsbeard) 28
trammel nets 200
Trapa natans (water chestnut) 142, 144–5
   mode of introduction 145
traps 176–8
trawling, in control of marine species 220–1
tributylin (TBT) 209
Trichosurus vulpecula (brushtail possum) 51
   control tools 183
   poisoning 180, 181
3-trifluoromethyl-4-nitrophenol (TFM) 194, 197
triploid fish 198–9
tropical fire ant, eradication 99
trout, eradication in Australia 195

tumbleweed (*Bassia scoparia*), eradication 53
cost 54
turkeys (*Melaegris gallopavo*), eradication 52
Tyroglyphus phylloxerae 79

Uist islands, hedgehog eradication programme 49
UN Convention on the Law of the Sea (UNCLOS) 115
Undaria pinnatifida
chemical control 225, 226
eradication 55
heat treatment 224
physical removal 215, 220
wrapping 222
United Nations Environment Programme (UNEP) Regional Seas Programme 116
Urocyon littoralis (grey fox), hyperpredation 237–8
USA
economic costs of invasive species 20
invasive species 20–1
prevention of introductions 4
user ID/ tracking, databases 42
utilization of aquatic invasive plants 146

value systems vii
Vatu-i-Ra, Pacific rat eradication 100–1
vegetative reproduction, aquatic plants 144
viruses
potential quarantine issues 17
use in biological control 82, 181–2, 198
Viwa Island, eradication of Pacific rates 96–7
volunteer groups, utilization in wetland and riparian areas 103–4
Vulpes vulpes (red fox), control tools 184

'wall of death' technique 179
watch lists 39–40, 45
water chestnut (*Trapa natans*) 142, 144–5
mode of introduction 145
water hyacinth (*Eichhornia crassipes*) vi, 108, 142, 144, 146
biological control 149, 150
eradication costs 54
mechanical control 147
mode of spread 145
use of herbicides 148

water lettuce (*Pistia stratiotes*) 142, 144
biological control 150, 151
water soldier (*Stratiotes aloides*) 142
Weedbusters 97–8, 104, 105
weed risk assessment system, Australia 8
weeds
abundance, relationship to impact 66–7
classification 126–7
containment, feasibility 64
definition 127
eradication 52–3, 58
costs 53–4
seed banks 57
see also aquatic plants; terrestrial plants
weevils (*Rhinocyllus conicus*), use in biological control 87–8
western goatsbeard (*Tragapogon dubius*) 28
West Nile virus 32
predictive model for introduction to Galápagos Islands 21
wetland areas
Convention on Wetlands (Ramsar 1971) 113–14
public participation 103–4
wharf piles, wrapping 222
wildlife trade regulations (EU) 114
windows of opportunity vii
World Summit on Sustainable Development (WSSD) 121
World Trade Organization (WTO), Agreement on the Application of Sanitary and Phytosanitary Measures 119–20
wrapping and encapsulation, marine species 218–19, 221–4, 222, 230

yellow crazy ant (*Anoplolepis gracilipes*) 155–6
invasion of Christmas Island 156–7, 158
aerial control campaign 161–7, 170–1
interim response 157–61
yellow waterlily (*Nymphaea mexicana*) 142
mode of introduction 145

zebra mussel (*Dreissena polymorpha*), 'gravity model' 21
zoonotic diseases, deliberate introduction 9
Zygina species 104